STAR WARE

The Amateur Astronomer's Guide to Choosing, Buying, and Using Telescopes and Accessories

FOURTH EDITION

PHILIP S. HARRINGTON

BICENTENNIAL
1807
WILEY
2007
BICENTENNIAL

John Wiley & Sons, Inc.

For my wife, Wendy, and our daughter, Helen,
the centers of my universe
and in memory of my parents,
Frank and Dorothy Harrington

Published by John Wiley & Sons, Inc., Hoboken, New Jersey
Published simultaneously in Canada

Wiley Bicentennial Logo: Richard J. Pacifico

For general information about our other products and services, please contact our Customer Care Department within the United States at (800) 762-2974, outside the United States at (317) 572-3993 or fax (317) 572-4002.

Wiley also publishes its books in a variety of electronic formats. Some content that appears in print may not be available in electronic books. For more information about Wiley products, visit our Web site at www.wiley.com.

Library of Congress Cataloging-in-Publication Data:
Harrington, Philip S.
   Star ware : the amateur astronomer's guide to choosing, buying, and using telescopes and accessories / Philip S. Harrington. — 4th ed.
      p. cm.
   Includes bibliographical references and index.
   ISBN 978-0-471-75063-5 (pbk.)
   1. Telescopes—Purchasing—Guidebooks.   2. Telescopes—Amateurs' manuals.   I. Title.
   QB88.H37 2007
   522'.2—dc22   2006025134

Printed in the United States of America

10 9 8 7 6 5 4 3 2 1

Contents

Preface to the Fourth Edition

> If the pure and elevated pleasure to be derived from the possession and use of a good telescope . . . were generally known, I am certain that no instrument of science would be more commonly found in the homes of intelligent people. There is only one way in which you can be sure of getting a good telescope. First, decide how large a glass you are to have, then go to a maker of established reputation, fix upon the price you are willing to pay—remembering that good work is never cheap—and finally see that the instrument furnished to you answers the proper tests for telescopes of its size. There are telescopes and there are telescopes.

With these words of advice, Garrett Serviss opened his classic work *Pleasures of the Telescope.* Upon its publication in 1901, this book inspired many an armchair astronomer to change from merely a spectator to a participant, actively observing the universe instead of just reading about it. In many ways, that book was an inspiration for the volume you hold before you.

The telescope market is radically different than it was in the days of Serviss. Back then, amateur astronomy was an activity of the wealthy. The selection of commercially made telescopes was restricted to only one type of instrument—the refractor—and sold for many times what their modern descendants cost today (after correcting for inflation).

By contrast, we live in an age that thrives on choice. Amateur astronomers must now wade through an ocean of literature and propaganda before being able to select a telescope intelligently. For many a budding astronomer, this chore appears overwhelming.

That is where this book comes in. You and I are going hunting for telescopes. After opening chapters that explain telescope jargon and history, today's astronomical marketplace is dissected and explored. Where is the best place to buy a telescope? Is there one telescope that does everything well? How should a telescope be cared for? What accessories are needed? The list of questions goes on and on.

Happily, so do the answers. Although there is no single set of answers that are right for everybody, all of the available options will be explored so that you can make an educated decision. All of the chapters that detail telescopes, binoculars, eyepieces, and accessories have been fully updated in this fourth edition to include dozens of new products. Reviews have also been expanded, based on my own experiences from testing equipment for *Astronomy* magazine as well as from hundreds of comments that I have received from readers around the world.

Unfortunately, so much astronomical equipment is now on the market that it is impossible to capture every product in these pages. That is why you and I will rely heavily on the supplemental online material available for each chapter. As you peruse the chapters, also visit their Internet counterparts, found in the *Star Ware* section of my Web site, www.philharrington.net.

Not all of the best astronomical equipment is available for sale, however; some of it has to be made at home. Ten new homemade projects are outlined further in the book. These range from simple accessories that can be made in less than an hour using common items that are probably lying around your basement or garage to advanced accessories requiring a good working knowledge of carpentry and electronics. All are very useful.

The book concludes with a discussion of how to assemble, care for, and use a telescope. All too often I hear from people who are frustrated with their telescopes. Not long ago, I was speaking with a friend of a friend who lamented that she didn't have a copy of this book before purchasing a small telescope from a megamart-type department store. She was frustrated that even after reading the instructions that came with her telescope, she couldn't get it to operate properly. We finally got the telescope to operate properly, no thanks to the inadequate instructions. Our back-and-forth exchange led me to add a section in this edition on how to assemble several typical of telescopes.

Yes, the telescope marketplace has certainly changed in the past century (even in the four years since the third edition of *Star Ware* was released), and so has the universe. The amateur astronomer has grown with these changes to explore the depths of space in ways that our ancestors could not have even imagined.

Acknowledgments

Putting together a book of this type would not have been possible were it not for the support of many other players. I would be an irresponsible author if I relied solely on my own humble opinions about astronomical equipment. To compile the telescope, eyepiece, and accessories reviews, I solicited input from amateur astronomers around the world. The responses I received were very revealing and immensely helpful. Unfortunately, space does not permit me to list the names of the hundreds of amateurs who contributed, but you all have my heartfelt thanks. I want to especially acknowledge the members of the "Talking Telescopes" e-mail discussion group that I established in 1999. Found online among Yahoo! Groups, TT is a great group that I encourage you to join. This book would be very different were it not for today's vast electronic communications network.

I also wish to acknowledge the contributions of the companies and dealers who provided me with their latest information, references, and other vital data. Tim Hagan from Helix Manufacturing deserves special recognition for allowing me to borrow and test equipment.

As you will see, chapter 8 is a selection of build-at-home projects for amateur astronomers. All were invented and constructed by amateur astronomers who were looking to enhance their enjoyment of the hobby. These amateurs were kind enough to supply me with information, drawings, and photographs so that I could pass their projects along to you. For their invaluable contributions, I wish to thank Ron Boe, Florian Boyd, Craig Colvin, Jim Dixon, Ed Hitchcock, Jack Kellythorne, and Craig Stark.

I wish to pass on my sincere appreciation to my proofreaders for this edition: Chris Adamson, John Bambury, Thom Bemus, Kevin Dixon, David Mitsky, Rod Mollise, John O'Hara, and Tom Trusock. I am very fortunate to have had this skilled set of veteran amateur astronomers—all among the most knowledgeable amateurs in the world—review the final manuscript. These guys know their stuff! Their thoughtful input and suggestions have been especially useful in a marketplace that is growing and changing as never before. Many thanks also to my editors Christel Winkler, Teryn Kendall, and Kimberly Monroe-Hill of John Wiley & Sons for their diligent guidance and help throughout the production phase of this book.

Finally, my deepest thanks, love, and appreciation go to my ever-patient family. My wife, Wendy, and daughter, Helen, have continually provided me with boundless love and encouragement over the years. Were it not for their understanding my need to go out at three in the morning or drive an hour or more from home just to look at the stars, this book would not exist. I love them both dearly for that.

You, dear reader, have a stake in all this, too. This book is not meant to be written, read, and forgotten about. It is meant to change, just as the hobby of astronomy changes. As you read through this occasionally opinionated book (did I say "occasionally"?), there may a passage or two to which you take exception. Or maybe you own a telescope or something else astronomical that you are either happy or unhappy with. If so, great! This book is meant to kindle emotion. Drop me a line and tell me about it. I want to know. Please address all correspondence to me in care of John Wiley & Sons, Inc., 111 River Street, Hoboken, New Jersey 07030. If you prefer, e-mail me at starware@ philharrington.net. And please check out additions and addenda in the *Star Ware* section of my Web site, www.philharrington.net. I shall try to answer all letters, but in case I miss yours, thank you in advance!

1

Parlez-Vous "Telescope"?

Before the telescope, ours was a mysterious universe. Events occurred nightly that struck both awe and dread into the hearts and minds of early stargazers. Was the firmament populated with powerful gods who looked down upon the pitiful Earth? Would the world be destroyed if one of these deities became displeased? Eons passed without an answer.

The invention of the telescope was the key that unlocked the vault of the cosmos. Although it is still rich with intrigue, the universe of today is no longer one to be feared. Instead, we sense that it is our destiny to study, explore, and embrace the heavens. From our backyards we are now able to spot incredibly distant phenomena that could not have been imagined just a generation ago. Such is the marvel of the modern telescope.

Today's amateur astronomers have a wide and varied selection of equipment from which to choose. To the novice stargazer, it all appears very enticing but also very complicated. One of the most confusing aspects of amateur astronomy is telescope vernacular—terms whose meanings seem shrouded in mystery. "Do astronomers speak a language all their own?" is the cry frequently echoed by newcomers to the hobby. The answer is yes, but it is a language that, unlike some foreign tongues, is easy to learn. Here is your first lesson.

Many different kinds of telescopes have been developed over the years. Even though their variations in design are great, all fall into one of three broad categories according to how they gather and focus light. *Refractors,* shown in Figure 1.1a, have a large lens (the *objective*) mounted in the front of the tube to perform this task, whereas *reflectors,* shown in Figure 1.1b, use a large mirror (the *primary mirror*) at the bottom of the tube. The third class of telescope, called *catadioptrics* (Figure 1.1c), places a lens (here called a *corrector plate*) in front of the primary mirror. In each instance, the telescope's *prime*

1

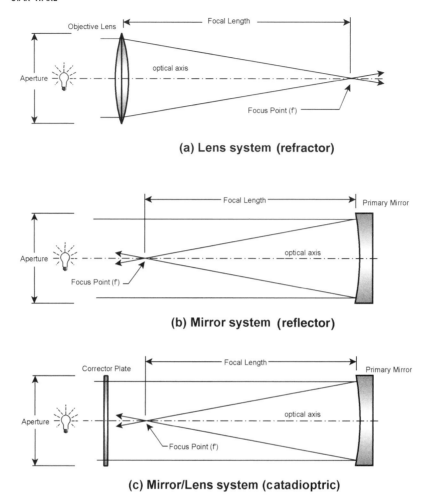

(a) Lens system (refractor)

(b) Mirror system (reflector)

(c) Mirror/Lens system (catadioptric)

Figure 1.1 *The basic principles of the telescope. Using a lens (a), a mirror (b), or (c) a combination, a telescope bends parallel rays of light to a focus point, or prime focus.*

optic (objective lens or primary mirror) brings the incoming light to a *focus* and then directs that light through an *eyepiece* to the observer's waiting eye. Although chapter 2 addresses the history and development of these grand instruments, we will begin here by exploring the many facets and terms that all telescopes share. As you read through the following discussion, be sure to pause and refer to the telescope diagrams found in chapter 2. This way, you can see how individual terms relate to the various types of telescopes.

Aperture

Let's begin with the basics. When we refer to the size of a telescope, we speak of its *aperture*. The aperture is simply the diameter (usually expressed in inches, centimeters, or millimeters) of the instrument's prime optic. In the case of a refractor, the diameter of the objective lens is cited, whereas in reflectors and catadioptric instruments, the diameters of their primary mirrors are specified. For instance, the objective lens in Galileo's first refractor was about 1.5 inches in diameter; it is therefore designated a 1.5-inch refractor. Sir Isaac Newton's first reflecting telescope employed a 1.3-inch mirror and would be referred to today as a 1.3-inch Newtonian reflector.

Many amateur astronomers consider aperture to be the most important criterion when selecting a telescope. In general (and there are exceptions to this rule, as pointed out in chapter 3), the larger a telescope's aperture, the brighter and clearer the image it will produce. And that is the name of the game: sharp, vivid views of the universe.

Focal Length

The *focal length* is the distance from the objective lens or primary mirror to the *focal point* or *prime focus,* which is where the light rays converge. In reflectors, this distance depends on the curvature of the telescope's mirrors, with a deeper curve resulting in a shorter focal length. The focal length of a refractor is dictated by the curves of the objective lens as well as by the type of glass used to manufacture the lens. In catadioptric telescopes, the focal length depends on the combined effect of the primary and secondary mirrors' curves.

As with aperture, focal length is commonly expressed in inches, centimeters, or millimeters.

Focal Ratio

When looking through astronomical books and magazines, it is not unusual to see a telescope specified as, say, an 8-inch f/10 or a 15-inch f/5. This f-number is the instrument's *focal ratio,* which is simply the focal length divided by the aperture. Therefore, an 8-inch telescope with a focal length of 56 inches would have a focal ratio of f/7, because $56 \div 8 = 7$. Likewise, by turning the expression around, we know that a 6-inch f/8 telescope has a focal length of 48 inches, because $6 \times 8 = 48$.

Readers familiar with photography may already be used to referring to lenses by their focal ratios. In the case of cameras, a lens with a faster focal ratio (that is, a smaller f-number) will produce brighter images on film, thereby allowing shorter exposures when shooting dimly lit subjects. The same is true for telescopes. Instruments with faster focal ratios will produce brighter images on film, thereby reducing the exposure times needed to record faint objects. However, a telescope with a fast focal ratio will *not* produce brighter images when used visually. The view of a particular object through, say, an 8-inch f/5 and an 8-inch f/10 will be identical when both are used at the same magnification. How bright an object appears to the eye depends only on telescope aperture and magnification.

Magnification

Many people, especially those new to telescopes, are under the false impression that the higher the magnification, the better the telescope. How wrong they are! It's true that as the power of a telescope increases, the apparent size of whatever is in view grows larger; but what most people fail to realize is that at the same time, the images become fainter and fuzzier. Finally, as the magnification climbs even higher, image quality becomes so poor that less detail will be seen than at lower powers.

It is easy to figure out the magnification of a telescope. If you look at the barrel of any eyepiece, you will notice a number followed by *mm*. It might be 25 mm, 12 mm, or 7 mm, among others; this is the focal length of that particular eyepiece expressed in millimeters. Magnification is calculated by dividing the telescope's focal length by the eyepiece's focal length. Remember to first convert the two focal lengths into the same units of measure—that is, both in inches or both in millimeters. (There are 25.4 millimeters in an inch.)

For example, let's figure out the magnification of an 8-inch f/10 telescope with a 25-mm eyepiece. The telescope's 80-inch focal length equals 2,032 mm ($80 \times 25.4 = 2{,}032$). Dividing 2,032 by the eyepiece's 25 mm focal length tells us that this telescope/eyepiece combination yields a magnification of 81× (read 81 *power*), because $2{,}032 \div 25 = 81$.

Most books and articles state that magnification should not exceed 60× per inch of aperture. This is true only under *ideal* conditions, something most observers rarely enjoy. Due to atmospheric turbulence (what astronomers call *poor seeing*), interference from artificial lighting, and other sources, many experienced observers seldom exceed 40× per inch. Some add the following caveat: never exceed 300× even if the telescope's aperture permits it. Others insist there is nothing wrong with using more than 60× per inch, as long as the sky conditions and optics are good enough. As you can see, the issue of magnification is always a hot topic of debate. My advice for the moment is to use the lowest magnification required to see what you want to see, but we are not done with the subject just yet. Magnification will be spoken of again in chapter 5.

Light-Gathering Ability

The human eye is a wondrous optical device, but its usefulness is severely limited in dim lighting conditions. When fully dilated under the darkest circumstances, the pupils of our eyes expand to about a quarter of an inch, or 7 mm, although this varies from person to person—the older you get, the less your pupils will dilate. In effect, we are born with a pair of quarter-inch refractors.

Telescopes effectively expand our pupils from fractions of an inch to many inches in diameter. The heavens now unfold with unexpected glory. A telescope's ability to reveal faint objects depends primarily on the area of its objective lens or primary mirror (in other words, its aperture), not on magnification; quite simply, the larger the aperture, the more light gathered. Recall from school that the area of a circle is equal to its radius squared multiplied by pi (approximately 3.14). For example, the prime optic in a 6-inch telescope has a light-gathering area of 28.3 square inches (since $3 \times 3 \times 3.14 = 28.3$). Doubling the aperture to 12 inches expands the light-gathering area to 113.1 square inches, an increase of 300%. Tripling it to 18 inches nets an increase of 800%, to 254.5 square inches.

A telescope's *limiting magnitude* is a measure of how faint a star the instrument will show. Table 1.1 lists the faintest stars that can be seen through some popular telescope sizes and is derived from the formula*:

$$\text{Limiting magnitude} = 9.1 + 5 \log D$$

where D = aperture.

Trying to quantify limiting magnitude, however, is anything but precise. Just because, say, an 18-inch telescope might see 15th-magnitude stars, it cannot see 15th-magnitude galaxies because of a galaxy's extended size. A deep-sky object's visibility is more dependent on its surface brightness, or magnitude per unit area, rather than on total integrated magnitude, as these numbers represent. Other factors affecting limiting magnitude include the quality of the telescope's optics, seeing conditions, light pollution, excessive magnification, and the observer's vision and experience. These numbers are conservative estimates; experienced observers under dark, crystalline skies can better these by half a magnitude or more.

Table 1.1 **Limiting Magnitudes**

Telescope Aperture		Faintest Magnitude
Inches	Millimeters	
2	51	10.6
3	76	11.5
4	102	12.1
6	152	13.0
8	203	13.6
10	254	14.1
12.5	318	14.6
14	356	14.8
16	406	15.1
18	457	15.4
20	508	15.6
24	610	16.0
30	762	16.5

*There are many formulas for calculating a telescope's limiting magnitude. In practice, I have found this one, from *Amateur Astronomer's Handbook* by J. B. Sidgwick, to be closest to reality. Sidgwick's formula is based on a naked-eye limiting magnitude of 6.5.

Resolving Power

A telescope's *resolving power* is its ability to see fine detail in whatever object at which it is aimed. Although resolving power plays a big part in everything we look at, it is especially important when viewing subtle planetary features, small surface markings on the Moon, or searching for close-set double stars.

A telescope's ability to resolve fine detail is always expressed in terms of *arc-seconds*. You may remember this term from high school geometry. Recall that in the sky there are 90° from horizon to the overhead point, or zenith, and 360° around the horizon. Each one of those degrees may be broken into 60 equal parts called *arc-minutes*. For example, the apparent diameter of the Moon in our sky may be referred to as either 0.5° or 30 arc-minutes, each one of which may be further broken down into 60 arc-seconds. Therefore, the Moon may also be sized as 1,800 arc-seconds.

Regardless of the size, quality, or location of a telescope, stars will never appear as perfectly sharp points. This is partially due to atmospheric interference and partially due to the fact that light is emitted in waves rather than mathematically straight lines. Even with perfect atmospheric conditions, what we see is a blob, technically called the *Airy disk*, which was named in honor of its discoverer, Sir George Airy, Britain's Astronomer Royal from 1835 to 1892.

Because light is composed of waves, rays from different parts of a telescope's prime optic (be it a mirror or a lens) alternately interfere with and enhance one another, producing a series of dark and bright concentric rings around the Airy disk (Figure 1.2a). The whole display is known as a *diffraction pattern*. Ideally, through a telescope without a central obstruction (that is, without a secondary mirror), 84% of the starlight remains concentrated in the central disk, 7% in the first bright ring, and 3% in the second bright ring, with the rest distributed among progressively fainter rings. Figure 1.2b graphically presents a typical diffraction pattern. The central peak represents the bright central disk, whereas the smaller humps show the successively fainter rings.

The apparent diameter of the Airy disk plays a direct role in determining an instrument's resolving power. This becomes especially critical for observations of close-set double stars. Just like determining a telescope's limiting magnitude, how close a pair of stars will be resolved in a given aperture depends on many variables, but especially on the optical quality of the telescope as well as on the sky. Based on the formula*:

$$\text{resolution} = 5.45 \div D$$

where D = aperture in inches.

Table 1.2 summarizes the results for most common amateur-size telescopes.

Although these values would appear to indicate the resolving power of the given apertures, some telescopes can actually exceed these bounds. The nineteenth-century English astronomer William Dawes found through experimentation that the closest a pair of 6th-magnitude yellow stars can be

*Also from *Amateur Astronomer's Handbook* by J. B. Sidgwick.

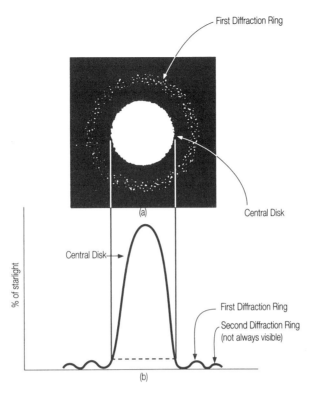

Figure 1.2 *The Airy disk (a) as it appears through a highly magnified telescope and (b) graphically showing the distribution of light.*

Table 1.2 **Resolving Power**

Telescope Aperture		Resolution Threshold
Inches	Millimeters	(theoretical) arc-seconds
2	51	2.7
3	76	1.8
4	102	1.4
6	152	0.91
8	203	0.68
10	254	0.55
12.5	318	0.44
14	356	0.39
16	406	0.34
18	457	0.60
20	508	0.27
24	610	0.23
30	762	0.18

to each other and still be distinguishable as two points can be estimated by the formula:

$$4.56 \div D$$

where D = aperture in inches.

This is called *Dawes' Limit* (Figure 1.3).

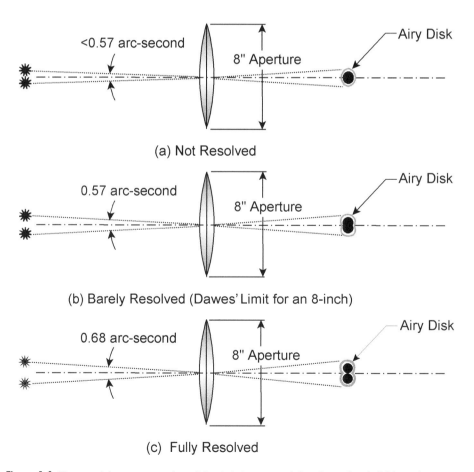

Figure 1.3 *The resolving power of an 8-inch telescope: (a) not resolved, (b) barely resolved, or the Dawes' Limit for the aperture, and (c) fully resolved.*

Table 1.3 **Dawes' Limit**

Telescope Aperture		Resolution Threshold
Inches	Millimeters	(theoretical) arc-seconds
2	51	2.3
3	76	1.5
4	102	1.1
6	152	0.76
8	203	0.57
10	254	0.46
12.5	318	0.36
14	356	0.33
16	406	0.29
18	457	0.25
20	508	0.23
24	610	0.19
30	762	0.15

Table 1.3 lists Dawes' Limit for some common telescope sizes.

When using telescopes less than 6 inches in aperture, some amateurs can readily exceed Dawes' Limit, while others will never reach it. Does this mean that they are doomed to be failures as observers? Not at all! Remember that Dawes' Limit was developed under very precise conditions that may have been far different than your own. Just as with limiting magnitude, reaching Dawes' Limit can be adversely affected by many factors, such as turbulence in our atmosphere, a great disparity in the test stars' colors and/or magnitudes, misaligned or poor quality optics, and the observer's visual acuity.

Rarely will a large aperture telescope—that is, one greater than about 10 inches—resolve to its Dawes' Limit. Even the largest backyard instruments can almost never show detail finer than between 0.5 arc-second (abbreviated 0.5″) and 1 arc-second (1″). In other words, a 16- to 18-inch telescope will offer little additional detail compared with an 8- to 10-inch telescope when used under most observing conditions—although the larger telescope will enhance an object's color. Interpret Dawes' Limit as a telescope's equivalent to the projected gas mileage of an automobile: "These are test results only—your actual numbers may vary."

We have just begun to digest some of the multitude of existing telescope terms. Others will be introduced in the succeeding chapters as they come along, but for now, the ones we have learned will provide enough of a foundation for us to begin our journey.

2

In the Beginning . . .

To appreciate the grandeur of the modern telescope, we must first understand its history and development. Since its invention, the telescope has captured the curiosity and commanded the respect of princes, paupers, scientists, and lay persons. Peering through a telescope renews the sense of wonder we all had as children. In short, it is a tool that sparks the imagination in us all.

Who is responsible for this marvelous creation? Ask this question of most people and they probably will answer, "Galileo." Galileo Galilei did, in fact, usher in the age of telescopic astronomy when he first turned his telescope, illustrated in Figure 2.1, toward the night sky. With it, he became the first person in human history to view craters on the Moon, the phases of Venus, four of the moons orbiting Jupiter, and many other hitherto unknown heavenly sights. Although he was ridiculed by his contemporaries and persecuted for heresy, Galileo's observations changed humankind's view of the universe as no single individual ever had before or has since. But he did not make the first telescope.

So who did? The truth is that no one knows for certain just who came up with the idea, or even when. Many historians claim that it was Jan Lippershey, a spectacle maker from Middelburg, Holland. Records indicate that in 1608 he first held two lenses in line with each other and noticed that they seemed to bring distant scenes closer. Subsequently, Lippershey sold many of his telescopes to his government, which recognized the military importance of such a tool. In fact, many of his instruments were sold in pairs, thus creating the first field glasses.

Other evidence may imply a much earlier origin. Archaeologists have unearthed glass in Egypt that dates to about 3500 B.C., while primitive lenses found in Turkey and Crete are thought to be 4,000 years old! In the third century B.C., the Greek mathematician Euclid wrote about the reflection and

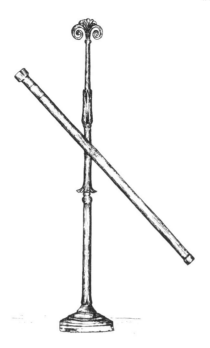

Figure 2.1 *Artist's rendition of Galileo's first telescope. Artwork by David Gallup.*

refraction of light. Four hundred years later, the Roman writer Seneca referred to the magnifying power of a glass sphere filled with water.

Although it is unknown whether any of these independent works led to the creation of a telescope, the English scientist Roger Bacon wrote of an amazing observation made in the thirteenth century: "Thus from an incredible distance we may read the smallest letters . . . the Sun, Moon and stars may be made to descend hither in appearance." Might he have been referring to the view through a telescope? We may never know.

Refracting Telescopes

Although its inventor may be lost to history, the earliest type of telescope is called the *Galilean* or *simple* refractor. A Galilean refractor consists of two lenses: a convex (curved outward) lens held in front of a concave (curved inward) lens a certain distance apart. The telescope's front lens is called the objective, while the other is referred to as the eyepiece, or *ocular*. The Galilean refractor places the concave eyepiece *before* the objective's prime focus, which produces an upright, extremely narrow field of view, much like today's inexpensive opera glasses.

Not long after Galileo made his first telescope, the German astronomer Johannes Kepler improved on the idea by simply swapping the concave eyepiece for a double convex lens and inserting it behind the prime focus. The *Keplerian refractor* proved to be far superior to Galileo's instrument. The modern refracting telescope continues to be based on Kepler's design. The fact that the view is upside down is of little consequence to astronomers because there is no up and down in space. (For terrestrial viewing, extra lenses may be added to flip the image a second time, reinverting the scene.)

Unfortunately, both the Galilean and the Keplerian designs have several optical deficiencies. Chief among these is *chromatic aberration* (Figure 2.2). When we look at any white-light source, we are not actually looking at a single wavelength (or color) of light but rather a collection of wavelengths mixed together. To prove this for yourself, shine sunlight through a prism. The light going in is refracted within the prism, exiting not as a unit but instead broken up, forming a rainbowlike spectrum. Each color of the spectrum has its own unique wavelength.

If you use a lens instead of a prism, each color will focus at a slightly different point. The net result is a zone of focus, rather than a point. Through this type of telescope, everything appears blurry and surrounded by halos of color. This effect is called chromatic aberration.

Another problem of simple refractors is *spherical aberration* (Figure 2.3). In this instance, the curvature of the objective lens causes the rays of light entering around its edges to focus at a slightly different point than those striking the center. Once again, the light focuses within a range rather than at a point, making the telescope incapable of producing a clear, razor-sharp image.

Modifying the inner and outer curves of the lens proved somewhat helpful. Experiments showed that both defects could be reduced (but not completely eliminated) by increasing the focal length—that is, decreasing the curvature—of the objective lens. And so, in an effort to improve image quality, the refractor became longer . . . and longer . . . and even longer! The longest refractor on record was constructed by Johannes Hevelius in Denmark during the latter part of the seventeenth century. His telescope measured about 150 feet from objective to eyepiece and required a complex sling system suspended high above the ground on a wooden mast to hold it in place! Can you imagine the effort it must have taken to swing around such a monster just to look at the Moon or a bright planet? Surely, there had to be a better way.

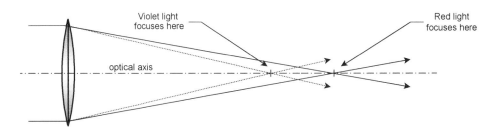

Figure 2.2 *Chromatic aberration is the result of a simple lens focusing different wavelengths of light at different distances.*

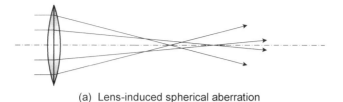

(a) Lens-induced spherical aberration

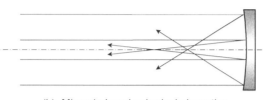

(b) Mirror-induced spherical aberration

Figure 2.3 *Spherical aberration. Both (a) lens-induced and (b) mirror-induced spherical aberration are caused by incorrectly figured optics.*

In an effort to combat these imperfections, the English mathematician Chester Hall developed a two-element *achromatic lens* in 1733. Hall learned that by using two matching lenses made of different types of glass, aberrations could be greatly reduced. In an achromatic lens, the outer element is usually made of crown glass, while the inner element is typically flint glass. Crown glass has a lower dispersion effect and therefore bends light rays less than flint glass, which has a higher dispersion. The convergence of light passing through the crown-glass lens is compensated by its divergence through the flint-glass lens, resulting in greatly dampened aberrations. Unfortunately, although Hall made several telescopes using this arrangement, the idea of an achromatic objective did not catch on for another quarter century.

In 1758, fellow Englishman John Dollond reacquainted the scientific community with Hall's idea when he was granted a patent for a two-element aberration-suppressing lens. Although quality glass was hard to come by for both of these pioneers, it appears that Dollond was more successful at producing a high-quality instrument. Perhaps that is why history records John Dollond, rather than Chester Hall, as the father of the modern refractor.

Regardless of who first devised it, this new and improved design has come to be called the *achromatic refractor* (Figure 2.4a), with the compound objective simply labeled an *achromat*. Although the methodology for improving the refractor was now known, the problem of getting high-quality glass (especially flint glass) persisted. In 1780, Pierre Louis Guinard, a Swiss bell maker, began experimenting with various casting techniques in an attempt to improve the glass-making process. It took him nearly twenty years, but Guinard's efforts

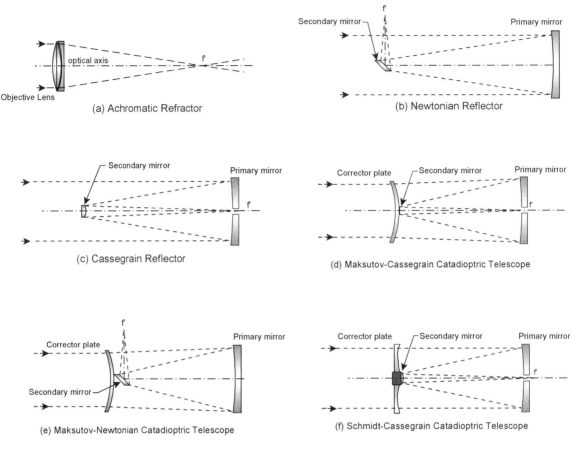

Figure 2.4 *Telescopes come in several different optical configurations: (a) achromatic refractor, (b) Newtonian reflector, (c) Cassegrain reflector, (d) Maksutov-Cassegrain telescope, (e) Maksutov-Newtonian telescope, and (f) Schmidt-Cassegrain telescope.*

ultimately paid off, because he learned the secret of producing flawless optical disks as large as nearly 6 inches in diameter.

Later, Guinard was to team up with Joseph von Fraunhofer, the inventor of the *spectroscope*. While studying under Guinard's guidance, Fraunhofer experimented by slightly modifying the lens curves suggested by Dollond, which resulted in the highest-quality objective yet created. In Fraunhofer's design, the front surface is strongly convex, the two central surfaces differ slightly from each other, requiring a narrow air space between the elements, while the innermost surface is almost perfectly flat. This lens system brings two wavelengths of light across the lens' full diameter to a common focus, thereby greatly reducing chromatic and spherical aberration.

The world's largest refractor is the 40-inch f/19 telescope at Yerkes Observatory in Williams Bay, Wisconsin.* This mighty instrument was constructed by Alvan Clark and Sons, Inc., the United States' premier telescope maker of the nineteenth century. Other examples of the Clarks's exceptional skill include the 36-inch at Lick Observatory in California, the 26-inch at the U.S. Naval Observatory in Washington, D.C., and many smaller refractors at universities and colleges worldwide. Even today, Clark refractors are considered to be among the finest available.

The most advanced modern refractors offer features that the Clarks could not have imagined. *Apochromatic refractors* effectively eliminate nearly all aberrations common to their Galilean, Keplerian, and achromatic cousins. The first apochromatic objective lens came from the genius of Ernst Abbe, a German mathematician and optical designer working for Carl Zeiss Optical. In 1868, two years after Zeiss had appointed Abbe as director of his company's research efforts, he devised a lens system to completely eliminate all traces of chromatic aberration and false color. Since this time, the design for the refracting telescope has hardly stood still—we will examine more modern designs, along with their designers, in chapter 3.

Reflecting Telescopes

The second type of telescope uses a large mirror, rather than a lens, to focus light to a point—not just any mirror, mind you, but a mirror with a precisely figured surface. To understand how a mirror-based telescope operates, we must first reflect on how mirrors work. Take a look at a mirror in your home. Chances are that it is flat, as shown in Figure 2.5a. Light that is cast onto the mirror's polished surface in parallel rays is reflected back in parallel rays. If the mirror is convex (Figure 2.5b), the light diverges after it strikes the surface. If the mirror is concave (Figure 2.5c), then the rays converge toward a common point, or focus. (It should be pointed out that household mirrors are *second-surface* mirrors—that is, their reflective coating is applied onto the back of the glass. Reflecting telescopes use *front-surface* mirrors, which are coated on the front.)

The first reflecting telescope was designed by James Gregory, from Aberdeen, Scotland, in 1663. His system centered around a concave mirror (called the *primary mirror*). The primary mirror reflected light to a smaller concave *secondary mirror,* which, in turn, bounced the light back through a central hole in the primary mirror and out to the eyepiece. This reflector, known as the Gregorian reflector, had the benefit of yielding an upright image, but its optical curves proved difficult for Gregory and his contemporaries to fabricate.

---

*Although the Yerkes 40-inch is the largest refractor ever used for regular astronomical observation, a 49.2-inch refractor with a focal length of 187 feet was shown at the Paris Universal Exhibition of 1900. It was mounted horizontally, with light reflected into the giant objective lens by a moving 79-inch flat mirror. Results were unimpressive, and the telescope was subsequently dismantled.

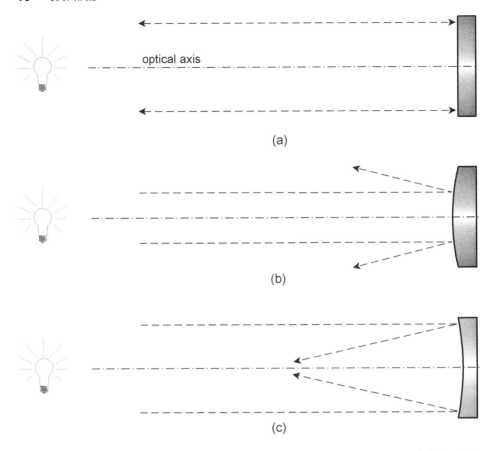

Figure 2.5 *Three mirrors, each with a different front-surface curve, reflect light differently. A flat mirror (a) reflects light straight back to the source, a convex mirror (b) causes light to diffuse, and a concave mirror (c) focuses light.*

A second design was later conceived by Sir Isaac Newton in 1672 (Figure 2.6). Like Gregory, Newton realized that a concave mirror would reflect and focus light back along the optical axis to a point called the prime focus. Here an observer could view a magnified image through an eyepiece. Quickly realizing that his head got in the way, Newton inserted a flat mirror at a 45° angle some distance in front of the primary mirror. The secondary, or *diagonal*, mirror acted to bounce the light 90° out through a hole in the side of the telescope's tube. This arrangement has since become known as the Newtonian reflector (Figure 2.4b).

The Newtonian reflector became the most popular design among amateur astronomers in the 1930s, when Vermonter Russell Porter wrote a series of articles for *Scientific American* magazine that popularized the idea of making your own telescope. A Newtonian reflector is relatively easy and inexpensive to make, giving amateurs the most bang for their buck. Although chromatic aberration is completely absent (as it is in all reflecting telescopes), the Newtonian is not without its faults. Coma, which turns pinpoint stars away from the center of view into tiny "comets," with their tails aimed outward from the center,

Figure 2.6 *Newton's first reflecting telescope. From* Great Astronomers *by Sir Robert S. Ball, London, 1912.*

is the biggest problem, which is exacerbated as the telescope's focal ratio drops. Optical alignment is also critical, especially in fast systems, and must be checked often.

The French sculptor Sieur Cassegrain also announced in 1672 a third variation of the reflecting telescope. The Cassegrainian reflector (yes, the telescope is correctly called a *Cassegrainian,* but since most other sources refer to it as a *Cassegrain,* I will from this point on, as well) is outwardly reminiscent of Gregory's original design. The biggest difference between a Cassegrain reflector (Figure 2.4c) and a Gregorian reflector is the curve of the secondary mirror's surface. The Gregorian uses a concave secondary mirror positioned outside the main focus, while Cassegrain uses a convex secondary mirror inside the main focus. The biggest plus to the Cassegrain is its compact design, which combines a large aperture inside a short tube. Optical problems include lower image contrast than a Newtonian, as well as strong curvature of field and coma, causing stars along the edges of the field to blur when those in the center are focused.

Both Newton and Cassegrain received acclaim for their independent inventions, but neither telescope saw further development for many years. One of the greatest difficulties to overcome was the lack of information on suitable materials for mirrors. Newton, for instance, made his mirrors out of bell metal whitened with arsenic. Others chose speculum metal, an amalgam of copper, tin, and arsenic. Both metals tarnished quickly.

Another complication faced by the makers of early reflecting telescopes was generating accurately figured mirrors. The first reflectors used spherically figured mirrors. In this case, rays striking the mirror's edge came to a different focus than the rays striking its center. The net result was spherical aberration. In order for all the light striking its surface to focus to a point, a primary mirror's concave surface must be a paraboloid accurately shaped to within a few millionths of an inch—a fraction of the wavelength of light. (A paraboloid is a three-dimensional version of a parabola, an open-ended curve with a single bend and two lines that never curve back and close. Instead, the two lines expand infinitely.) Neither design caught on at first, since good mirrors were just too difficult to come by.

The first reflector to use a parabolic mirror was constructed by the Englishman John Hadley in 1722. The primary mirror of his Newtonian measured about 6 inches across and had a focal length of 62.63 inches. But whereas Newton and the others had failed to generate mirrors with accurate parabolic concave curves, Hadley succeeded. Extensive tests were performed on Hadley's reflector after he presented it to the Royal Society. In direct comparison between it and the society's 123-foot-focal-length refractor of the same diameter, the reflector performed equally well and was immeasurably simpler to use.

A second success story for the early reflecting telescope was that of James Short, another English craftsman. Short created several fine Newtonian and Gregorian instruments in his optical shop from the 1730s through the 1760s. He placed many of his telescopes on a special type of support that permitted easier tracking of sky objects (what is today termed an *equatorial mount*—see chapter 3).

Sir William Herschel, a musician who became interested in astronomy when he was given a telescope in 1722, ground some of the finest mirrors of his day. As his interest in telescopes grew, Herschel continued to refine the reflector by devising his own system. The *Herschelian* design called for the primary mirror to be tilted slightly, thereby casting the reflection toward the front rim of the oversized tube, where the eyepiece would be mounted. The biggest advantage to this arrangement is that with no secondary mirror to block the incoming light, the telescope's aperture is unobstructed by a second mirror. Disadvantages included image distortion due to the tilted optics and heat from the observer's head. Herschel's largest telescope was completed in 1789. The metal speculum around which it was based measured 48 inches across and had a focal length of 40 feet. Records indicate that it weighed more than one ton.

Even this great instrument was to be eclipsed in 1845, when William Parsons, the third earl of Rosse, completed the largest speculum ever made. It measured 72 inches in diameter and weighed in at an incredible 8,380 pounds. This telescope (Figure 2.7), mounted in Parsonstown, Ireland, is famous in the annals of astronomical history as the first to reveal spiral structure in certain nebulae and are now known to be spiral galaxies.

The poor reflective qualities of speculum metal, coupled with its rapid tarnishing, made it imperative to develop a new mirror-making process. That evolutionary step was taken in the following decade. The first reflector to use a glass mirror instead of a metal speculum was constructed in 1856 by Dr. Karl Steinheil

Figure 2.7 *Lord Rosse's 72-inch reflecting telescope. From* Elements of Descriptive Astronomy *by Herbert A. Howe, New York, 1897.*

of Germany. The mirror, which measured 4 inches across, was coated with a very thin layer of silver; the procedure for chemically bonding silver to glass had been developed by Justus von Liebig about 1840. Although it apparently produced a very good image, Steinheil's attempt received little attention from the scientific community. The following year, Jean Foucault—creator of the Foucault pendulum and the Foucault mirror test procedure, among others—independently developed a silvered mirror for his astronomical telescope. He brought his instrument before the French Academy of Sciences, which immediately made his findings known to all. Foucault's methods of working glass and testing the results elevated the reflector to new heights of excellence and availability.

Although silver-on-glass mirrors proved to be far superior to the earlier metal versions, the method was still not without flaws. For one thing, silver tarnished rapidly, although not as fast as speculum metal. Experiments in the early twentieth century were aimed to remedy the situation, which ultimately led to the current process of evaporating a thin film of aluminum onto glass in a vacuum chamber. Even though aluminum is not as highly reflective as silver, its longer useful life span makes up for that slight difference. Indeed, modern enhanced coatings can reflect up to 99% of the light striking the mirror surface, assuming you are willing to pay the higher price. Most reflectors, however, use standard aluminum coatings that reflect between 85% and 89% of the light striking the mirror surface. For most of us, standard aluminizing is just fine, although for critical observing, such as searching for distant supernovae in other galaxies or hunting comets, it's worth the extra expense.

Although reflectors do not suffer from the refractor's chromatic aberration, they are anything but flawless. We have already seen how spherical aberration can destroy image integrity, but other problems must be dealt with as well. These include *coma,* which describes objects away from the center of view appearing like tiny comets, with their tails aimed outward from the center; *astigmatism,* resulting in star images that focus to crosses rather than points; and *light loss,* which is caused by obstruction by the secondary mirror and the fact that no reflective surface returns 100% of the light striking it.

Today, there exist many variations of the reflecting telescope's design. While the venerable Newtonian has remained popular among amateur astronomers, the Gregorian is all but forgotten. In addition to the Classical Cassegrain, we find two modified versions: the Dall-Kirkham and the Ritchey-Chrétien.

The Ritchey-Chrétien Cassegrain reflector design is based on hyperbolically curved primary and secondary mirrors, concave for the former and convex for the latter. Both mirrors proved very difficult and expensive to make (or, at least, make *well*). The design was developed in the early 1910s by the American optician George Ritchey and Henri Chrétien, an optical designer from France. Although some minor field curvature plagues the Ritchey-Chrétien design, it is completely free of coma, astigmatism, and spherical aberration. Interestingly, Ritchey, who had made optics for some of the instruments used at Mount Wilson Observatory in California, so severely criticized his boss George Ellery Hale's then-new 100-inch Hooker Telescope for not using the Ritchey-Chrétien design that he was fired and, ultimately, ostracized from U.S. astronomy. Later, when Hale was conceiving the 200-inch Hale reflector for Mount Palomar, he refused to use the Ritchey-Chrétien design, because it had Ritchey's name on it. Ultimately, however, history has proven that Ritchey was right, because nearly every new large telescope built or designed since the 200-inch has used the Ritchey-Chrétien design.

Invented by U.S. optician Horace Dall in 1928, the Dall-Kirkham variant of the Cassegrain reflector didn't catch on until Allan Kirkham, an amateur astronomer from Oregon, mentioned the concept to Albert Ingalls, then the editor of *Scientific American* magazine. Ingalls subsequently published an article that referred to the design as the Dall-Kirkham, and the name stuck. Dall-Kirkham Cassegrains use modified ellipsoidal primary mirrors and spherical secondary mirrors. Although easier and less costly to make, Dall-Kirkhams are plagued by strong field curvature and coma at focal ratios less than f/15. As such, the design is best for narrow-field studies, like planetary work.

Finally, for the true student of the reflector, there are several lesser-known instruments, such as the Wright, Houghton, and Tri-Schiefspiegler (a three-mirror telescope with tilted optics). Not easily available, most of these instruments must be obtained from custom manufacturers.

Like the refractor, today's reflectors enjoy the benefit of advanced materials and optical coatings. Although they are a far cry from the early telescopes of Newton, Gregory, and Cassegrain, we must still pause a moment to consider how different our understanding of the universe might be if it were not for early optical pioneers.

Catadioptric Telescopes

Some comparatively new designs in the twentieth century launched a whole new breed of telescope: the *catadioptric*. These telescopes combine attributes of both refractors and reflectors into one instrument. They can produce wide fields with few aberrations. Many declare that this genre is (at least, potentially) the perfect telescope; others see it as a collection of compromises.

The first catadioptric was devised in 1930 by the German astronomer Bernhard Schmidt. The Schmidt telescope passes starlight through a corrector plate *before* it strikes the spherical primary mirror. The curves of the corrector plate eliminate the spherical aberration that would result if the mirror was used alone. One of the chief advantages of the Schmidt is its fast f-ratio, typically f/1.5 or less. However, because of its fast optics, the Schmidt's prime focus point is inaccessible to an eyepiece, restricting the instrument to photographic applications only. To photograph through a Schmidt, film is placed in a special curved holder (to accommodate a slightly curved focal plane) at the instrument's prime focus, not far in front of the main mirror.

The second type of catadioptric instrument to be developed was the Maksutov telescope. By rights, the Maksutov telescope should probably be called the Bouwers telescope, after A. Bouwers of Amsterdam, Holland. Bouwers developed the idea for a photovisual catadioptric telescope in February 1941. Eight months later, Dimitri Maksutov, an optical scientist working independently in Moscow, came up with the exact same design. Like the Schmidt, the Maksutov combines features of both refractors and reflectors. The most distinctive trait of the Maksutov is its deep-dish front corrector plate, or *meniscus*, which is placed inside the spherical primary mirror's radius of curvature. (If the radius is the distance between the center of a circle and its edge, then a mirror's radius of curvature is the distance, or radius, from the mirror's curved surface and the center of that curve.) Light passes through the corrector plate to the primary mirror and then to a convex secondary mirror.

Most Maksutovs resemble a Cassegrain in design and are therefore referred to as Maksutov-Cassegrains (Figure 2.4d). In these telescopes, the secondary mirror returns the light toward the primary mirror, passing through a central hole out to the eyepiece. This layout allows a long focal length to be crammed into the shortest tube possible.

In 1957, John Gregory, an optical engineer working for Perkin-Elmer Corporation in Connecticut, modified the original Maksutov-Cassegrain scheme to improve its overall performance. The main difference in the Gregory-Maksutov telescope is that instead of a separate secondary mirror, a small central spot on the interior of the corrector is aluminized to reflect light to the eyepiece.

Although not as common, a Maksutov telescope may also be constructed in a Newtonian configuration (Figure 2.4e). In this scheme, the secondary mirror is tilted at 45°. As in the classical Newtonian reflector, light from the target passes through a hole in the side of the telescope's tube to the waiting eyepiece. The greatest advantage of the Maksutov-Newtonian over the traditional Newtonian is the availability of a short focal length (and therefore a wide field of view) with greatly reduced coma and astigmatism.

Two hybrids of the Schmidt camera have also been developed: the Schmidt-Newtonian and the Schmidt-Cassegrain (Figure 2.4f). Since its commercial introduction in the 1960s by Celestron International, the Schmidt-Cassegrain has grown to become one of the most popular telescopes sold today. It combines a short-focal-length spherical concave mirror with a spherical convex secondary mirror and a Schmidt-like corrector plate. The net result is a large-aperture telescope that fits into a comparatively small package. Is the Schmidt-Cassegrain the right telescope for you? Only you can answer that question—with a little help from chapter 3.

Finally, the Schmidt-Newtonian hybrid combines a Schmidt front corrector plate, a spherical primary mirror, and flat secondary mirror tilted at 45°—with the top-mounted focuser of a traditional Newtonian. The Schmidt-Newtonian is an ideal instrument for observers craving sharp, wide-field views of expansive star fields and broad nebulae.

The telescope has certainly come a long way in its four-hundred-year history, but that history is by no means finished. The age of orbiting observatories, such as the Hubble Space Telescope, has opened untold possibilities. Back here on the ground, new giant telescopes, like the Keck reflectors in Hawaii, using segmented mirrors—and even some whose exact curves are controlled and varied by computers to compensate for atmospheric conditions (so-called *adaptive optics*)—are now being aimed toward the universe. New materials, construction techniques, and accessories are coming into use. All of this means that the future will see even more diversity in this already diverse field. Stay tuned!

3

So You Want to Buy a Telescope!

So, you want to buy a telescope? That's wonderful! A telescope will let you visit places that most people are not even aware exist. With it, you can soar over the stark surface of the Moon, travel to the other worlds in our solar system, and plunge into the dark void of deep space to survey clusters of jewel-like stars, huge interstellar clouds, and remote galaxies. You will witness firsthand exciting celestial objects that were unknown to astronomers only a generation ago. You can become a citizen of the universe without ever leaving your backyard.

Just as a pilot needs the right aircraft to fly from one place to another, so, too, must an amateur astronomer have the right instrument to journey into the cosmos. As we have seen already, many different types of telescopes have been devised in the past four centuries. Some remain popular today, while others are of interest from a historical viewpoint only.

Which telescope is right for you? Had I written this book back in the 1950s or 1960s, there would have been only one answer: a 6-inch f/8 Newtonian reflector. Just about every amateur astronomer either owned one or knew someone who did. Although many different companies made this type of instrument, the most popular one was the RV-6 Dynascope by Criterion Manufacturing Company of Hartford, Connecticut, which for years retailed for $194.95. The RV-6 was to telescopes what the old Volkswagen Beetle was to cars of that era—a triumph of simplicity and durability at a great price.

Times have changed, the world has grown more complicated, and the hobby of amateur astronomy has become more complex. The venerable RV-6 is no longer manufactured, although some can still be found in classified advertisements. Today, looking through astronomical product literature, we find sophisticated Schmidt-Cassegrains, mammoth Newtonian reflectors, and state-of-the-art refractors. With such a variety from which to choose, it is hard to know where to begin.

Optical Quality

Before examining specific types of telescopes, a few terms used to rate the caliber of telescope lenses and mirrors must be defined and discussed. In the everyday world, when we want to express the accuracy of something, we usually write it in fractions of an inch, centimeter, or millimeter. For instance, when building a house, a carpenter might call for a piece of wood that is, say, 4 feet long plus or minus $1/8$ inch. In other words, as long as the piece of wood is within $1/8$ inch of 4 feet, it is close enough to be used.

In the optical world, however, close is not always close enough. Because the curves of a lens or a mirror must be made to such tight tolerances, it is not practical to refer to optical quality in everyday units of measurement. Instead, optical quality is usually expressed in fractions of the wavelength of light. Because each color in the spectrum has a different wavelength, opticians use the color that the human eye is most sensitive to: yellow-green. Yellow-green, in the middle of the visible spectrum, has a wavelength of 550 nanometers, which is 0.00055 millimeter or 0.00002 inch.

For a lens or a mirror to be accurate to, say, $1/8$ wave (a value frequently quoted by telescope manufacturers), its surface shape cannot deviate from perfection by more than 0.000069 mm, or 0.000003 inch! This means that none of the little irregularities (commonly called *hills* and *valleys*) on the optical surface exceed a height or depth of $1/8$ of the wavelength of yellow-green light. As you can see, the smaller the fraction, the better the optics. Given the same aperture and conditions, telescope A with a $1/8$-wave prime optic (lens or mirror) should outperform telescope B with a $1/4$-wave lens or mirror, while both should be exceeded by telescope C with a $1/20$-wave prime optic.

Stop right there! Companies are quick to boast about the quality of their primary mirrors and objective lenses, but in reality, we should be concerned with the *final wavefront* reaching the observer's eye, which is double the wave error of the prime optic alone. For instance, a reflecting telescope with a $1/8$-wave mirror has a final wavefront of $1/4$ wave. This value is known as *Rayleigh's Criterion* and is usually considered the lowest-quality level that will produce acceptable images. Clearly, an instrument with a $1/8$ to $1/10$ final wavefront is very good. However, even these figures must be taken loosely because there is no industry-wide method of testing.

Because of increasing consumer dissatisfaction with the quality of commercial telescopes, both *Sky & Telescope* and *Astronomy* magazines often test equipment and subsequently publish the results. Talk about a shot heard around the world! Both organizations quickly found out that the claims made by some manufacturers (particularly a few producers of Newtonian reflectors and Schmidt-Cassegrains) were a bit, shall I say, inflated.

In light of this shake-up, many companies have dropped claims of their optics' wavefront, referring to them instead as *diffraction limited,* which means that the optics are so good that performance is limited only by the wave properties of light itself and not by any flaws in optical accuracy. In general, to be diffraction limited, an instrument's final wavefront must be at least $1/4$ wave, the Rayleigh Criterion. Once again, however, this can prove to be a subjective claim.

Some manufacturers of premium telescope optics state the quality of their products in terms of the *Strehl ratio*. Back in chapter 1, we discussed how a telescope focuses the light from a star not to a perfect point but rather into a tiny disk called the Airy disk, which is surrounded by a series of diffraction rings. Using mathematics that are far too cumbersome to contribute to the explanation here, optical engineers can calculate the distribution of light in the Airy disk produced by a theoretically perfect telescope. Not surprisingly, the brightest part of the Airy disk is in its center. The Strehl ratio is the amount of light a real-world telescope focuses into the center of an Airy disk divided by the amount of light focused there in a "perfect" telescope of the same aperture and focal length. Therefore, an unattainably perfect telescope would have a Strehl ratio of 1.00. Although I refrain from quoting published values for specific telescope models in later chapters, in general we want to look for values as close to 1.00 as possible (or 100, if quoting as a percentage) when judging a telescope's excellence based on its Strehl ratio. But as with fractional wavelength values, it pays to question a manufacturer about claims of its telescope's Strehl ratio by asking how it was determined and what is the accuracy of those measurements. Statistics that are not verifiable are worthless.

Telescope Point-Counterpoint

So which telescope would I recommend for you? None of them . . . or all of them! Actually, there is no one answer anymore. It all depends on what you want to use the telescope for, how much money you can afford to spend, and many other considerations. To help sort all this out, you and I are about to go telescope hunting together. We will begin by looking at each type of telescope that is commercially available. The chapter's second section will examine the many different mounting systems used to hold a telescope in place. Finally, all considerations will be weighed to let *you* decide which telescope is right for you.

Binoculars

Binoculars may be thought of as two low-power refracting telescopes that happen to be strapped together. Light from whatever target is in view enters a pair of objective lenses, bounces through two identical sets of prisms, and exits through the eyepieces. Unlike astronomical telescopes, which usually flip the image either upside down or left to right, binoculars are designed to produce an upright image. This, coupled with their wide-field views, adds to the appeal, especially for beginners.

Just as the inventor of the telescope is not clearly known, the identity of whoever came up with the first pair of binoculars is also lost to history. The first person to patent the concept of fastening two telescopes together to make binoculars was none other than Jan Lippershey, who was also the first person to patent the telescope. His instrument was unveiled in December 1609. As is the case today, the most popular use of these early binoculars was for terrestrial, rather than celestial, viewing. But as you can well imagine, trying to support these cumbersome contraptions was difficult, at best. Not only were the long tubes awkward but keeping them parallel was a tough task. More often than not, early binoculars were plagued by double images.

Dr. Ernst Abbe, a German physicist and mathematician working for Carl Zeiss Optics, devised an ingenious "binocular prismatic telescope," as he called it, in 1894. This signaled two very important advances in binocular design. First, since the light path through the instrument now ricocheted within a set of prisms, the binoculars' physical length could be reduced while maintaining its optical focal length. A second, side benefit was that the prisms flipped the image right-side up, for a correct view of the world. Although Abbe designed the prisms used in his early binoculars, he could not patent the prisms because it was discovered that they had actually been invented in 1854 by the Italian optician Paolo Ignazio Porro. As a result, Abbe's creation became known as Porro prism binoculars.

A second prismatic binocular design was invented a few years later. In 1897, Hensoldt AG, an optical company founded in 1852 by Moritz Carl Hensoldt in Wetzlar, Germany, introduced the first roof prism binocular. The Penta Model A, a small 7 × 29 model, led to a rivalry between Hensoldt AG and Carl Zeiss Optics that would last for the next three decades, until the two companies merged in 1928.

All Porro prism and roof prism binoculars (Figure 3.1) are labeled with two numbers, such as 7 × 35 (pronounced "seven by thirty-five") or 10 × 50. The first number refers to the pair's magnification, while the second number

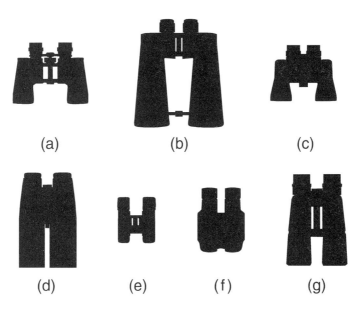

(a)   (b)   (c)

(d)   (e)   (f)   (g)

Figure 3.1 *Binoculars come in many different shapes and sizes, but all share one of two basic designs. Silhouettes (a) through (c) show typical Porro prism binoculars, while (d) through (g) illustrate the roof prism design.*

specifies the diameter (in millimeters) of the two front lenses. Typically, values range from seven power (7×) to twenty-five power (25×), with objectives measuring between 35 mm (1.5 inches) and 150 mm (6 inches). Chapter 4 will give you the full story of binoculars, hopefully helping you to choose a pair that is best suited for your needs.

The ultimate in portability, binoculars offer unparalleled views of rich Milky Way star fields thanks to their low power and wide fields of view. As much as this is an advantage to the deep-sky observer, it is a serious drawback to those interested in looking for fine detail on the planets, where higher powers are required. In these cases, the hobbyist has no choice but to purchase a telescope.

Refracting Telescopes

After many years of being all but ignored by the amateur community, the astronomical refractor (Figure 3.2) is making a strong comeback. Hobbyists are rediscovering the exquisite images seen through well-made refractors. Crisp views of the Moon, razor-sharp planetary vistas, and pinpoint stars are all possible through the refracting telescope.

Achromatic Refractors. As mentioned in chapter 2, many refractors of yesteryear were plagued with a wide and varied assortment of aberrations and image imperfections. The most difficult of these faults to correct are chromatic aberration and spherical aberration.

If we look in a dictionary, we will find the following definition of the word *achromatic*:

> ach·ro·mat·ic (ăk′rə-măt/ik) adj. 1. Designating color perceived to have zero saturation and therefore no hue, such as neutral grays, white, or black. 2. Refracting light without spectral color separation.

Figure 3.2 *When most people picture a telescope, they usually conjure up images of a refractor, such as this 4.7-inch SkyView Pro 120 achromatic instrument from Orion Telescopes. Photo courtesy of Orion Telescopes.*

Achromatic objective lenses, in which a convex crown lens is paired with a concave flint element, go a long way in suppressing chromatic aberration. These are designed to bring red and blue—the colors at opposite ends of the visible spectrum—to the same focus, with the remaining colors in between converging to nearly the same focus point. It works, too! Indeed, at f/15 or greater, chromatic aberration is effectively eliminated in backyard-size refractors. Even at focal ratios down to f/10, chromatic aberration is frequently not too offensive if the objective elements are well made. High-quality achromatic refractors sold today range in size from 2.6 inches (66 mm) up to 11 inches (279 mm). Even though most of the chromatic aberration can be dealt with effectively, a lingering bluish or purplish glow frequently can be seen around brighter stars and planets. This glow is known as *residual chromatic aberration* and is almost always present in achromatic refractors.

How apparent this false color will be depends on the refractor's aperture and focal ratio; it will become more intrusive as aperture increases. In his book *Amateur Astronomer's Handbook,* author John Sidgwick offers the following formula as guidance:

$$\text{focal length} \geq 2.88D^2$$

What does this mean? Simply put, residual chromatic aberration will not be a factor as long as an achromatic refractor's focal length (abbreviated f.l.) is greater than 2.88 times its aperture (D, in inches) squared. So a 4-inch refractor will exhibit some false color around brighter objects if its focal length is less than 46 inches (f/11.5), while a 6-inch refractor will show some false color if shorter than 103 inches (f/17.3).

In *Telescope Optics: Evaluation and Design,* authors Harrie Rutten and Martin van Venrooij suggest an even more conservative value. They suggest that the minimum focal ratio be governed by the formula:

$$\text{f/ratio}_{(min)} \geq 0.122D$$

In this case, the aperture D is measured in millimeters. Therefore, a 4-inch (100-mm) achromatic refractor needs to be no less than f/12.2 to eliminate chromatic aberration, while a 6-inch must be at least f/18.3.

Chromatic aberration does more to damage image quality than just add false color. The light that should be contributing to the image that our eye perceives is instead being scattered over a wider area, causing image brightness to lag. Image contrast also suffers due to the scattering of light. Specifically, darker areas are washed out by overlapping fringes of brighter surrounding areas (as an example, think of the belts of Jupiter or the dark areas on the surface of Mars).

Why do I raise this issue? Recently, the telescope market has been flooded with 4- to 6-inch f/8 to f/10 achromatic refractors from China and Taiwan. Their promise of large apertures (for refractors, that is) and sharp images have attracted a wide following. Despite their low prices, these telescopes have proven to be surprisingly good. Yes, there is some false color evident around the Moon, planets, and brighter stars, but for the most part, it is not terribly objectionable. There are a few, however, that are not quite ready for prime time. Each is addressed in chapter 5, but don't skip ahead just yet—there is more to the story.

A strong point in favor of the refractor is that its aperture is a *clear* aperture. That is, nothing blocks any part of the light as it travels from the objective to the eyepiece. As you can tell from looking at the diagrams in chapter 2, this is not the case for reflectors and catadioptric instruments. As soon as a secondary mirror interferes with the path of the light, some loss of contrast and image degradation are inevitable.

In addition to sharp images, smaller achromatic refractors—that is, up to perhaps 5 inches aperture—are also famous for their portability and ruggedness. If constructed properly, a refractor should deliver years of service without its optics needing to be realigned (recollimated). The sealed-tube design means that dust and dirt are prevented from infiltrating the optical system, while contaminants can be kept off the objective's exterior simply by using a lens cap.

On the minus side of the achromatic refractor is its small aperture. Although this is of less concern to lunar, solar, and planetary observers, the instrument's small light-gathering area means that faint objects such as nebulae and galaxies will appear dimmer than they do in larger-but-cheaper reflectors. In addition, the long tubes of larger achromatic refractors can make them difficult to store and transport to dark, rural skies.

Another problem common to refractors is their inability to provide comfortable viewing angles at all elevations above the horizon. The long tube and short tripod typically provided can work against the observer in some cases. For instance, if the mounting is set at the right height to view near the zenith, the eyepiece will swing high off the ground as soon as the telescope is aimed toward the horizon. This disadvantage can be partially offset by inserting a star diagonal between the telescope's drawtube and eyepiece, but this has disadvantages of its own. Using either a mirror or a prism, a star diagonal bends the light either 45° or 90° to make viewing more comfortable. Most astronomical refractors include 90° star diagonals, because they are better for high-angle viewing, although some instruments that double as terrestrial spotting scopes include a 45° diagonal instead. The latter usually flips everything right side up but dims the view slightly in the process.

One of today's most popular styles of achromatic refractors are the so-called short tubes. Ranging in aperture from 66 mm to 150 mm, these telescopes all share low focal ratios (low f-ratios), short focal lengths, and compact tubes. Although larger telescopes can require considerable effort to set up, short-tubed refractors are ideal as "grab-and-go" instruments that can be taken outside at a moment's notice to enjoy a midweek clear night. They are designed primarily as wide-field instruments, ideal for scanning the Milky Way or enjoying the view of larger, brighter sky objects. Because they are not intended for high-magnification viewing, these instruments are not as suitable for viewing the planets as other telescope designs. Also keep in mind that, based on the previous discussion, these telescopes still obey the laws of optics and, as such, suffer from false color and residual chromatic aberration. Their performance cannot compare with apochromatic refractors that share similar apertures and focal lengths (described in the next section). Of course, they don't share their exorbitant price tags, either. Personally, I feel that short-tubed refractors are ideal *second* telescopes, but if this is your first and only, I would recommend other designs, unless you live on the top floor of a walk-up apartment building and need their portability.

Apochromatic Refractors. If *achromatic* means "no color," then *apochromatic* might be defined as "really, really no color, and I mean it this time." Even though an achromat brings two wavelengths of light from opposite ends of the spectrum to a common focus, it still leaves residual chromatic aberration along the optical axis. Although not as distracting as chromatic aberration from a single lens, residual chromatic aberration can still contaminate critical viewing and photography.

Apochromats—or apos, as they are affectionately known to many owners—further eliminate residual chromatic aberration, allowing manufacturers to increase aperture and decrease focal length. First popularized in the 1980s, apochromatic refractors use either two-, three-, or four-element objective lenses with one or more elements of an unusual glass type—often fluorite, SD (*special-dispersion*), or ED (*extra-low-dispersion*) glass. Apochromats minimize the dispersion of light by bringing three wavelengths of light (and, in theory, all others in between) to the same focus, reducing residual chromatic aberration dramatically, and thereby permitting shorter, more manageable focal lengths.

Much has been written about the pros and cons of fluorite (monocrystalline calcium fluoride) lenses. Fluorite has been called the most colorful mineral in the world, often displaying intense shades of purple, blue, green, yellow, reddish orange, pink, white, and brown. Unlike ordinary optical glass, which is made primarily of silica, fluorite is characterized by low-refraction and low-dispersion characteristics, which are perfect for suppressing chromatic aberration to undetectable levels. Unfortunately, it is very difficult to obtain fluorite large enough for a lens, and so telescope objectives (and camera lenses, etc.) are made from fluorite "grown" through an artificial crystallization technology.

In optics, the most popular myth about fluorite lenses is that they do not stand the test of time. Some "experts" claim that fluorite absorbs moisture and/or fractures more easily than other types of glass. This is simply not true under normal conditions. Fluorite objectives work very well and are just as durable as conventional lenses. Like all lenses, they will last a lifetime if given a little care.

While durability is not a problem, there are a couple of hitches to using fluorite refractors. One problem with fluorite that is not popularly known is its high thermal expansion. This means that the fluorite element will require more time to adjust to ambient temperature than a nonfluorite lens. Telescope optics change shape slightly as they cool or warm, and this characteristic is more pronounced in fluorite than in other materials.

Other hindrances are shared by all apos. Like most commercially sold achromatic refractors, apochromatic refractors are limited to smaller apertures, usually somewhere in the range of 3 to 7 inches or so. This limitation is not because of unleashed aberrations at larger apertures; it is simply a question of economics, which brings us to their second (and biggest) stumbling block: apochromats are not cheap! When we compare dollars per inch of aperture, it soon becomes apparent that apochromatic refractors are the most expensive telescopes. Given the same type of mounting and accessories, an apochromatic refractor can retail for more than twice the price of a comparable achromat. That's a big difference, but the difference in image quality can be even bigger. For a first telescope, an achromatic refractor is just fine, but if this is going to be your ultimate "dream" telescope, then you ought to consider an apochromat.

Reflecting Telescopes

Reflectors offer an alternative to the small apertures and big prices of refractors. Let's compare. Each of the two or more elements in a refractor's objective lens must be accurately figured and made of high-quality, homogeneous glass. By contrast, the single optical surfaces of a reflector's primary and secondary mirrors favor construction of large apertures at comparatively modest prices.

Another big advantage that a reflector enjoys over a refractor is complete freedom from chromatic aberration. (Chromatic aberration is a property of light refraction but not reflection.) This means that only the true colors of whatever a reflecting telescope is aiming at will come shining through. Of course, the eyepieces used to magnify the image for our eyes use lenses, so we are not completely out of the woods.

These two important pluses are frequently enough to sway amateurs in favor of a reflector. They feel that although there are drawbacks to the design, these are outweighed by the many strong points. But just what are the problems of reflecting telescopes? Some are peculiar to certain breeds, while others affect them all.

One shortcoming common to all telescopes of this genre is the simple fact that mirrors do not reflect all the light that strikes them. Just how much light is lost depends on the kind of reflective coating used. For instance, most telescope mirrors are coated with a thin layer of aluminum and overcoated with a clear layer of silicon monoxide for added protection against scratches and pitting. This combination reflects about 89% of visible light. This sounds good, but for primary and secondary mirrors with standard aluminum coatings, the combined reflectivity is only 79% of the light striking the primary! That's why special enhanced coatings have become popular in recent years. Enhanced coatings increase overall system reflectivity to between 90% and 96%. Some say, however, that enhanced coatings scatter light, in turn decreasing image contrast. If so, the decrease is minimal.

Reflectors also lose some light and, especially, image contrast because of obstruction by the secondary mirror. Just how much light is blocked and contrast lost depends on the size of the secondary, which in turn depends on the focal length of the primary mirror. Both must be matched correctly to maximize light throughput to the eyepiece. A too-large secondary mirror needlessly blocks a telescope's full aperture, thereby damaging image contrast, while a secondary mirror that is too small reduces the telescope's effective aperture. Generally speaking, the shorter the focal length of the primary, the larger its secondary must be to bounce all of the light toward the eyepiece.

Tradition has it, however, that central obstruction is expressed as a percentage of the aperture's diameter, not area. Therefore, a reflector with a 10% central obstruction by area is referred to as having a 16% obstruction by diameter. Both refer to an obstruction that measures 1.27 inches across. Primary mirrors with very fast focal ratios can have central obstructions in the neighborhood of 20%, even 25%. The only reflectors that do not suffer from this ailment are referred to as *off-axis* reflectors. Classic off-axis designs include the Herschelian and members of the Schiefspiegler family of instruments.

Since the idea of a telescope that uses mirrors to focus light was first conceived in 1663, different designs have come and gone. Today, two designs continue to stand the test of time: the Newtonian reflector and the Cassegrain reflector. Each reflector shall be examined separately.

Newtonian Reflectors. For sheer brute-force light-gathering ability, Newtonian reflectors (Figure 3.3) rate a best buy. No other type of telescope will give you as large an aperture for the money. Given a similar style mounting, you could buy an 8-inch Newtonian reflector for the same amount of money needed for a 4-inch achromatic refractor. Commercial models range from 3 inches to more than 2 feet in diameter, with focal ratios stretching from f/3.5 to roughly f/10. (Of course, not all apertures are available at all focal ratios. Can you imagine climbing more than 20 feet to the eyepiece of a 24-inch f/10!)

For the sake of discussion, I have divided Newtonians into two groups based on focal ratio. Those instruments with focal ratios less than f/6 have

Figure 3.3 *This 6-inch f/8 Newtonian reflector on a basic Dobsonian mount would be an ideal first instrument for any new stargazer.*

very deeply curved mirrors, and so are referred to here as *deep-dish* Newtonians. Reflectors with focal ratios of f/6 and greater will be called *shallow-dish* telescopes.

Pardon my bias, but shallow-dish reflectors have always been my favorite type of telescope. They are capable of delivering clear views of the Moon, the Sun, and other members of the solar system, as well as thousands of deep-sky objects. Shallow-dish reflectors with apertures between 3 inches (80 mm) and 8 inches (203 mm) are usually small enough to be moved from home to observing site and quickly set up with little trouble. Once the viewing starts, most amateurs happily find that looking through both the eyepiece and the small finderscope is effortless because the telescope's height closely matches the eye level of the observer.

A 6-inch f/8 Newtonian is still one of the best all-around telescopes for those new to astronomy. It is compact enough so as not to be a burden to transport and assemble, yet it is large enough to provide years of fascination—and all at a reasonable price. Better yet is an 8-inch f/6 to f/9 Newtonian. The increased aperture permits even finer views of nighttime targets. Keep in mind, however, that as aperture grows, so grows a telescope's size and weight. This means that unless you live in the country and can store your telescope where it is easily accessible, a shallow-dish reflector larger than an 8-inch might be difficult to manage.

Most experienced observers agree that shallow-dish reflectors are tough to beat. In fact, an optimized Newtonian reflector can deliver views of the Moon and the planets that eclipse those through a catadioptric telescope and compare favorably with a refractor of similar size, but at a fraction of the refractor's cost. Although the commercial telescope market now offers a wide range of superb refractors, it has yet to embrace the long-focus reflector fully.

Although shallow-dish reflectors provide fine views of the planets and the Moon, they grow to monumental lengths as their apertures increase (we oldtimers still remember 12.5-inch f/8s!). The eyepiece of an 18-inch f/4.5 reflector is about 78 inches off the ground, while an 18-inch f/8 would tower at nearly 12 feet! That's why most 10-inch and larger Newtonians have comparatively fast focal ratios, falling into the deep-dish category.

Many of these large-aperture Newtonians use thin-section primary mirrors. Traditionally, primary mirrors have a diameter-to-thickness ratio of 6:1. This means that a 12-inch mirror measures a full 2 inches thick. That is one heavy piece of glass to support. Thin-section mirrors cut this ratio to about 12:1, slashing the weight by 50%. This sounds good at first, but practice shows that large, thin mirrors tend to sag under their own weight (thicker mirrors are more rigid), thereby distorting the parabolic curve, when held in a conventional three-point mirror cell. To prevent mirror sag, a better support system is needed to support the primary at nine (or more) evenly spaced points across its back surface. These cells are frequently called *mirror flotation systems*, as they do not clamp down around the mirror's rim, thereby preventing possible edge distortions due to pinching. Many premium Dobsonian-mounted reflectors use primary cells that support the edge of their mirrors with woven straps or slings that resemble those used on folding lawn chairs. Others have more elaborate (though not necessarily better) support systems that hold the mirror securely, while not actually physically gripping it.

If big aperture means bright images, why not buy the biggest aperture available? Actually, there are several reasons not to do so. For one thing, unless they are made very well, Newtonians (especially those with short focal lengths) are susceptible to several optical irregularities, including spherical aberration, astigmatism, and coma.

Spherical aberration results when light rays from the edge of an improperly made mirror (or lens) focus to a slightly different point than those from the optic's center. In general, the faster a primary mirror's focal ratio, the greater the need for an accurate parabolic curve. Slower focal ratios are more forgiving. In fact, a parabolic mirror may be not required at all. Some Newtonian reflectors will work perfectly well with spherical mirrors, which are considerably easier (and, therefore, cheaper) to fabricate. In his classic work *How to Make a Telescope,* author Jean Texereau recommends a formula derived by André Couder that calculates the minimum focal lengths for given apertures that a spherical primary will satisfy Rayleigh's Criterion, and, therefore, produce satisfactory images. The formula reads:

$$f.l.^3 = 88.6A^4$$

where A is the telescope aperture and f.l. is the focal length, both expressed in inches. Table 3.1 tabulates the results for some common apertures.

As you can see, as aperture grows, the minimum acceptable focal length also grows, and quite quickly. Some manufacturers (notably from Taiwan and China) supply some of their telescopes with spherical mirrors. Although they seem to work well enough in general, they are all quite close to the minimum values cited in Table 3.1. I must also point out that many optical purists believe that Texereau's values are low, that they should all be increased by 30% or more to produce a well-functioning Newtonian telescope based on a spherical mirror. Therefore, approach with wariness any telescope with a focal ratio below the values in Table 3.1 that claims to use a spherical mirror.

Astigmatism is caused by a mirror or lens that was not symmetrically ground around its center or by misaligned or even pinched optics. Many reflectors show strong astigmatism simply because their secondary mirrors are not

Table 3.1 **Minimum Focal Length for Spherical Mirrors to Satisfy Rayleigh's Criterion**

Aperture		Minimum Focal Length (inches)		
Inches	Millimeters	Inches	Millimeters	Focal Ratio (f.l./A)
3.0	76	19	483	6.3
4.5	114	33	838	7.3
6.0	152	49	1245	8.2
8.0	203	71	1803	9.0
10.0	254	96	2438	9.6

CONSUMER CAVEAT: Aperture vs. Light Pollution

There is an old myth that if you observe from a light-polluted area, large apertures may produce results inferior to instruments with smaller apertures. This belief was based on the premise that because larger mirrors gather more starlight, they must also gather more sky glow, which must wash out the field of view. After conducting my own tests using an 18-inch Newtonian reflector under a variety of conditions, I found that this just is not the case. It is true, however, that larger apertures are more sensitive to heat currents and turbulent atmospheric conditions. There have been nights when I couldn't focus my 18-inch reflector, but images through my 4-inch refractor and 8-inch reflector were tack sharp, but that has nothing to do with ambient lighting.

aligned correctly. The result: elongated star images that appear to flip their orientation by 90° when the eyepiece is brought from one side of the focus point to the other. Coma, especially apparent in deep-dish telescopes, is evident when stars near the edge of the field of view distort into tiny blobs resembling comets, while stars at the center appear as sharp points. With any or all of these imperfections present, resolution suffers greatly. (Note that coma can be eliminated by using a coma corrector—see chapter 6.)

Both shallow-dish and deep-dish Newtonians share many other pitfalls as well. One of the more troublesome is that of all the different types of telescopes, Newtonians are among the most susceptible to collimation problems. If either or both of the mirrors are not aligned correctly, image quality will suffer greatly, possibly to the point of making the telescope unusable. Sadly, many commercial reflectors are delivered with misaligned mirrors. The new owner, perhaps not knowing better, immediately condemns his or her telescope's poor performance as a case of bad optics. In reality, however, the optics may be fine, just a little out of alignment. Chapter 9 details how to examine and adjust a telescope's collimation, a procedure that should be repeated frequently. The need for precise alignment grows more critical as the primary's focal ratio shrinks, making it especially important to double-check collimation at the start of every observing session if your telescope is f/6 or less. Some observers even recheck collimation during an observing session, particularly if they are using a large scope in an area subject to a major temperature drop during the night.

There are cases where no matter how well aligned the optics are, image quality is still lacking. Here, the fault undoubtedly lies with one or both of the mirrors themselves. As the saying goes, you get what you pay for, and that is as true with telescopes as with anything else. Clearly, manufacturers of low-cost models must cut their expenses somewhere in order to underbid their competition. These cuts are usually found in the nominal-quality standard equipment supplied with the instrument but sometimes may also affect optical testing procedures and quality control.

Cassegrain Reflectors. Although they have never attained the widespread following among amateur astronomers that Newtonians continue to enjoy, Cassegrain reflectors (Figure 3.4) have always been considered highly competent instruments. Cassegrains are characterized by long focal lengths, making them ideally suited for high-power, high-resolution applications such as solar, lunar, and planetary studies. Although Newtonians also may be constructed with these focal ratios, observers would have to go to great lengths to reach their eyepieces! Not so with the Cassegrain, where the eyepiece is conveniently located behind the primary mirror.

The Cassegrain's long focal length is created not by the primary mirror (which typically ranges around f/4) but by the convex, hyperbolic secondary mirror. As it reflects the light from the primary mirror back toward the eyepiece, the convex secondary mirror actually magnifies the image, thereby stretching the telescope's effective focal ratio to between f/10 and f/20. The net result is a telescope that is much more compact and easier to manage than a Newtonian of equivalent aperture and focal length.

Unfortunately, while the convex secondary mirror gives the Cassegrain its great compactness, it also contributes to many of the telescope's biggest disadvantages. First, in order to reflect all the light from the primary mirror back toward the eyepiece, the secondary mirror must be placed quite close to the primary. This forces its diameter to be noticeably larger than the flat diagonal of a Newtonian. With the secondary blocking more light, image sharpness and contrast suffer. Second, the convex secondary mirror combined with the short-focus primary mirror make optical alignment critical, while at the same time causing the telescope to be more difficult to collimate than a similar Newton-

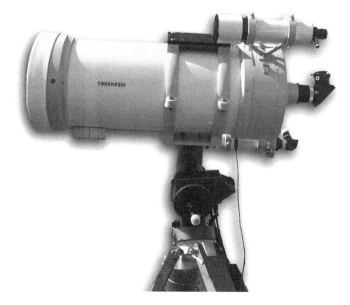

Figure 3.4 *Owning a state-of-the-art Cassegrain reflector like this one from Takahashi is a lifelong dream of many amateurs.*

ian. Finally, Cassegrains are prone to coma just like deep-dish Newtonians, making it impossible to achieve sharp focus around the edge of the field of view.

The eyepiece's placement along the optical axis also can work against the instrument's performance. The most obvious objection will become painfully apparent the first time an observer aims a Cassegrain near the zenith and tries to look through the eyepiece. That can be a real pain in the neck, although the use of a star diagonal will help alleviate the problem. Another problem that may not be quite as apparent involves a localized case of light pollution, caused by extraneous light passing around the secondary mirror and flooding the field of view. To combat this problem, manufacturers invariably install a long baffle tube protruding in front of the primary mirror. The size of the baffle is critical, because it must shield the eyepiece field from all sources of incidental light while allowing all of the light from the target to shine through.

As you thumb through the "Cassegrain Reflector" section in chapter 5, you'll see that there are three different types of reflectors listed. In addition to the Classical Cassegrain described here, there are the Ritchey-Chrétien and Dall-Kirkham variants. The Classical Cassegrain is beleaguered by coma toward the edge of the field, proving especially taxing for astrophotography. In his pioneering work during the 1910s, American optical designer George Ritchey, with assistance from French optical scientist Henri Chrétien, found that by changing the concave curve in the primary mirror from a paraboloid to a hyperboloid, and modifying the hyperbolic secondary mirror to suit, he was able to significantly reduce coma. This adjustment also let Ritchey lower the telescope's focal ratio, from around f/15 for a Classical Cassegrain, to around f/8 or f/9.

A scan through Appendix A shows that Ritchey-Chrétien reflectors are a rare breed among amateur astronomers, owing to the expense of creating their complex aspherical optics. In 2005, Meade Instruments introduced a line of RCX400 reflectors that are advertised as Ritchey-Chretiens. We'll talk more about these scopes in chapter 5, but for now, let's just say that they are not true Ritchey-Chrétiens. Instead, they are hybrid telescopes that combine a spherical primary mirror with a specially designed corrector plate that collectively warp the light in such a way that by the time it reaches the hyperbolic secondary mirror, it is effectively attenuated as it would be in a pure Ritchey-Chrétien (which does not require a corrector plate). You will find more on the RCX line under the "Exotic Catadioptrics" section in chapter 5.

Dall-Kirkham Cassegrains use modified ellipsoidal primary mirrors and spherical secondary mirrors. These curves prove far easier and cheaper to make than Ritchey-Chrétiens. If properly designed, a Dall-Kirkham reflector can yield wonderfully sharp images but only over a small field. Move too far off-axis and coma takes over. Just how strong the coma appears depends on the focal length of the primary mirror, but in general, the shorter the primary's focal length, the greater the off-axis coma. Overall focal ratios are usually no less than f/11, with the best results achieved as the focal ratio increases.

Although Cassegrains remain the most common type of telescope in professional observatories, their popularity among today's amateur astronomers is low. So it should come as no surprise to find that so few companies offer complete Cassegrain systems for the hobbyist.

Catadioptric Telescopes

Most amateur astronomers looking for a compact telescope now favor hybrid designs that combine some of the best attributes of both the reflector and the refractor, creating a completely different kind of beast: the catadioptric. Catadioptric telescopes (also known as *compound telescopes*) are comparative Johnny-come-latelies on the amateur scene. Yet in only a few decades, they have developed a loyal following of backyard astronomers who staunchly defend them as the ultimate telescopes.

Most lovers of catadioptrics fall into one, two, or possibly all three of the following categories:

1. They are urban or suburban astronomers who prefer to travel to remote observing sites.
2. They enjoy astrophotography (or aspire to at least try it).
3. They just love gadgets.

If any or all of these profiles fit you, then a catadioptric telescope just might be your perfect telescope.

Catadioptric telescopes for visual use may be constructed in either Newtonian or Cassegrain configurations. Four catadioptric designs have made a lasting impact on the world of amateur astronomy: the Schmidt-Cassegrain, the Schmidt-Newtonian, the Maksutov-Cassegrain, and the Maksutov-Newtonian. Cassegrain-style catadioptric telescopes are immediately recognizable by the position of their focusers behind their primary mirrors, like a Cassegrain reflector. Newtonian-based catadioptrics place their focusers toward the front of the tube, like traditional Newtonian reflectors. For our purposes here, the discussion will be confined to these designs.

Schmidt-Cassegrain Telescopes. Take a look through just about any astronomy magazine published just about anywhere in the world and you are bound to find at least one advertisement for a Schmidt-Cassegrain telescope (also known as a Schmidt-Cas or an SCT). As your eyes digest the ads chock-full of mouth-watering celestial photographs that have been taken through these instruments, you suddenly get the irresistible urge to run right out and buy one. You would not be the first person to find these telescopes so appealing. Although Schmidt-Cassegrain telescopes are available in apertures from 5 inches to 16 inches, the favorite size is the 8-inch model.

Is the Schmidt-Cassegrain (Figure 3.5) the perfect telescope? Admittedly, it can be attractive. By far, its greatest asset has to be the compact design. No other telescope can fit as large an aperture and as long a focal length into such a short tube assembly as a Schmidt-Cas (they are usually only about twice as long as the aperture). If storing and transporting your telescope are big concerns, then this will be an especially important benefit.

Here is another point in their favor. Nothing can end an observing session quicker than a fatigued observer. For instance, using a Newtonian reflector, with its eyepiece positioned at the front end of the tube, usually means that the observer must remain standing—sometimes even on a stool or a ladder—just to take a peek. Compare this to a Schmidt-Cassegrain telescope, where the

Figure 3.5 *Schmidt-Cassegrain telescopes, like the Celestron CPC shown here, remain very popular among today's amateur astronomers. Photo by Kevin Dixon.*

observer can enjoy a comfortable, seated viewing of just about all points in the sky. Your back and legs will certainly thank you! The eyepiece is only difficult to reach when the telescope is aimed close to the zenith. As with a refractor and a Cassegrain, a right-angle star diagonal placed in the focuser before the eyepiece will help a little, but these have their drawbacks, too. Most annoying of all is that a diagonal will flip everything right to left, creating a mirror image that makes the view difficult to compare with star charts.

Another big plus of the Schmidt-Cassegrain is its sealed tube. The front corrector plate acts as a shield to keep dirt, dust, and other foreign contaminants off the primary and secondary mirrors. This is especially handy if you travel a lot with your telescope and are constantly taking it in and out of its carrying case. A sealed tube also can help to extend the useful life of the mirrors' aluminized coatings by sealing well against the elements. (Always make sure the mirrors are dry before storing the telescope to prevent the onset of mold and mildew.)

Although the corrector protects the two mirrors against dust contamination, it slows a telescope from reaching thermal equilibrium with the night air and can also act as a dew collector. Depending on local weather conditions, correctors can fog over in a matter of hours or even minutes, or they may remain clear all night. To help fight the onslaught of dew, manufacturers sell *dew caps* or *dew shields*. Dew caps are a must-have accessory for all Cassegrain-based catadioptrics. (Consult chapter 7 for more information.)

Sounds good so far, right? Well, what about the telescope's optical performance? Here is where the Schmidt-Cassegrain telescope begins to teeter.

Because of the comparatively large secondary mirrors required to reflect light back toward their eyepieces, SCTs produce images that have less contrast than some other telescope designs of the same aperture. A typical 8-inch SCT has a secondary mirror mounting that measures about 2.75 inches across. That's a whopping 34% central obstruction! This can prove especially critical when searching for fine planetary detail or when hunting for faint deep-sky objects at the threshold of visibility. The enhanced optical coatings available on all popular models improve light transmission and reduce scattering. These coatings can make the difference between seeing a marginally visible object and missing it, and they are an absolute must for all Schmidt-Cassegrains. Still, planet watching, which demands sharp, high-contrast images, suffers in Schmidt-Cassegrains.

The image sharpness in a Schmidt-Cassegrain is often not as precise as that obtained through a refractor or a reflector. Perhaps this is due to the loss of contrast mentioned earlier or because of optical misalignment, another frequent problem of the Schmidt-Cassegrain. In any telescope, optical misalignment will play havoc with image quality. What should you do if the optics of a Schmidt-Cassegrain are out of alignment? If only the secondary mirror is off, you may follow the procedure outlined in chapter 9, but if the primary mirror is out, you must return the telescope to the manufacturer for service. Remember—although just about anyone can take a telescope apart, not everyone can put it back together!

In general, a Schmidt-Cassegrain telescope represents a very good value for the money. In fact, taking inflation into account, an SCT costs less today than it did when first introduced in 1970. Schmidt-Cassegrains offer acceptable views of the Sun, the Moon, the planets, and deep-sky objects, and they work reasonably well for astrophotography (although stars often appear bloated in long exposures). But for exacting views of celestial objects, SCTs are outperformed by other types of telescopes. For observations of solar system members, it is hard to beat a shallow-dish Newtonian (especially f/8 or higher) or a good refractor, while the myriad faint deep-sky objects are seen best with large-aperture, deep-dish Newtonians. Schmidt-Cassegrains are jack-of-all-trades-but-master-of-none telescopes.

Schmidt-Newtonian Telescopes. These catadioptric telescopes look just like conventional Newtonians at first glance. But looking a little more closely shows that Schmidt-Newtonian telescopes use spherical primaries and front corrector plates, rather than the paraboloidal primaries found in high-quality traditional Newtonian telescopes, to gather light and bring it into sharp focus. The result is a fast-focal-ratio telescope that shows much less astigmatism and edge-of-field coma than a traditional Newtonian of the same aperture and focal length. This combination makes this model ideal for wide-field, through-the-telescope astrophotography and for visual observing.

Many of the advantages and disadvantages of Schmidt-Cassegrains translate directly to Schmidt-Newtonians. Like an SCT, the sealed tube has the advantage of keeping dust and dirt off the optics, which is always a good thing.

At the same time, however, it also slows cool-down time when the telescope is first brought outside from a warm house. The corrector plate also necessitates some form of active dew prevention so that it doesn't dew over on damp evenings.

Optically, Schmidt-Newtonians are designed to produce wide fields of view. As a result, these instruments favor low- to medium-power deep-sky observing, comet hunting, and astrophotography. High-magnification applications, such as detailed study of the Moon and the planets, are less satisfying with this type of instrument. The optional enhanced coatings (strongly recommended) produce brilliant images, but contrast is lessened by the large central obstructions typically required of the design. But keep in mind that collimation, especially of the secondary mirror, is critical in a Schmidt-Newtonian. If the collimation is not spot-on, astigmatism and coma will soften star images and make for less-than-satisfying views.

Maksutov Telescopes.   The final stop on our telescope world tour is the Maksutov catadioptric. Many people feel that Maksutovs are the finest telescopes of all. And why not? Maks provide views of the Moon, the Sun, and the planets that rival those of the best refractors and long-focus reflectors, and they are easily adaptable for astrophotography (although their high focal ratios mean longer exposures than faster telescopes of similar aperture). Smaller models are also a breeze to travel with.

Maksutov telescopes come in both Newtonian and Cassegrain configurations, the latter being more common. Typically, Maksutov-Cassegrains have focal ratios greater than f/10, and so are more appropriate for high-powered planet watching as well as hunting for small deep-sky objects. Like SCTs, their compact design makes Maksutov-Cassegrains very popular among observers who want the convenience of a short-tubed instrument capable of delivering refractor-sharp views of the Moon and the planets. Like an SCT, the comfort of viewing from a seated position is also likely behind their increasing popularity.

Maksutov-Newtonians, with their faster focal ratios typically around f/6, are excellent general-purpose instruments. While they excel at deep-sky observing as well as guided astrophotography, make no mistake—a well-made Mak-Newt will also show stunning planetary views.

Is there a downside to the Maksutov? Unfortunately, yes. Because of the thick corrector plate in front, Maks take longer to acclimate to the cool night air than any other common type of telescope. A telescope will not perform at its best until it has equalized with the outdoor air temperature. To help speed things along, some companies install fans, but the optics are still slow to reach thermal equilibrium. Some Maks—notably the Questar instruments—also cost more per inch of aperture than nearly any other type of telescope. Also, there is the problem of the corrector plate, which like any exposed optic, can act as a dew magnet if it is not protected by some sort of anti-dew system.

To help digest all this information, take a look at Table 3.2, which summarizes all the pros and cons mentioned here. Use it to compare the good points and the bad among the more popular types of telescopes sold today.

Table 3.2 **Telescope Point–Counterpoint: A Summary**

	Point	Counterpoint
A. Binoculars Typically 1.4″ (35 mm) to 3.9″ (100 mm) apertures	• Most are comparatively inexpensive • Extremely portable • Wide field makes them ideal for scanning	• Low power makes them unsuitable for objects requiring high magnification • Small aperture restricts magnitude limit
B. Achromatic Refractors Typically 2.6″ (66 mm) to 6″ (152 mm), f/5 and above	• Portable in smaller apertures • Sharp images • Moderate price vs. aperture • Good for Moon, Sun, planets, double stars, and bright astrophotography	• Small apertures • Mounts may be shaky (attention, department-store shoppers!) • Potential for chromatic aberration
C. Apochromatic Refractors Typically 2.4″ (60 mm) to 7″ (178 mm) aperture, f/5 and above	• Portable in smaller apertures • Very sharp, contrasty images • Excellent for Moon, Sun, planets, double stars, and astrophotography	• Very high cost vs. aperture • Small apertures (for the price) limit magnitude penetration
D. Shallow-Dish Newtonian Reflectors Typically 4″ (102 mm) to 8″ (203 mm) aperture, f/6 and above	• Low cost vs. aperture • Easy to collimate • Very good for Moon, Sun, planets (especially f/8 and above), deep-sky objects, and astrophotography	• Bulky/heavy, over 8″ • Collimation must be checked often • Open tube end permits dirt and dust contamination • Excessively long focal lengths may require a ladder to reach the eyepiece • Dobsonian mounts do not track the stars (some have this option, however)
E. Deep-Dish Newtonian Reflectors Typically 4″ (102 mm) and larger, below f/6	• Low cost vs. aperture • Wide fields of view • Large apertures mean maximum magnitude penetration • Easy to collimate • Excellent for both bright and faint deep-sky objects (solar system objects okay, but usually inferior to same size shallow-dish reflectors though not always)	• Larger apertures can be heavy/bulky, especially those with solid tubes • Even with fast focal ratio, larger apertures may still require a ladder to reach the eyepiece • Very low-cost versions may indicate compromise in quality • Dobsonian mounts do not track the stars (some have this option, however) • Collimation is critical (must be checked before each use) • Open tube ends permit dirt and dust contamination • Large apertures very sensitive to seeing conditions

	Point	Counterpoint
F. Cassegrain Reflector Typically 6″ (152 mm) and larger, f/12 and above	• Portability • Convenient eyepiece position • Good for Moon, Sun, planets, and smaller deep-sky objects (e.g., double stars, planetary nebulae)	• Large secondary mirror • Moderate-to-high price vs. aperture • Narrow fields • Offered by few companies
G. Schmidt-Cassegrain (Catadioptric) Typically 5″ (127 mm) to 16″ (406 mm) apertures, f/10	• Moderate cost vs. aperture • Portability • Convenient eyepiece position • Wide range of accessories • Easily adaptable to astrophotography • Good for viewing Moon, Sun, planets, bright deep-sky objects, and astrophotography	• Large secondary mirror reduces contrast • Image quality not as good as refractors or reflectors • Bloated star images in long exposure photos • Slow f-ratio means longer exposures than faster Newtonians and refractors • Corrector plate prone to dewing over • Potentially difficult to find objects, especially near the zenith (fork mounts only) • Mirror shift during focusing can cause images to jump
H. Schmidt-Newtonian (Catadioptric) Typically 6″ (152 mm) to 10″ (254 mm), f/4 to f/5	• Moderate cost vs. aperture Portability • Fast focal ratio means shorter exposures during astrophotography • Wide field of view	• Large secondary mirror reduces contrast • Image quality not as good as refractors or reflectors • Corrector plate prone to dewing over • Collimation is critical
H. Maksutov-Newtonian (Catadioptric) Typically 5″ (127 mm) to 10″ (254 mm), typically f/6	• Sharp images, approaching refractor quality • Convenient size for quick setup • Excellent for Moon, planets, Sun, bright deep-sky objects	• Corrector plates prone to dewing over • Costlier than like-size Newtonians
I. Maksutov-Cassegrain (Catadioptric) Typically 3.5″ (90 mm) and larger, f/12 and higher	• Sharp images, approaching refractor quality • Convenient eyepiece position • Easily adaptable to astrophotography • Excellent for Moon, planets, Sun; good for bright, small deep-sky objects	• Some are very expensive vs. aperture • Some have finderscopes that are inconveniently positioned • Some models use threaded eyepieces, making an adapter necessary for other brand oculars • Difficult to collimate, if needed • Slow f-ratio means longer exposures and narrow fields of coverage

Support Your Local Telescope

The telescope itself is only half of the story. Can you imagine trying to hold a telescope by hand while struggling to look through it? If the instrument's weight did not get you first, surely every little shake would be magnified into a visual earthquake! To use a true astronomical telescope, we have no choice but to support it on some kind of external mounting. For small spotting scopes, this might be a simple tabletop tripod, while the most elaborate telescopes come equipped with equally elaborate support systems.

Selecting the proper mount is just as important as picking the right telescope. A good mount must be strong enough to support the telescope's weight while minimizing any vibrations induced by the observer (such as during focusing) and the environment (from wind gusts or even nearby road traffic). Indeed, without a sturdy mount to support the telescope, even the finest instrument will produce only blurry, wobbly images. A mounting also must provide smooth motions when moving the telescope from one object to the next and allow easy access to any part of the sky.

All telescope mounting systems fall into one of two broad categories based on their construction: altitude-azimuth or equatorial. Figure 3.6 shows examples of both.

Altitude-Azimuth Mounts

Altitude-azimuth mounts (frequently shortened to either alt-azimuth or alt-az) are the simplest types of telescope support available. As their name implies, alt-az systems move both in azimuth (horizontally) and in altitude (vertically). All camera tripod heads, for instance, are alt-az systems.

Figure 3.6 *Three of today's most popular telescope mounts, from left to right: a Newtonian reflector on a Dobsonian altitude-azimuth mount; a Schmidt-Cassegrain telescope on a German equatorial mount with computerized GoTo control; and a Schmidt-Cassegrain telescope on a computerized, fork-style altitude-azimuth mount.*

This type of mounting is most frequently supplied with smaller, less-expensive refractors and Newtonian reflectors. It allows the instrument to be aimed with ease toward any part of the sky. Once pointed in the proper direction, the mount's two axes (that is, the altitude axis and the azimuth axis) can be locked in place. Some alt-azimuth mounts are outfitted with slow-motion controls, one for each axis. By twisting one or both of the control knobs, the observer can fine-tune the telescope's aim as well as keep up with the Earth's rotation.

In the past twenty years, a variation of the alt-azimuth mount called the Dobsonian has become extremely popular among hobbyists. Dobsonian mounts are named for John Dobson, an amateur telescope maker and astronomy popularizer from the San Francisco area. Back in the 1970s, Dobson began to build large-aperture Newtonian reflectors to see the "real universe." With the optical assembly complete, he faced the difficult challenge of designing a mount strong enough to support the instrument's girth yet simple enough to be constructed from common materials using hand tools. What resulted was an off-shoot of the alt-az mount. Using plywood and some basic materials and tools, Dobson devised a telescope mount that was capable of holding steady his huge Newtonians. Plywood is an ideal material for a telescope mount, because it has incredible strength as well as a terrific vibration-damping ability. Formica and Teflon together create smooth bearing surfaces, allowing the telescope to flow across the sky. No wonder Dobsonian mounts have become so popular.

Although both traditional alt-az mounts as well as Dobsonian mounts are wonderfully simple to use, they also possess some drawbacks. Perhaps the most obvious is caused not by the mount but by the Earth itself! If an alt-azimuth mount is used to support a terrestrial spotting scope, then the fact that it moves horizontally and vertically plays in its favor. However, the sky is always moving due to the Earth's rotation. Therefore, to study or photograph celestial objects for extended periods of time without interruption, our telescopes must move right along with them. If we were located exactly at either the North or South Pole, the stars would appear to trace arcs parallel to the horizon as the Earth spins. Tracking the stars would simply require aiming the telescope at the desired target, locking the altitude axis in place, and slowly moving the azimuth axis with the sky.

Once we leave the poles, the tilt of the Earth's axis causes the stars to follow long, curved paths in the sky, causing most to rise diagonally in the east and to set diagonally in the west. With an alt-azimuth mount, it becomes necessary to nudge the telescope both horizontally and vertically in a steplike fashion to keep up with the sky. This is decidedly less convenient than the single motion enjoyed by an equatorial mount, the second way of supporting a telescope.

Equatorial Mounts

"If you can't raise the bridge, lower the river," so the saying goes. This is the philosophy of the equatorial mount. Since nothing can be done about the stars' apparent motion across the sky, the telescope's mounting method must accommodate it. An equatorial mount may be thought of as an altitude-azimuth mount tilted at an angle that matches your observing site's latitude.

Like its simpler sibling, an equatorial mount is made up of two perpendicular axes, but instead of referring to them as altitude and azimuth, we use the terms *right-ascension* (or *polar*) *axis* and *declination axis.* In order for an equatorial mount to track the stars properly, its polar axis must be aligned with the celestial pole, a procedure detailed in chapter 10.

There are many benefits to using an equatorial mount. The greatest is the ability to attach a motor drive onto the right-ascension axis so the telescope follows the sky automatically and (almost) effortlessly. Another point favoring an equatorial mount is that once aligned to the pole, it can make finding objects in the sky much easier by simplifying hopping from one object to the next using a star chart, as well as by permitting the use of setting circles.

On the minus side of equatorial mounts, however, is that they are almost always larger, heavier, more expensive, and more cumbersome than alt-azimuth mounts. This is why the simple Dobsonian alt-az design is so popular for supporting large Newtonians. An equatorial large enough to support, say, a 12- to 15-inch f/4 reflector probably would tip the scales at close to 200 pounds, while a plywood Dobsonian mount would weigh under 50 pounds.

Just as there are many kinds of telescopes, so, too, are there many kinds of equatorial mounts. Some are quite extravagant, while others are simple to use and understand. We will examine the two most common styles.

German Equatorial Mounts. For years, this was the most popular type of mount among amateur astronomers. The German equatorial is shaped like a tilted letter T with the polar axis representing the long leg and the declination axis marking the letter's crossbar. The telescope is mounted to one end of the cross bar, while a weight for counterbalance is secured to the opposite end.

The simplicity and sturdiness of German equatorials have made them a perennial favorite for supporting all types of telescopes. They allow free access to just about any part of the sky (as with all equatorial mounts, things get a little tough around the poles), are easily outfitted with a motor-driven clock

---

CONSUMER CAVEAT: Rap Test

Regardless of the type used, a mounting must support its telescope steadily to be of any value. To test a mount's rigidity, I recommend doing the rap test. Hit the telescope tube toward the top end with the ball of your palm while looking through the telescope at a target, either terrestrial or celestial. Don't really whack the telescope, but hit it with just enough force to make it shake noticeably. Then, count how long it takes for the image to settle down. To me, less than 3 seconds is excellent, while 3 to 5 seconds is good. I consider 5 to 10 seconds only fair, while anything more than 10 seconds is poor.

drive, and may be held by either a tripod or a pedestal base. To help make polar alignment easier, some German equatorials have small alignment scopes built right into their right ascension axes—a big hit among astrophotographers.

Of course, as with everything in life, there are some flaws in the German mount as well. One strike against the design is that it cannot sweep continuously from east to west. Instead, when the telescope nears the meridian, the user must move it away from whatever was in view, swing the instrument around to the other side of the mounting, and re-aim it back to the target. Inconvenient as this is for the visual observer, it is disastrous for the astrophotographer caught in the middle of a long exposure because as the telescope is spun around, the orientation of the field of view is also rotated.

A second burden to the German-style mount is a heavy one to bear: they can weigh a lot, especially for telescopes larger than 8 inches. Most of their weight comes from the axes (typically made of solid steel) as well as the counterweight used to offset the telescope. At the same time, I must quickly point out that weight does not necessarily beget sturdiness. For instance, some heavy German equatorial mounts are so poorly designed that they could not even steadily support telescopes half as large as those they are sold with. (More about checking a mount's rigidity later in this chapter.)

If you are looking at a telescope that comes with a German mount, pay especially close attention to the diameter of the right-ascension and declination shafts. On well-designed mounts, each shaft is at least $1/8$ of the telescope's aperture. For additional solidity, better mounts use ball bearings in their axes, while inferior versions use cheap bushings.

Finally, if you must travel with your telescope to a dark-sky site, moving a large German equatorial mount can be a tiring exercise. First, the telescope must be disconnected from the mounting. Next, depending on how heavy everything is, the equatorial mount (or *head*) might need to be separated from its tripod or pedestal. Last, all three pieces (along with all eyepieces, charts, and other accessories) must be carefully stored away. The reverse sequence occurs when setting up at the site, and the whole thing happens all over again when it is time to go home.

Fork Equatorial Mounts. While German mounts are preferred for telescopes with long tubes, fork equatorial mounts are popular with compact instruments, such as Schmidt-Cassegrains and Maksutov-Cassegrains. Fork mounts support their telescopes on bearings set between two fork *tines,* or *prongs,* that permit full movement in declination. The forks typically extend from a rotatable circular base which, in turn, acts as the right ascension axis when tilted at the proper angle.

Perhaps the biggest plus to the fork mount is its light weight. Unlike its Bavarian cousin, a fork equatorial usually does not require counterweighting to achieve balance. Instead, the telescope is balanced by placing its center of gravity within the prongs, like a seesaw. This is an especially nice feature for Cassegrain-style telescopes, because it permits convenient access to the eyepiece regardless of where the telescope is aimed—that is, except when it is aimed near the celestial pole. In this position, the eyepiece can be notoriously difficult to reach, usually requiring the observer to lean over the mounting without bumping into it.

The fork mounts that come with most SCTs and Maksutov-Cassegrain telescopes are designed for maximum convenience and portability. They are compact enough to remain attached to their telescopes, and together fit into their cases for easy transporting. Once at the observing site, the fork quickly secures to its tripod by using thumbscrews. It can't get much better than that, especially when compared with the German alternative.

Fork mounts quickly become impractical, however, for long-tubed telescopes such as Newtonians and refractors. In order for a fork-mounted telescope to be able to point toward any spot in the sky, the mount's two prongs must be long enough to let the ends of the instrument swing through without colliding with any other part of the mounting. To satisfy this requirement, the prongs must grow in length as the telescope becomes longer. At the same time, the fork tines also must grow in girth to maintain rigidity. Otherwise, if the fork arms are undersized, they will transmit every little vibration to the telescope. (In those cases, maybe they ought to be called tuning fork mounts!)

One way around the need for longer fork prongs is to shift the telescope's center of gravity by adding counterweights onto the tube. Either way, however, the total weight will increase. In fact, in the end the fork-mounted telescope might weigh more than if it was held on an equally strong German equatorial.

GoTo Mounts. Technically, the GoTo mount is not a separate mounting design, but rather, the marriage of a telescope mount to a built-in computer that can move the telescope to a selected object. Many GoTo mounts are actually altitude-azimuth fork mounts, making polar aligning unnecessary. (To use altitude-azimuth GoTo mounts for long-exposure guided astrophotography, an equatorial wedge must be used to tilt the telescope to match the observing site's latitude and the mount polar aligned as with a standard equatorial mount.) Instead, the onboard computer compensates automatically for the observer's location and then controls the mount's drive motors to track the sky accurately.

How does this magic work? In general, the user simply enters the current time and date into a small, handheld control box and selects his or her location from the computer's city database (those with built-in GPS technology do these automatically), and then hits the align button. The telescope's computerized aiming system then slews toward two of its memorized alignment stars. It should stop in the general location of the first alignment star, but it is up to the user to fine-tune the instrument by using the hand controller until the first alignment star is centered in the field of view. Then the align button is pressed to store that position. After repeating the process with the second alignment star, the LCD displays whether the alignment was successful or not. If so, the telescope is initialized and ready to go for the evening; if not, the process must be repeated.

There is no denying that GoTo telescopes have great appeal. Many who purchase one of the cheaper GoTo instruments that seem to be proliferating these days, however, find them to be disappointing. Inexpensive GoTo telescopes are often difficult to set up, especially when it comes to initializing the built-in brain. Here's an all-too-typical experience. A few years ago, a young girl in one of my undergraduate astronomy classes told me she had been given an inexpensive GoTo telescope as a gift. She was a smart girl, getting As in all her classes, but she just couldn't get the telescope's GoTo to go anywhere. After we were finally able to get it working, she was disappointed by two things. First, the drive system drained a set of batteries after only a few hours, which meant she would either have to replace them after an evening out or use an external power source. Either alternative compromised the telescope's attractiveness. Second, she eventually realized the smaller aperture's limitations.

Bear in mind that for a given investment, you can choose a larger, non-GoTo telescope for the same price as a small, computerized instrument. In the latter, you are dividing your dollar, with part going for the optics and part for the electronics. This is a compromise at best. Weigh all your options before making a decision, but I think you know which I would recommend.

Your Tel-O-Scope

Astrologers have their horoscopes, but amateur astronomers have "tel-o-scopes." What telescope is in your future? Perhaps the stars can tell us. To be perfect partners, you and your telescope have to be matched both physically and emotionally. Without this spiritual link, dire consequences can result. Here are eight questions to help you focus in on which telescope is best matched to your profile. Answer each question as honestly and realistically as you can; there is no right or wrong answer.

Once completed, add up the scores that are listed in brackets after each response. By comparing your total score with those found in Table 3.3 at the end of your tel-o-scope session, you will get a good idea of which telescopes are best suited for your needs, but use the results only as a guide, not as an absolute. And no peeking at your neighbor's answers!

1. Which statement best describes your level of astronomical expertise?
   a. Casual observer [1]
   b. Enthusiastic beginner [4]
   c. Intermediate space cadet [6]
   d. Advanced amateur [10]
2. Will this be your first telescope or binoculars?
   a. Yes [4]
   b. No [8]
3. If not, what other instrument(s) do you already own? (If you own more than one, select only the one that you use most often.)
   a. Binoculars [1]
   b. Achromatic refractor [4]
   c. Apochromatic refractor [11]
   d. Newtonian reflector (4" to 5" aperture) [4]
   e. Newtonian reflector (6" to 10" aperture on equatorial mount) [6]

f. Newtonian reflector (>10" aperture on equatorial mount) [11]
g. Newtonian reflector (<12" aperture on Dobsonian mount) [5]
h. Newtonian reflector (12" to 15" aperture on Dobsonian mount) [7]
i. Newtonian reflector (16"+ aperture on Dobsonian mount) [10]
j. Cassegrain reflector (<10" aperture) [7]
k. Cassegrain reflector (10" and larger aperture) [11]
l. Schmidt-Cassegrain catadioptric [9]
m. Inexpensive Maksutov catadioptric (e.g., Meade ETX) [5]
n. Expensive Maksutov catadioptric (e.g., Questar) [11]

4. What do you want to use the telescope for primarily? Choose only one.
a. Casual scan of the sky [1]
b. Informal lunar/solar/planetary observing [4]
c. Estimating magnitudes of variable stars [6]
d. Comet hunting [5]
e. Detailed study of solar system objects [10]
f. Bright deep-sky objects (star clusters, nebulae, galaxies) [6]
g. Faint deep-sky objects [8]
h. Astrophotography of bright objects (Moon, Sun, etc.) [4]
i. Astrophotography of faint objects (deep-sky, etc.) [9]

5. How much money can you afford to spend on this telescope? (Be conservative; remember that you might want to buy some accessories for it—see chapters 6 and 7.)
a. $100 or less [1]
b. $200 to $400 [3]
c. $400 to $800 [5]
d. $800 to $1,200 [7]
e. $1,200 to $1,600 [9]
f. $1,600 to $2,000 [11]
g. $2,000 to $3,000 [13]
h. As much as it takes (Are you looking to adopt an older son? I am available.) [15]

6. Which of the following scenarios best describes your particular situation?
a. I live in the city and will use my telescope in the city. [1]
b. I live in the city but will use my telescope in the suburbs. [6]
c. I live in the city but will use my telescope in the country. [8]
d. I live in the suburbs and will use my telescope in the suburbs. [9]
e. I live in the suburbs but will use my telescope in the country. [10]
f. I live in the country and will use my telescope in the country. [12]
g. I live in the country but will use my telescope in the city. [–5] (Just kidding!)

7. Which of the following best describes your observing site?
a. A beach [4]
b. A rural open field or meadow away from any body of water [8]
c. A suburban park near a lake or a river [5]
d. A suburban site with a few trees and a few lights but away from any water [6]
e. An urban yard with a few trees and a lot of lights [4]
f. A rural hilltop far from all civilization [13]
g. A desert [11]
h. A rural yard with a few trees and no lights [9]

8. Where will you store your telescope?
   a. In a room on the ground floor of my house [6]
   b. In a room in my ground-floor apartment/co-op/condominium [5]
   c. Upstairs [4]
   d. In my (sometimes damp) basement [4]
   e. In a closet on the ground floor [5]
   f. I'm not sure; I have very little extra room [4]
   g. In a garden/tool shed outside [9]
   h. In my garage (protected from car exhaust and other potential damage) [7]
   i. In an observatory [11]

Now add up your results and compare them to Table 3.3.

Table 3.3 **The Results Are In . . .**

If your score is between . . .	Then a good telescope for you might be . . .
17 and 30	Binoculars
25 and 45	Achromatic refractor (3.1-inch aperture on a sturdy alt-azimuth mount)
35 and 55	"Econo-Dob" Newtonian reflector (4- to 8-inch aperture on a Dobsonian mount) -OR- Achromatic refractor (3- to 4-inch aperture on an alt-azimuth or an equatorial mount) -OR- "Econo-Mak" (e.g., Meade ETX series) for those who absolutely must have a small telescope (e.g., for airline travel)
45 and 65	Newtonian reflector (6- to 8-inch aperture on an equatorial mount) -OR- Achromatic refractor (4- to 5-inch aperture on an alt-azimuth or an equatorial mount) -OR- Maksutov (4- to 6-inch aperture, mount often sold separately)
55 and 75	Newtonian reflector (12- to 15-inch aperture on a Dobsonian mount) -OR- Achromatic refractor (5- to 6-inch aperture on an equatorial mount) -OR- Small apochromatic refractor (2.7- to 3.1-inch aperture) if this will be a second telescope -OR- Schmidt-Cassegrain (8-inch aperture)—for astronomers who own small cars and must travel to dark skies -OR- Maksutov (6-inch or larger aperture, mount often sold separately)

(continued)

Table 3.3 **(continued)**

If your score is between . . .	Then a good telescope for you might be . . .
65 and 85	Schmidt-Cassegrain (8- to 11-inch aperture)—for astronomers who own small cars and must travel to dark skies -OR- Newtonian reflector (10-inch or larger aperture on an equatorial mount)
75 and 90	Cassegrain reflector (8-inch or larger aperture) -OR- Apochromatic refractor (4-inch or larger aperture) -OR- Newtonian reflector (16-inch or larger aperture on a Dobsonian or an equatorial mount) -OR- Maksutov (premium 3.5-inch or larger aperture)

The results of this test are based on buying a new telescope from a retail outlet and should be used only as a guide. The range of choices for each score is purposely broad to give the reader the greatest selection. For instance, if your score was 58, then you can select from either the 45 to 65 or the 55 to 75 score ranges. These indicate that good telescopes for you to consider include a 4- to 5-inch achromatic refractor or a 6- to 8-inch Newtonian reflector on an equatorial mount, a 12- to 15-inch Newtonian on a Dobsonian mount, or an 8-inch SCT. You must then look at your particular situation to see which type is best for you based on what you have read up to now. If astrophotography is an interest, then a good choice would be an 8-inch SCT. If your primary interest is observing faint deep-sky objects, then I would suggest the Dobsonian-mounted Newtonian, while someone interested in viewing the planets would do better with the refractor. If money (or, rather, lack of) is your strongest concern, then choose either the 6- or 8-inch Newtonian. Of course, you might also consider selecting from a lower-score category if one of those suggestions better fits your needs.

Some readers may find the final result inconsistent with their responses. For instance, if you answer that you live and observe in the country, you already own a Schmidt-Cassegrain telescope, and you want a telescope to photograph faint deep-sky objects but are willing to spend $100 or less, then your total score would correspond to, perhaps, a Newtonian reflector. That would be the right answer, except that the price range is inconsistent with the answers, because such an instrument would cost anywhere from $600 on up. Sorry, that's just simple economics, but there are always alternatives.

Just what are your alternatives? Who makes the best instruments for the money? Chapter 4 continues with a survey of today's astronomical binoculars, while chapter 5 evaluates today's telescope marketplace.

4

Two Eyes Are Better Than One

It might strike some readers as odd that I chose to begin the discussion of astronomical equipment with binoculars rather than telescopes. After all, why use "just binoculars" when there are so many telescopes to be had? The answer is simple. I strongly recommend that if you are limited in budget or are just starting out in the hobby, do *not* even consider buying an inexpensive telescope. Those telescopes are inevitably disappointing. Spend your money wisely by purchasing a good pair of binoculars, maybe a tripod and binocular mount, plus a star atlas and few of the guidebooks listed in chapter 7. Not only are they great for astronomy, binoculars are perfect for more down-to-Earth activities, such as viewing sporting events, birding, and what I jokingly call recreational voyeurism. There will always be time to buy a telescope later.

The fact of the matter is that binoculars remain the best way for someone who is new to astronomy to start exploring the universe that lies beyond the naked eye. Their natural, two-eyed view seems far more instinctive than squinting through a cyclopean telescope. Increasing the comfort level of an observer not only makes observing more pleasant but also enhances visual performance. Image contrast, resolution, and the ability to detect faint objects are all improved just by using two eyes instead of one.

Perhaps you are not a novice stargazer, but rather an advanced amateur who has been touring the universe for years, even decades. Great! We share that common bond. Every amateur astronomer should own a pair of good-quality binoculars regardless of his or her other telescopic equipment.

Quick Tips

When choosing a pair of binoculars, the first decision to make is magnification. Low-power binoculars are great for wide views of the Milky Way, but higher-power glasses are preferred for specific objects. Keep in mind that as magnification increases, so does the binoculars' weight. This is especially important if you plan on holding the glasses in your hands. Most people can support up to 10×50 binoculars with minimal strain and fatigue, but if either the magnification or aperture is increased by much more, hand support becomes impractical.

Many manufacturers offer zoom binoculars that can instantly double or even triple their magnification by pressing a thumb lever. Unfortunately, nothing in this life is free, and zoom binoculars pay heavily. To perform this feat, zoom models require far more complex optical systems than fixed-power glasses. More optical elements increase the risk for imperfections and inferior performance. Additionally, image brightness suffers terribly at the high end of their magnification range, causing faint objects to vanish. A good rule of thumb is to stick with fixed-, or single-power binoculars.

Another critical choice is the binoculars' objective lens diameter. When it comes to selecting a telescope, amateur astronomers often assume that bigger is better. Although this is generally true for telescopes, it is not always the case for binoculars. To be perfectly matched for stargazing, the diameter of the beam of light leaving the binoculars' eyepieces (the *exit pupil*) should match the diameter of the observer's pupils.

Although the diameter of each person's pupil can vary depending on the person's age, ambient lighting, and other factors, values typically range from about 2.5 mm in broad daylight to about 7 mm in total darkness. If the binoculars' exit pupil is much smaller than the observer's pupils at the moment of observation, then we lose the *rich-field* viewing capability. On the other hand, too large an exit pupil will waste some of the light the binoculars collect. (For a complete discussion on what exactly an exit pupil is, fast-forward to the discussion about eyepieces in chapter 6. Go ahead, I'll wait.)

The exit pupil in millimeters for any pair of binoculars is found by dividing their aperture (in millimeters) by their magnification. If you are young and plan on doing most of your observing from a rural setting, you would do best with a pair of binoculars that yield a 7-mm exit pupil, such as 7×50 or 10×70. However, if you are older or are a captive of a light-polluted city or suburb, you may do better with binoculars yielding a 4-mm or 5-mm exit pupil (8×40, 10×50, etc.).

Rather than define and discuss exit pupils, some manufacturers specify a binocular's *twilight factor*, which is a numerical value that is supposed to rate performance in dim light. The twilight factor is calculated by multiplying the aperture by the magnification, then taking the square root of the product. For example, a pair of 7×50 binoculars has an exit pupil of 7.1-mm and a twilight factor of 18.7, while a pair of 10×50 binoculars has a 5-mm exit pupil and a twilight factor of 22.4. By definition, the higher the twilight factor, supposedly the better the dim-light images.

Does this mean that the 10 × 50s are always better for astronomical viewing than 7 × 50s? Maybe, maybe not. I have never put a lot of weight in the twilight factor, because it does not factor in local sky conditions. In my experience, the exit pupil is what really tells the tale of how well a pair of binoculars will show the night sky, not the twilight factor. Of course, we must also take into account a wide range of other mitigating factors, such as lens and prism quality and the types of coatings used. So, that brings us to the second rule of thumb: dismiss twilight factors.

Modern binoculars use one of two basic prism styles: compact roof prism binoculars and heavier Porro prism binoculars, both shown in Figure 4.1. Which should you consider? All other things being equal, Porro prism binoculars will yield brighter, sharper, more contrasty images than roof prism glasses. Why? Roof prism glasses, which only started to become popular back in the 1960s, do not reflect light as efficiently as Porro prisms. By design, roof prisms require that one prism surface be aluminized to bounce light through to the eyepieces. As the light strikes the aluminizing, a bit of it is lost, resulting in a dimmed image. Porro prisms (at least those made from high-quality glass) reflect all of the incoming light without the need for a mirrored surface, yielding a brighter final image.

This is not meant to condemn all roof prism binoculars. Many of the stars in the binocular industry, including Zeiss and Canon, make some remarkable

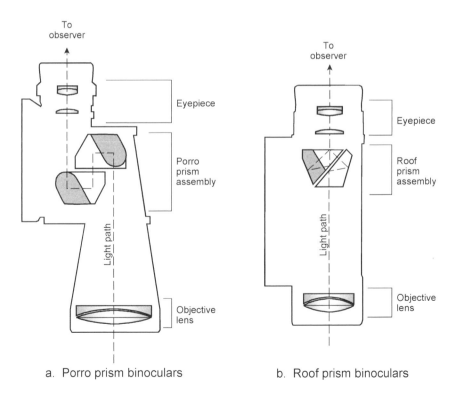

a. Porro prism binoculars   b. Roof prism binoculars

Figure 4.1 *Cross sections through (a) Porro prism binoculars and (b) roof prism binoculars.*

roof prism models. The finest use the Abbe-König style of roof prism, which has better light transmission qualities than the more common Schmidt-Pechan design found in lesser models. But be prepared to pay for this quality, which brings us to the second disadvantage of roof prism binoculars: their high cost. Part of the higher expense is the result of the more complex light path that demands greater optical precision in manufacturing. Unless constructed with great care, the image quality of roof prism binoculars will be unacceptable. Even still, assuming all other things are equal, most Schmidt-Pechan roof prism glasses are inferior to Porro prism models for viewing the night sky. Having said that, there are still several star-worthy roof binoculars out there.

If you are considering roof prism binoculars, you will likely come across the phrase *phase coating*. Phase coatings are used to correct for light loss inherent in the roof prism design. As an image is bounced through a roof prism assembly, it is split into two pathways, involving different internal reflections that are perpendicular to each other. The horizontal plane is passed slightly out of phase to the vertical plane, causing some dimming and loss of contrast. The best roof prism binoculars remedy this so-called phase error by applying a special coating to the prism surface. Lesser roof prism glasses, without phase coatings, suffer from inferior image contrast and resolution.

Porro prisms, because of their design, are immune from phase error, but their quality is affected by the type of glass used. Better binoculars use prisms made from BaK-4 (barium crown) glass, while less-expensive binoculars use prisms of BK-7 (borosilicate). BaK-4 prisms transmit brighter, sharper images because their design passes practically all of the light that enters (what optical experts call *total internal reflection*). Because of their optical properties, BK-7

Figure 4.2 *A binocular's tail often tells the tale. Here, we see that these astro-savvy binoculars have the right stuff, including BaK-4 prisms and fully multicoated optics.*

prisms suffer from light falloff, and consequently, somewhat dimmer images. The problem lies in BK-7's *index of refraction,* which is a measure of how much a beam of light will bend, or refract, when it passes through a medium. Beyond the objective lens, light strikes the wedge-shaped prism face at too steep an angle (referred to as the *angle of incidence*) to be reflected completely through the prism; instead, some is lost. More light is lost as it exits the prism, and again as the light enters and exits the second Porro prism in the set. The higher index of refraction of BaK-4 permits total internal reflection at a steeper angle of incidence, so that all of the image-forming light is fully reflected.

Most manufacturers will state that their binoculars have BaK-4 prisms right on the glasses (Figure 4.2), but if not, you can always check for yourself. Hold the binoculars at arms length and look at the circle of light floating behind the eyepieces. This is the previously mentioned exit pupil. The exit pupils will appear perfectly circular if the prisms are made from BaK-4 glass, but diamond-shaped with gray edge shadows (because of the light falloff) with BK-7 prisms.

Shop Till You Drop

Besides the type of prism used, there are many other factors to take into account when shopping for a pair of astronomical binoculars. If possible, it is always best to look at several brands and models side by side.

Let's start with coatings. Does the manufacturer state that the lenses are coated? Optical coatings improve light transmission and reduce scattering. An uncoated lens scatters about 4% of the light hitting it. By applying a thin layer of magnesium fluoride onto both surfaces of the lens, scattering is reduced to 1.5%. A lens that has been coated with magnesium fluoride will have a purplish tint when held at a narrow angle toward a light. If the coating is too thin, it will look pinkish; if it is too thick, then a greenish tint will be cast. Uncoated lenses have a whitish glint.

Just about all binoculars sold today have optics that are coated with magnesium fluoride, but that does not necessarily imply quality. The words *coated optics* on some inexpensive binoculars usually mean that only the exposed faces of the objectives and eye lenses are coated but the internal optics are probably

CONSUMER CAVEAT: Ruby Coatings

Are ruby-coated lenses any good? Ruby, or red, coatings are quite the rage in certain circles, although documented evidence of their real benefits remains rather dubious. Manufacturers tell us that they are best at reducing glare in very bright light, such as sunlit snow or sand. Whether this is true, I will leave for others to debate, but ruby-coated binoculars offer no benefit to astronomical viewing. Indeed, they may actually work against stargazers.

not. For the best views, we need all optical surfaces coated, so fully coated optics are preferred. But wait, there's more! It has been found that while a single coating on each optical surface is good, building up that coating in several thin layers is even better at enhancing performance. So-called multicoated lenses reduce reflection to less than 0.5%. These show a greenish reflection when turned toward a light. Combining the best with the best, binoculars that have fully multicoated optics, which means that every optical surface has been coated with several microscopically thin layers of magnesium fluoride, are the cream of the crop for astronomical viewing.

A well-designed pair of binoculars will also have good eye relief. Eye relief is the distance that the binoculars must be held away from the observer's eye to see the entire field of view at once. If eye relief is too short, the binoculars need to be held uncomfortably close to the observer, a problem to consider especially if you wear glasses. At the same time, if eye relief is excessive, then the binoculars become difficult to center and to hold steadily over the observer's eyes. Most observers favor binoculars with eye relief between about 14 mm and 20 mm. In general, the value tends to shrink as magnification increases.

How about focusing? Most binoculars have a knurled focusing wheel between the barrels, which is called *center focusing.* Some *individual-focusing* binoculars require that each eyepiece be focused separately. Regardless of the method used, the focusing device should move smoothly. Avoid the less-expensive binoculars (none listed here, incidentally) that use a *fast-focus* thumb lever instead of a wheel. This latter approach permits rapid, coarse focusing, which may be fine for moving objects, but it is not accurate enough for the fine focusing required when stargazing. Even worse are *fixed-focus* models that do not have any focusing mechanism at all. They are set at the factory to focus about 100 meters away from the observer. The stars are a lot farther away!

At the same time, recognizing that most people have one good eye and one bad eye, most manufacturers design their center-focus binoculars so that one eyepiece can be focused differently from the other. This *diopter adjustment,* as it is called, is usually located in the right eyepiece. First, close your right eye and focus the view for your left eye using the center-focusing wheel. Then, using only your right eye, focus the view by twisting the right eyepiece's barrel. With both eyes open, the scene should appear equally sharp.

Binoculars over 10× and larger than 60 mm in aperture are classified as giant binoculars. These binoculars are especially useful for comet hunting, lunar examination, and any application that requires a little more magnification than traditional glasses. If you have your heart set on purchasing a high-power pair, then some sort of external support (typically a photographic tripod) is a must. But make certain that the binoculars are tripod adaptable. Usually, that means there is a 0.25–20 threaded socket located in the hinge between the barrels, which makes an L-shaped tripod adapter a necessity. You will also need a sturdy, tall tripod. Most tripods are geared toward terrestrial viewing, not celestial, and as such, do not extend very high. Chapter 7 discusses about what to look for in a camera tripod and reviews a number of useful binocular mounts.

When evaluating binoculars, make sure they are comfortable to use. To check the fit, flex the binoculars back and forth around their central hinge. Notice how the distance between the eyepieces changes as the barrels are pivoted? That distance, called the *interpupillary distance,* must match the distance between the observer's two eyes. The average adult has an interpupillary distance between 2.3 inches (58 mm) and 2.8 inches (71 mm). For binoculars to work properly, the centers of their eyepieces must be able to spread between those two values. If the eyepieces are even slightly out of adjustment, the observer will see not one but two separate and distinct off-center circles. Unless both sets of eyes and eyepieces line up exactly, the most perfect optics, the finest coatings, and the best prisms are worthless.

Once the mechanics have been checked, move on to the optics. Just like telescopes, binoculars can suffer from several different optical flaws. *Field curvature* occurs when the light rays passing through the edge of a lens do not focus in the same plane as those passing through its center. As a result, images near the periphery of the field are out of focus, while images at the center are sharp—and vice versa. Distortion bends and contorts objects geometrically, changing straight lines into arcs toward the edge of the field of view. With positive, or *pincushion,* distortion, lines bend outward at their ends (concave), while with negative distortion, they bow inward (convex). Astigmatism elongates images on either side of focus. On one side, images will stretch horizontally, while on the other, the elongation will flip 90°, extending them vertically. Distortion and field curvature are best checked during the day by aiming the binoculars at a distant pole or other vertical reference, while astigmatism is easiest to detect by viewing the stars at night.

Equipment Reviews

There are so many binoculars currently available on the market, how can one tell which is better than the rest? Given such a wide variety from which to choose, it should come as no surprise that the answer is not an easy one. Here are capsule reviews of more than 200 different models from more than two dozen companies. Consumers should note that some of the companies listed manufacture their own products, while others simply rebadge goods from other companies, typically located in the Far East.

Even among the companies that make suitable astronomical binoculars, there are so many models that it is impossible to capture them all in the pages here. As a result, additional reviews of many fine binoculars can also be found in the *"Star Ware* Extras" section of my Web site, www.philharrington.net. The binoculars found there are also listed in Appendix 1 in the back of this book, which are annotated with a note directing readers to the Web site.

Apogee. Although better known for its imported telescopes and eyepieces, Apogee also imports several lines of binoculars. For instance, its line of Astro-Vue binoculars ranges in size from 10×60s to a huge 20×100 colossus. All include BaK-4 prisms and multicoated objective lens. Mechanical construction is typical for inexpensive binoculars, with adequate center focusing and decent ergonomics. What really draws positive attention are the two small thumbwheels located adjacent to the eyepieces. Astro-Vues come with built-in broadband nebula filters that can be rotated into the light path by turning each thumbwheel a quarter turn. (Chapter 7 discusses the usefulness of broadband filters, which I'll let you review at your convenience.) But the question to answer here is: great or gimmick? Are broadband filters useful with binoculars? Although broadband filters darken the background sky to make the view more aesthetically pleasing, they do little to bring out more detail in objects. The objects that stand to gain the most from these filters are emission and planetary nebulae. The problem is that almost all planetary nebulae are so small that they will appear starlike in binoculars. Most emission nebulae also need more magnification and light-gathering ability than these binoculars can muster. So, in truth, the number of objects actually enhanced by these binoculars is very small.

Apogee's RA-88-SA binoculars (Figure 4.3) are built around a pair of 88-mm rich-field refractors. Triple-element objective lenses promise better false-color suppression than the more common, two-element achromatic objectives. And unlike most large binoculars, these glasses come with 90° star diagonals built right into the binocular body, making them more comfortable to use when aimed high in the sky.

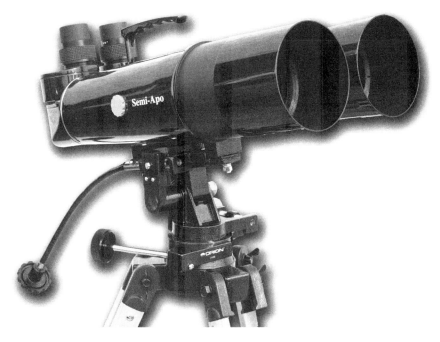

Figure 4.3 *The RA-88 giant binoculars from Apogee, with interchangeable eyepieces, deliver impressive performance at a surprisingly low price, but a sturdy mount is a necessity.*

But how well do they work? I'm happy to report that they work very well. Color suppression is better than I had expected, given the price, while images are sharp across the inner two-thirds of the field, with good contrast. Some owners have complained collimation is lost when adjusting interpupillary spacing. Although there is some play in the adjustment rod because of its coarse threading, I haven't found collimation to be a concern with the models I tested. Apogee sells two other sets of eyepieces for the RA-88, producing 26× and 32×, but it should be noted that the binoculars' focusers are not designed to accept standard 1.25-inch eyepieces.

Recently, Apogee introduced smaller RA-70-SA binoculars that include the same features as the larger 88s. Although this model is too new to discuss here, it looks promising.

Astromeccanica. Astromeccanica binocular telescopes are assembled in Italy from refractors manufactured in China, and then imported into North America by Oberwerk. The result of this international effort is an impressive pair of instruments that come mounted on alt-azimuth fork mounts perched atop steel pedestals. The mounts, made from custom-machined aluminum, move smoothly horizontally and vertically, although personally, I feel the pedestals are too short for comfortable viewing near the zenith, even from a seated position.

Both the Astromeccanica Ziel 120 and Astromeccanica Helios 150 binoscopes feature 1.25-inch helical focusers built into right-angle star diagonals. Unfortunately, three aluminized mirrors are used to bounce the light around to the eyepieces rather than prisms, dimming the images in the process. Still, a true binocular telescope will produce superior views to a similar, single telescope outfitted with a binoviewer eyepiece attachment. Images through the Astromeccanica binoculars are right side up but mirror-imaged left-right.

In testing both models, I found chromatic aberration to be no more intolerable than through a single refractor. The 150-mm model, made from a pair of 6-inch f/5 refractors, shows the greater amount of false color, but I didn't find it terribly offensive unless viewing brighter stars and planets or the Moon. False color was also clearly seen through the smaller 120-mm f/5 model, but again it was not objectionable in most instances. Terrestrial viewing, however, was an interesting experience. For reasons that I still can't explain, views were *depth reversed*. In other words, when looking toward a horizon of hills beyond closer trees, the trees appeared farther away than the hills. A very odd sensation indeed!

Barska. Barska is a small company that imports inexpensive binoculars from China. In fact, some of its X-Trail binoculars are the least expensive found in this book. Can binoculars that retail for between $25 and $65 be any good? Sure, as long as you keep your expectations realistic. Images will be reasonably sharp in the center half of the field, but fall off toward the edge. The fully coated optics and BK-7 prisms yield contrast and image brightness that are good enough to view the Moon and brighter deep-sky objects, but these binoculars will probably miss objects that are challenging through similarly sized, premium-quality binoculars. I would avoid the 20 × 50s, however, because they have ruby-coated objective lenses. The other models have standard coatings.

Barska's 70-mm and 80-mm X-Trail binoculars deserve special mention as being the lowest-priced pairs of giant binoculars sold today. Like their smaller brethren, the 15 × 70, 20 × 80, and 30 × 80 Barskas include fully coated optics but add BaK-4 prisms for brighter images than would be possible with BK-7 glass. Images are reasonably good, although the greater light-gathering ability tends to accentuate chromatic aberration around brighter objects. Not surprisingly, stars along the outer third of the field are softened from distortion. Images are acceptable but cannot compare with higher-priced giant binoculars, including those seen through Barska's own Cosmos line.

Barska Cosmos binoculars come in three sizes: 15 × 60, 20 × 80, and 25 × 100. All include multicoated optics and BaK-4 prisms, individually focused eyepieces, and aluminum barrels. The 80-mm and 100-mm models include a sliding mount on a reinforcing bar, while the 15 × 60s have a traditional tripod socket in the central hinge. The most-tempting feature of Barska binoculars is their exceptionally low price. Can a pair of 80-mm or 100-mm binoculars that cost far less than some 50-mm binoculars be any good? Let me answer that question this way. If you have always wanted to own a pair of 4-inch binoculars, but just didn't have the capital to invest in a more-expensive pair, then the Barska 25 × 100s are certainly worth a look. No, they will not equal the more-expensive competition—you're not getting something for nothing—but they are certainly worth the investment if money is too tight for other 100s.

Bushnell. Bushnell offers dozens of different binocular models, yet for all of this diversity, only a few pass the astronomical litmus test. The company's Legacy binoculars range in size from 8 × 22 to 10 × 50, including two zoom variations, but only the 8 × 40 and 10 × 50 models offer useful combinations of aperture and magnification. Each model has multicoated optics, BaK-4 prisms, and a built-in tripod socket. Weight is typical for their sizes, although many users will find the short 9-mm eye relief uncomfortable, especially if wearing eyeglasses. In testing the 10 × 50 Legacy binoculars in the April 2005 issue of *Astronomy* magazine, I found images to be every bit as bright as the best low-end binoculars. These binoculars had slightly less image contrast and slightly more distortion and astigmatism than my top-rated Scenix binoculars from Orion, however.

The best of Bushnell's bunch are its Legend 10 × 50 and 12 × 50 Porro prism models. Although centered star images are bright and sharp, some pincushion distortion can be seen toward the outer 25% of the field, which is to be expected from binoculars in this price range. But thanks to fully multicoated optics, BaK-4 prisms, reasonably good eye relief and field size, sculpted barrels, and an integrated tripod socket, Legends are sure to please all but the most critical binocularists. I would just recommend sticking to the 10 × 50s. Although the 12 × 50s weigh the same, that extra 2× is enough to magnify hand-supported jitters.

Canon. Canon, famous among photographers for its fine cameras, shook the binocular world a few years back when it introduced its *image-stabilized* (IS) binoculars. These remarkable binoculars contain no ordinary Porro prisms, mind you; rather, they are mounted in a flexible, oil-filled housing. Inside each of the barrels is a microprocessor that detects all motions of the observer and sends a controlling signal to each of the prism assemblies. The prisms, which Canon calls Vari-Angle prisms, then flex to compensate for the movement. Powered by two AA batteries, the stabilization effect is activated by the push of a button on the top of the right-hand barrel. It works like a charm, even in temperatures as low as +14°F (−10°C) and as high as +112°F (+44°C).

Of course, this technology doesn't come cheaply. Even the smallest Canon IS binoculars, only 8 × 25 (not listed in this book because of the small aperture), typically sell for about $250, while the 18 × 50s go for about five times that price. Are these binoculars worth the price? Many owners state emphatically that they are, using such flowery language as "wonderful," "amazing," and simply "wow!" And despite their comparatively small apertures, there is no denying that the views through Canon IS binoculars are often superior to those through many larger conventional models. Canon, also renowned for its fine photographic lenses, uses fully multicoated optics to produce what is arguably the sharpest set of binocular optics sold today at any price. Stars remain pinpoints across the entire field through all of the IS glasses. Some claim that the fine optics combined with the stable images is like adding 20 mm to the aperture.

Don't get me wrong. The technology is terrific, as are the optics. Images are sharp across the entire field of any of the Canon IS glasses. Personally, however, when I engage the IS feature by pushing a button, the view strikes me as odd, taking on a strange quality that is difficult to describe. The image moves as I move, but seems to do so just slightly out of sync with my motions. In fact, the effect almost makes me feel slightly seasick, although I am prone to that sort of thing. Also, the binoculars still have to be lowered every time you want to look back at a star chart. Expect to lower them fairly often just to rest, because they are on the heavy side for their respective apertures. The 10 × 42 and 18 × 50 models come with a tripod socket for securing the binoculars to an external support, although that would seem to defeat the purpose (and expense) of image stabilization. So, although images are exceptionally sharp through every Canon IS binocular, I still feel that consumers would be much better off spending the same amount of money on a larger pair of conventional binoculars and a sturdy tripod.

Carton Optical. Based in Japan with a North American sales office in Canada, Carton Optical's Adlerblick line of binoculars includes full multicoatings on all optical surfaces, conventional center focus, and prisms of BaK-4 glass for bright, clear images across their full fields. Although all models have very good optics, the 7 × 50s impress me as the best in overall image quality. Images are sharp across the inner 80% of the field, beyond which distortion begins to blur the view. By comparison, the 10 × 50s only have sharp images across the inner 60% or so of the field. Although a tripod socket is built into the central hinge, Adlerblick binoculars are exceptionally light, making them easy to support by hand.

Celestron International. Celestron International currently offers five lines of binoculars suitable for astronomical viewing, with its UpClose binoculars the least expensive of the bunch. UpClose binoculars are easily hand-holdable, although eye relief is rather short for eyeglass wearers. Each is also tripod adaptable, a feature not found in many sub-$100 binoculars. They produce sharp images in the center of the field but lack the image contrast and brightness of more expensive glasses because of their BK-7 prisms. Still, they make good first binoculars for budding stargazers.

If you have a little more money to invest in your binoculars, Celestron's Outland LX 8 × 40 and 10 × 50 models are well worth considering. Each includes fully multicoated optics and BaK-4 Porro prisms for bright images, while the air-tight, nitrogen-filled barrels ensure that those images will remain clear even in damp conditions. Twist-up eyecups help to keep stray light from creeping around the edge of the eyepiece and into the observer's eyes. Focusing is smooth across the entire range, and the rubber-coated barrels are easy to grip. Considering that this model is available for less than $100, Celestron's Outland LX binoculars are an excellent value.

Celestron's OptiView LPR binoculars come with built-in broadband nebula filters that the manufacturer claims help diminish the ill effects brought on by light pollution. By turning small thumbwheels at the base of the prism housings, the filters can swing in and out. As I reported in the April 2005 issue of *Astronomy* magazine, I didn't see a positive difference during either suburban or rural observing. Urban stargazers might enjoy some aesthetic benefit, although the improvement will not be very dramatic.

The Ultima DX series of 8 × 56, 9 × 63, and 10 × 50 binoculars, redesigned in 2006, feature center focus, premium components, and sealed barrels filled with dry nitrogen to prevent fogging. These binoculars are also tripod adaptable. The Ultima DXs produce an exceptionally sharp image. Eye relief is also very good, making for comfortable viewing. Although the redesigned models are a few ounces heavier than earlier models, their increased fields of view allow for a more panoramic effect. Also on the plus side, the new Ultimas have twist-up eyecups as well as waterproof bodies that are covered with a shock-resistant rubber covering. Much like the predecessor Ultima models, the new Ultima DXs are fine, moderately priced binoculars.

Last but not least, Celestron's four giant SkyMaster binoculars offer very good value for the money. The 12 × 60 and 15 × 70 SkyMasters have surprisingly good image brightness, quality, and contrast considering their price. Although images are a little soft across the outer 25% of the field, the overall visual effect through these binoculars is impressive. And while these binoculars should be mounted on a support to get the most out of them, their relatively light weight makes it possible to hold them by hand long enough to enjoy views of many favorite sky objects. If you are in the market for giant econobinoculars, the 15 × 70s are highly recommended.

Celestron's 20 × 80 and 25 × 100 SkyMasters offer equal performance and equal value in their respective size ranges. Views are relatively free of distortion, although images soften toward the outer edge of the field. There is also a fair amount of false color visible around brighter objects, which is typical of many giant binoculars. These binoculars are far too large and heavy to hold by hand, however, so some sort of external support is essential. A reinforcing bar bridges the objectives and prisms for added stability, while a permanently attached tripod adapter slides back and forth along a rod for easy mounting and balancing. Although a good idea if the binoculars are mounted to a photographic tripod, it might cause a problem attaching them to some of the parallelogram binocular mounts discussed in chapter 7.

Fujinon. Fujinon, the name that has come to be known among photographers for excellent film and camera products, is also synonymous with the finest binoculars as well. Fujinon binoculars are famous for their sharp, clear images that snap into focus. They suffer from little or no astigmatism, residual chromatic aberration, or other optical faults that plague glasses of lesser design. Binoculars don't come much better than these.

Although not all Fujinon binoculars are designed for astronomical viewing, the Polaris series (also known as the FMT-SX series, see Figure 4.4) is made with the stargazer in mind. Ranging from 6×30 (too small for astronomical viewing) to 16×70, all models are waterproofed and sealed, purged of air, and filled with dry nitrogen to eliminate internal lens fogging in high-humidity environments. To maintain their airtight integrity, eyepieces must be individually focused, which is an inconvenience when supporting by hand. Another inconvenience is their high price, but many people are willing to pay for the outstanding optical quality. The FMT-SX 7×50s and 10×70s are especially popular among amateur astronomers, and I personally find the 16×70s to be

Figure 4.4 *The author's daughter, Helen, viewing through a pair of Fujinon 16×70 FMT-SX binoculars. Note how she is supporting the ends of barrels, rather than gripping them around their prisms. Although still a tiring exercise, this is the best way to support giant binoculars by hand.*

outstanding performers from my suburban backyard as well as from darker, rural surroundings. I fondly recall a view from the Stellafane convention in Vermont a few years ago of M6 and M7, two open clusters in the tail of Scorpius, as they just glided over some distant pine trees. That view painted a three-dimensional portrait in my mind that has proven far more lasting and meaningful than any photograph.

Fujinon recently unveiled a redesigned version of the FMT-SX that, apart from some cosmetic changes, appear the same on paper as the models it replaced. In critical testing, however, star images near the edge of the field are not quite as sharp as those in the previous style. Although most people will not notice the difference, it is apparent when a new pair is tested against an older pair.

If you're working within a tighter budget but still want Fujinon quality, consider the company's Poseidon MT-SX series. Fujinon MT-SX binoculars are available in 7×50 and 10×50 models, retailing for about $100 less than their FMT-SX counterparts. Like their more-expensive brethren, MT-SX binoculars are waterproof and nitrogen filled, again requiring individually focused eyepieces. So what's the difference? For one, eye relief is considerably shorter, only 12 mm in each, which some users might find uncomfortable. FMT-SX binoculars also use flat-field optics to reduce edge-of-field distortion, while MT-SX models do not. But the amount of distortion present in any Fujinon binoculars is so small that few complain.

Fujinon also makes giant-giant binoculars called Large-Size Binoculars. That's an understatement, if ever I heard one! Featuring the same quality construction as the Polaris line, including BaK-4 Porro prisms and fully multicoated optics throughout (referred to as Electron Beam Coating by the manufacturer), Fujinon's Large-Size binoculars are available in either 25×150 or 40×150 versions. The 25×150s come in three models: the least-expensive MT-SX, featuring straight-through eyepieces; EDMT-SX, utilizing ED (extra-low dispersion) glass in their objective lenses for better color correction; and EM-SX models, including ED objectives and eyepieces mounted in 45° prisms. The 40×150 EDMT-SX binoculars also have ED objectives but conventional, straight-through eyepieces. Fujinon is justifiably proud of its Large-Size binoculars, and rightfully so. After all, the late Japanese amateur Yuji Hyakutake used 25×150 MTs to discover Comet Hyakutake (C/1996 B2)—one of the twentieth century's most spectacular comets.

In 1999, Fujinon joined the image-stabilized world with its 14×40 Techno-Stabi binoculars. I mentioned previously when discussing the Canon IS binoculars that if you are sensitive to motion sickness, you might feel a bit queasy using those. Well, the Fujinons really take to the high seas, with a much larger compensation angle (5° versus 1° for the Canons). Images tend to really float around, making them difficult to center after panning. Although this may be ideal for seafarers, it is less desirable for landlubber astronomers.

Optically, although the images are as sharp as Canon's line, the Techno-Stabi binoculars seem dimmer for reasons that escape me, though it likely has something to do with the roof prisms and coatings. Some slight residual chromatic aberration is also evident around brighter objects, but flare and ghosting are well suppressed. Perhaps the most annoying feature of the Fujinons is their two-button activation process. Push the button that is nearest the eyepieces to turn the power on initially; depress it again to stabilize the image; then press the other button to turn the whole thing off. It's very easy to confuse one button for the other in the dark. The stabilization circuit stays on by itself for one minute after the button is released, after which time, it goes into standby mode. The larger Canons stay on for five minutes before standing by. I must admit, however, that I like the Fujinon's off-center center-focusing knob, which is adjacent to the right eyepiece.

Jim's Mobile, Incorporated. Jim's Mobile introduced a new, innovative product called the RB-66 binoscope a few years ago. The RB stands for *reverse binoculars,* which is something that the amateur market had never quite seen the likes of before. More recently, the manufacturer has expanded the line with its RB-10 and RB-16 monster binoculars.

The RB-66 is, in effect, tandem 6-inch f/5 Newtonian reflectors that create a pair of giant binoculars. The RB-10 binoculars are built around a pair of 10-inch f/4.7 reflectors, while the RB-16 uses 16-inch f/4.8 reflectors. Each instrument comes mounted on an alt-azimuth mount. Light enters the two tubes, bounces off the primary mirrors back to the secondary mirrors, and finally exits out through focusers that are facing each other. A pair of star diagonals then turns the light path toward the front of the instrument, where the observer can view from a seated position. JMI uses modified focusers that slide back and forth parallel to the tubes.

Unlike conventional binoculars, which are ready to go immediately, the RB binoscopes require some initial tweaking whenever they are set up. First, the optical collimation must be checked. To gain access to the three adjustment screws behind each mirror, you must disassemble the plastic clamshell covers of the RB-66 and RB-10, which is done by removing the instrument's carrying handle and pulling away the cover's top half. The RB-16's primaries can be accessed directly. Each side is independently collimated like a conventional Newtonian.

Once collimation is complete, both telescopes must be aimed parallel to each other. This proves easier to check and confirm if you aim at a known object, such as the Moon or a bright star or planet. Center the target in one of the eyepieces and then look through the other. By flipping two rocker switches back and forth on the control panel, a pair of electric motors moves the right-hand telescope horizontally and the left telescope vertically.

Although setup can be annoyingly time-consuming, the views through the JMI RBs are quite amazing. The crescent Moon through the RB-66, for instance, creates a surreal, faux 3-D effect that I find quite stunning. Although there are better telescopes for the planets, the RB-66 is in a league of its own when it comes to widespread deep-sky objects like the Orion Nebula, Lagoon Nebula, Veil Nebula, and the Andromeda Galaxy. Object contrast, brightness, and aesthetics are significantly enhanced by using both eyes. There is also slight improvement in stellar magnitude penetration, typically between 0.1 and 0.2 magnitude.

Kowa. A well-known name in the spotting-scope industry, Kowa also markets a number of binoculars, including its distinctive High Lander binoculars. The silver barrels feature a built-in carrying handle and removable eyepieces that are tilted 45°. Each dry-nitrogen-filled barrel also has a threaded dust cap and an extendable dew cap that slides out in front of the objective lens.

Kowa High Landers are available with either achromatic objective lenses or multielement apochromatic objective lenses highlighted by a fluorite element, the latter being called High Lander Prominars. Both versions include fully multicoated optics and come standard with a pair of wide-field 14-mm eyepieces, yielding 32× in these f/5.48 instruments. The eyepieces can be easily removed and swapped for an optional pair of 9-mm eyepieces to raise magnification to 50×. Although they have 1.25-inch barrels, they are specially designed to plug securely into the eyepiece holders. Standard 1.25-inch eyepieces also fit but cannot be locked in place.

The Prominars have some of the finest optics of any binoculars that I have ever used. They give no hint of distortion at all, producing a textbook-perfect flat field. Images are also sharp and show excellent contrast with no sign of purplish or yellowish tinges, even when viewing the Moon and bright stars, like Vega. Star images were absolutely pinpoint sharp with exceptional contrast. The Pleiades, the Double Cluster, and the star clouds through central Cygnus were breathtaking, each sparkling like diamonds.

Meade Instruments. Although Meade's 9 × 63 Astronomy binoculars use a pair of roof prisms rather than Porro prisms, they represent an excellent value. For around $200, these binoculars include fully coated optics, a central tripod socket, and have 22 mm of eye relief, making them ideal for eyeglass wearers. Images are bright and contrasty, with no evidence of stray flaring around demanding objects like the Moon or Venus. True, there is some minor chromatic aberration and some expected softening around the edge of the field, but both are easily ignored while appreciating the view.

Meade also sells several other lines of binoculars. Several models in its inexpensive TravelView product line are ideal for beginners. You will find them discussed in detail in the *Star Ware* section of my Web site, www.philharrington.net.

Miyauchi. Not exactly a brand name that rolls off the tongue easily, Miyauchi binoculars display high-quality lenses and prisms packed into one-of-a-kind streamlined, gray barrels. The individually focused eyepieces and interpupillary adjustment are also very smooth. Some have inferred that the latter may be too smooth, causing the binoculars to go out of adjustment easily, but I have never found this to be a problem with the models that I have tested.

The smallest Miyauchi is the recently introduced 7 × 50 Binon. If I gave an award for unusual looks, these would win, thanks to their machined aluminum bodies that are trimmed with maroon leather. They may look a little bizarre, but inside they are a thing of beauty. Fully multicoated triplet objective lenses work wonders at suppressing false color, while BaK-4 Porro prisms produce high-contrast images that are sharp nearly to the edge of the 9.5° fields in the individually focusing Erfle eyepieces. Focusing may also take a little getting used to because of the eyepieces' opposite threading. In other words, if the left eyepiece has to be turned clockwise to bring the image into focus, the right eyepiece will need to be turned counterclockwise. But that's just another quirk for a strange looking pair of binoculars that happen to work very well.

Miyauchi's 22 × 60 Bs-iC Pleiades and 22 × 71 NBA-71 Saturn II binoculars are very well constructed, although they have comparatively small exit pupils. Each one-piece aluminum die-cast binocular barrel is filled with dry nitrogen to prevent internal fogging. The eyepieces are tilted at 45° angles, which is advertised as being easier for looking skyward. Personally, I find the angle quite uncomfortable, so you should try before you buy, if possible. Both focus individually, with click stops for precise setting. Optically, both models feature long focal-length objective lenses for excellent color correction. Star images are sharp and clear, with little distortion detectable toward the edge of the field of view. The eyepieces of the Saturn IIs can be replaced by optional 40× and 115× eyepiece sets, making them good for high-power lunar and planetary viewing as well as deep-sky viewing. Miyauchi includes a side-mounted finder for easier aiming.

Miyauchi added 40 × 100 Saturn III binoculars to its line-up in early 2006. Like the smaller Saturn IIs, the long focal ratio of their objective lenses (in this case, f/7.5) gives these binoculars better false-color suppression than more traditional, shorter focal length objectives. Featuring the Miyauchi's trademark aluminum body, the Saturn IIIs have interchangeable eyepieces, a nicely integrated carrying handle, and nitrogen-filled barrels. Optional eyepiece sets can decrease magnification to 33× or increase it to 75×. Because these binoculars come with prisms that are tilted at 45° like the smaller Saturn IIs, my advice about trying before buying applies here as well. Also be aware that a robust mount is going to be needed to hold these 13-pound monsters steadily. To that end, Miyauchi offers a specially designed fork mount, although I prefer one of the mounts from Universal Astronomics that are discussed in chapter 7.

Miyauchi's 26 × 100 Galaxy Bj-iCE binoculars feature five-element ED objectives to virtually eliminate all signs of false color around the edges of bright objects. Just like the Saturn series, the Galaxy eyepieces can be replaced with a 37× set, if desired. Note, however, that the Saturn eyepieces will not fit the Galaxy binoculars, and vice versa. Whether you buy the extra eyepieces is up to you, but the optional 3× finderscope, which screws on in place of the removable carrying handle, is highly recommended.

Then, there are the Miyauchi BR-141 25 × 141 apochromatic binoculars. One of the world's most expensive binoculars, BR-141s feature a pair of five-element, 5.6-inch objective lenses, each with a false-color-suppressing fluorite element. What a pair! The eyepieces are removable, so that standard 25× magnification can be changed to several magnifications up to 45× with auxiliary eyepieces from Miyauchi. Scanning the skies through a pair of 5.6-inch apochromatic refractors is an amazing trip—if you can afford the first-class ticket! The BR-141s come standard with 25× eyepieces, a side-mounted 3× finderscope, a padded carrying case, and a mount/tripod combination; if preferred, the body can also be purchased alone.

Nikon. The name Nikon may be more familiar to photographers than astronomers, but that is changing fast. Nikon produces several lines of outstanding binoculars.

I have found Nikon's low-end Action binoculars to be a little disappointing, however. Although they are manufactured with BaK-4 prisms and are stated to include aspheric lenses to reduce optical distortions, image contrast and edge correction are still lacking. The binoculars have very good ergonomics, however, with a nice, sculpted feel that makes them easy to hold. Their bayonet-style eyecups twist out rather than fold like many other models, which also makes for comfortable viewing. Nikon Action binoculars range from a modest 7×35 pair to 16×50s.

Nikon's Action EX All-Terrain Binoculars (ATB) bear more than a passing resemblance to the Action line, and also include sealed, waterproofed barrels filled with dry nitrogen. Like the Action line, the Action EX ATBs have a very nice feel. Although they are easy to grip and support by hand, the 12×50 and 16×50 models should still be supported by a tripod because of their relatively high magnifications. Looking critically at their performance, apart from some minor edge distortion, images through the Action EXs are sharp and clear. Even though Action EX binoculars are very good, I still believe there are better models available that cost less, including Orion's Scenix. The Nikons are more ruggedly made, however, and will likely hold their collimation better if bumped around.

Owners of Nikon's Superior E glasses swear by this series. Although designed more for birders than astronomers, these 10×42 and 12×50 Porro prism binoculars are just as adept looking skyward at night as they are during the day. Construction quality is second to none, as anyone can tell with a turn of the ultra-smooth focusing knob. The body, constructed of magnesium, is covered with a rubberized material that is a pleasure to grip. And the optics are, well, Nikon. That means images are bright and sharp across the field, and contrast is excellent. Some users have noted, however, that the eye relief causes some blacking out of the view unless the binoculars are held just so, which can lead to increased fatigue after prolonged use.

To me, the finest Nikon binoculars wear the Criterion badge. For handheld use, Nikon's 7×50 Prostars have almost all the right stuff. Not surprisingly, optical performance is outstanding thanks to fully multicoated objective lenses made of extra-low-dispersion (ED) glass, which go a long way in eliminating chromatic and spherical aberrations as well as coma. Like Fujinon Polaris FMT binoculars, Nikon Prostars come with individually focused eyepieces and are sealed and waterproofed to prevent internal fogging. Their field of view is comparable to the Fujinons, but eye relief is less.

Nikon's 10×70 and 18×70 Astroluxes share the same features as the 7×50 Prostars and are remarkable giant binoculars. Images are tack sharp for all models except for the outer 5% or 10% or so of the field, with very good contrast. I feel, however, that they are slightly inferior in terms of image sharpness and contrast to similar-size Fujinon FMT-SXs when compared side-by-side. Still, the wide apparent fields of view and excellent focus of the eyepieces certainly carry them high in the overall ratings.

Be aware that all Nikon Series E, Prostar, and Astroluxe binoculars do not include a standard tripod socket. Instead, they require a special tripod adapter that clamps onto the post between the binocular barrels. Although this works well, it costs much more than the L-shaped brackets used by most other binoculars. Also, because most camera tripods are not tall enough for viewing the sky comfortably, many amateurs favor parallelogram binocular mounts. Unfortunately, the design of the Nikon tripod adapter makes it impossible to attach the binoculars to many of these mounts without a second adapter plate. If you are planning to attach these Nikons to one of these mounts, contact the mount's manufacturer beforehand for recommendations.

Finally, there is Nikon's 20 × 120 Bino-Telescope. The name is a bit misleading, because unlike other binocular telescopes discussed in this chapter, the Nikons do not have interchangeable eyepieces. Still, it goes without saying that a view spanning 3° is striking. The Bino-Telescope includes 21 mm of eye relief, fully multicoated optics, BaK-4 prisms, and an integral fork mount. Customized tripods are available separately. These instruments were designed originally for seafaring use and are waterproofed to prevent fogging. Unfortunately, the fork mount was also designed primarily for horizontal viewing. As a result, the binoculars are limited in vertical travel to about 70° above the horizon, at which point they hit the mount.

Oberwerk. Although Oberwerk is not a name with high recognition value, this company stands out as offering some giant binoculars at small-binocular prices.

Oberwerk's 8 × 56 and 11 × 56 binoculars include BaK-4 prisms, fully multicoated optics, and built-in tripod sockets, a great combination in their price range. Images in both are very sharp, although I saw some minor ghosting around brighter objects in the 11 × 56s that I tested. The distortion was likely an artifact caused by irregularities in the paint inside the barrels of the particular pair I tested. Several spots around the internal field stop looked as though they had been touched up by hand with less-than-flat black paint. Still, I prefer them to the 8 × 56 model, which suffers from tunnel vision. Both have good image contrast and sharp stars across 80% of their fields. Eye relief is excellent in both cases, making these binoculars strong contenders for those forced to wear eyeglasses when observing.

Oberwerk 60-mm FMC Mini-Giants are available in a choice of 9×, 12×, 15×, or 20× magnification. All are housed in the same rubber-coated body and include BaK-4 prisms as well as optical coatings on all lens surfaces. Images are good considering what you pay for binoculars, just over $100, but they do not compare with the likes of Fujinons or Nikons. Stars in the center of the field are sharp, but become blunted as the eye moves toward the field edge. Eye relief, unfortunately, is quite short in all but the 8× and 9× models, with only 10 mm in the 20× glasses. All models include a standard tripod socket built into the central hinge, an especially important feature because these glasses are no lightweights.

Moving up, Oberwerk 11 × 70 and 15 × 70 Giant binoculars have received high praise from many users. Images are excellent in the center and reasonably sharp to about the outer third of the field. They are light enough to be supported by hand for brief periods, but a tripod or other support is strongly recommended for extended observing. The Oberwerks have a smooth, black vinyl coating, center-focusing eyepieces, a built-in tripod socket, and a soft carrying case. The knurled center-focusing knob turns smoothly and precisely, with no detectable binding or image shift.

Oberwerk's 80-, 90-, and 100-mm Giant models always garner a lot of interest because they seem to offer so much binocular for little money. Are they too good to be true? The three 20 × 80 models all have appealing attributes, but if it were up to me, I'd save up the extra money and wait until I could afford the Deluxe II version. One look at their features and it's easy to tell why. For around $300, the Deluxe IIs pack in triple-element objective lenses for better-than-average false-color suppression, fully multicoated lenses and BaK-4 prisms, and waterproof barrels filled with dry nitrogen. Images are bright and clear nearly all the way to the edge. I offer only one caveat. The Deluxe IIs are very good optically, but their heavy weight may prove too taxing for many binocular mounts. Unless you already own one of these models or are prepared to invest in one soon, consider the Standard 20 × 80s, which weigh 2.5 pounds less. Images aren't quite up to Deluxe II criteria, but they are clear and crisp over the inner 70% of the field. The 20 × 80 LW, short for *lightweight*, weigh a pound less still and work well on most binocular mounts. They impress me as being the equal of the 20 × 80 Standards in every respect except for eye relief, which is a critical 5 mm shorter. Optically, both are the equal of Orion's more expensive 20 × 80 GiantViews.

Then, there are Oberwerk's 20 × 90, 22 × 100, and 25 × 100IF Supergiants. The 20 × 90s are especially nice performers, with very good image sharpness, contrast, and freedom from flaring. Part of the pleasure of using these binoculars is their relatively wide field of view, which is just big enough to squeeze in the three stars in Orion's belt. Another part comes from ergonomics, like their very comfortable rubber eyecups.

I am not as enthusiastic about the 22 × 100s, however. Despite their high-quality construction, images strike me as a little soft in the center of the field despite repeated focusing; however, they are fun to use. For instance, the visual impression I enjoyed through these binoculars one recent summer of M8, the Lagoon Nebula in Sagittarius, was something that even the finest photographs can't possibly convey. I just feel that Oberwerk's 20 × 90 binoculars are a better choice.

There is no question that Oberwerk's 25 × 100IF monsters offer amazing views, but their girth can take a little getting used to. At 9 pounds, they require a very sturdy tripod or other mount to hold them steady. Even then, their mass can cause them to loosen unless they are securely tightened in place. But once mounted properly, binoculars of this size can be a thrilling experience. Deep-sky objects are no longer ill-defined smudges as with smaller binoculars, but rather, resolved patches of nebulosity or stardust. The fully multicoated optics do a good job suppressing flare in all but the most extreme cases. A gibbous or full Moon, for instance, will be fringed by false color and flaring around its edges.

Finally, we come to Oberwerk's pièce de résistance, its 25/40 × 100 Long-Range Observation Binoculars (sold by other retailers as Border Hawk binoculars). Built originally for the Chinese military, these are *big* binoculars, weighing more than 26 pounds. Throw in the optional wooden tripod, and you're toting around more than 42 pounds! For that, you get multicoated optics and BaK-4 prisms, plus the flexibility of two magnifications. A pair of rotating turrets lets users select between two individually focusing Erfle eyepieces that produce either 25× or 40×. Eye relief is tight, however, especially at 40× (a paltry 8 mm). Focus is crisp across the inner half of the view, but stars tend to become distended toward the outer edge. There is little residual chromatic aberration and no detectable flaring or ghosting. Even at 25×, these gargantuans magnify enough to show remarkable detail in deep-sky objects—detail that is missed in lower-power, though costlier, binoculars. The 40× eyepieces offer enough extra oomph to reveal the rings of Saturn and, just barely, the two equatorial belts of Jupiter—that is, assuming you can find the planet in the first place. Aiming these binoculars is, at best, challenging. Many users have also added their own unity finder, but the deluxe version includes a side-mounted 7 × 50 finderscope.

Orion Telescopes and Binoculars. Few astronomy-related companies are into binoculars in as big a way as California-based Orion Telescopes and Binoculars, which imports several different lines. Its least-expensive astro binoculars are the 8 × 40, 10 × 50, and 12 × 50 WorldViews. Images are sharp across the inner 75% of their fields of view and have acceptable levels of image contrast and sharpness. I don't find them as comfortable to hold, however, as Orion's Scenix binoculars.

In fact, if I had to crown a line of sub-$100 binoculars as best in breed, Orion's Scenix would win. Available in 7 × 50, 8 × 40, and 10 × 50 combinations, Scenix feature fully coated optics enhanced by multicoated objectives. In a test of inexpensive 10 × 50 binoculars that I conducted in *Astronomy* magazine's April 2005 issue, the Orion Scenix 10 × 50s were judged a cut above the rest in terms of image contrast and brightness. Focusing was smooth and precise, and images were sharp and clear. Distortion and field curvature were evident around the outer 25% of the field, which is less than most models in this price range. The view compared favorably with Orion's 10 × 50 UltraViews, which cost about $70 more. The only downside was eye relief, which is just over half that of the UltraViews and could make viewing uncomfortable when wearing eyeglasses.

Orion's UltraView series is the next rung up on the quality and price ladder. I have owned a pair of 10 × 50 UltraViews for several years and am very happy with them. Focusing is smooth, with no binding or rough spots. Eyeglass wearers should especially enjoy the UltraViews' exceptional eye relief of 22 mm. Images are great across about 80% of the field, with stars only beginning to distort toward the outermost edge. The latest UltraViews also include unique twist-and-lock eyecups in place of the more traditional folding eyecups. I am undecided as to whether I like them, only because the eyecups are not as soft and pliable as those on my older pair. But regardless, Orion UltraViews represent exceptional values.

Orion Vista 7 × 50, 8 × 42, and 10 × 50 binoculars, shown in Figure 4.5, compare favorably to other moderately priced models. Images are clearer and crisper than through the UltraViews, with little or no detectable edge distortion. Eye relief is good and focusing is very smooth. The only real drawback to Vistas is their slightly narrower-than-average fields of view, which is why I chose the wider-field UltraViews mentioned above instead. But then, I like wide-field binoculars. The Vistas, however, have the sharper optics.

Orion also markets 8 × 56, 9 × 63, 12 × 63, and 15 × 63 Mini Giant binoculars. These are ideal for stargazers who need a little extra magnification but still want the option of handheld binoculars for a quick nighttime view out the back door before bedtime. The 9 × 63s are a good value, with optics that are nearly as sharp as the Celestron 9 × 63 Ultimas but have better eye relief (26 mm, great for eyeglass wearers but actually too long for others) and a cheaper price tag. The 8 × 56 Orion Mini Giants, however, do not quite hold up to the challenge of the Celestron 8 × 56 Ultimas in terms of overall optical and mechanical quality. The 12 × 63s and 15 × 63s are fine for higher-power viewing, with decent image sharpness and contrast but require a tripod or other external support to be useful.

Looking for *big* binoculars? Orion sells 11 × 70 and 20 × 70 Little Giant IIs; 15 × 70, 20 × 80, and 25 × 100 GiantViews; and 20 × 80 and 30 × 80 MegaViews. Images in the centers of their fields are sharp, but like many giant binoculars, all suffer from edge curvature to some degree. In the ever-expanding field of low-priced giant binoculars, Orion's Little Giant IIs are some of the best, with very good image brightness, contrast, and sharpness. They are also amazingly lightweight for their size. Although a tripod is still needed for long-term use, Little Giant IIs are light enough for quick hand-supported stargazing.

Orion's 15 × 70 GiantViews are also light enough for a quick view by hand, but a tripod is necessary for even a quick glance skyward through the larger 80-mm and 100-mm GiantViews. Like the Little Giants, the Orion GiantViews

Figure 4.5 *Orion Telescope's Vista binoculars. Photo courtesy of Orion Telescopes.*

deliver sharp images on-center but soften off-center. Image brightness is good, however, and the individually focusing eyepieces move quite smoothly.

Orion's 20× and 30× MegaView 80-mm binoculars exceed the GiantViews in terms of image sharpness and contrast. Eye relief is also very good, which makes viewing more comfortable. Slight amounts of chromatic aberration around brighter objects and negative distortion are evident, but neither condition is objectionable. Like the 80-mm and 100-mm GiantViews discussed earlier, the MegaViews are bridged by a metal bar with a captive tripod mounting post. Although the built-in post makes a separate tripod L-adapter unnecessary, it makes attaching them to some parallelogram binocular mounts difficult.

Pentax. Best known for its cameras, Pentax offers some fine Porro prism binoculars that are well-suited for astronomical viewing. Although relatively inexpensive, Pentax XCF models produce bright, sharp images. Distortion and astigmatism, however, are a bit more pronounced than through the comparably priced Orion Scenix and Nikon Action EX lines, blurring the outer third of each field. XCF binoculars are easy to hold and have smooth focusing. Eye relief is a little short, so eyeglass wearers may have a problem taking in the full field of view.

The new Pentax PCF WP II Porro prism binoculars replace the PCF V and PCF WP series from a few years ago. Each of the four waterproof models (8×40, 10×50, 12×50, and 20×60) has nitrogen-filled barrels, a center-focus thumbwheel with a locking mechanism, and a click stop diopter adjustment on the right eyepiece. Their rubberized barrels, combined with excellent ergonomics, make the PCF WP IIs a pleasure to hold by hand.

All Pentax PCF WP II binoculars are a pleasure to look through as well. Their fully multicoated optics produce good-quality views of the night sky. Centralized star images are tack sharp but tend to soften toward the outer third of the field, as with most binoculars in this price range. This model also has generous eye relief.

Stellarvue. Stellarvue is a name more familiar for its refractors than for binoculars, yet the two models in its lineup have garnered their own share of loyal fans. Both have fully multicoated optics, BaK-4 prisms, individually focusing eyepieces, and tripod sockets, and come with fitted, hard carrying cases. Both use oversized objective lenses that are then masked down to the stated apertures. The 20×85 B85s, for instance, have 90-mm objectives, but an internal aperture stop cuts the working aperture to 85 mm.

The stop-down technique works well, judging by the sharp images I've seen through a pair. Often, stars look a little bloated through high-power binoculars, but not through these. Likewise, deep-sky objects stand out nicely against their surroundings. Edge distortion started to become evident about two-thirds of the way out from the center of the field.

Vixen Company, Ltd. Trying to understand the binocular offerings from Vixen takes a little detective work, because not all of their binoculars are available in all countries by the same name. For instance, the Vixen Geoma binoculars sold in North America appear to be identical to the Vixen Ultima line sold elsewhere. My philosophy is that "a rose is a rose." The saddest part is that only four Geoma models are currently imported by Vixen North America: 7×50, 8×56, 9×63, and 10×50. All have long eye relief, fully multicoated optics, built-in tripod sockets, and hard carrying cases. Construction is solid, so expect your investment to last for years. Optically, images are as good as we have come to expect from all Vixen products, with only minimal edge distortion and false color. Unfortunately for me, finding the rest of the Ultima line would require a trip across the pond to Europe, Asia, Australia, or beyond. Stargazers in those areas can also select from 7×42, 8×42, and 10×42 models, as well as 6.5×44 and 9.5×44 apochromatic binoculars that include ED lens elements in their objectives to reduce residual chromatic aberration. Eye relief is very good in the 42-mm models but only a scant 13 mm in the two ED editions.

Vixen also makes some uniquely designed supergiant binoculars that stand out from the crowd as excellent values. All models have 45° prisms, which prove to be a mixed blessing when binoculars are tilted skyward. Although fine for viewing up to about a 45° elevation, the angle quickly becomes a pain in the neck as the binoculars approach the zenith.

First up is the 80-mm Vixen BT80M-A, a pair of 3.1-inch f/11.25 refractors that come with a pair of 25-mm orthoscopic eyepieces, generating 36×. But what makes these truly unusual is that the removable eyepieces can be swapped for any 1.25-inch-diameter eyepiece, for an almost unlimited range of magnifications. With a set of high-quality eyepieces, such as those supplied, the image quality through these binoculars is exceptional, with little evidence of residual chromatic aberration or field curvature. Swapping the supplied 25-mm orthoscopics for a pair of Nagler Type 6s or other super-mega-ultra eyepieces detailed in chapter 7 gives you quite the starship!

To support the BT80M-A, Vixen recommends its Custom D alt-azimuth mount on an aluminum tripod. The Custom D mount includes smoothly operating manual slow-motion controls on both axes and a counterweight to help offset the balance disparity when the binoculars are in place. A dovetail plate for attaching them to the Custom D mount is permanently mounted to the barrels, but a 0.25–20 tapped hole is also provided for connecting them to a standard photographic tripod. Although the Custom D mount is heavily constructed, one owner complained that it still shakes under the weight of the binocular telescope. Another inconvenience is that because the tripod does not have a center post to adjust height, the only way to change the height of the binoculars is to extend the tripod's legs.

Vixen also makes two very impressive pairs of HFT-A Giant 125-mm binoculars. One is supplied with a pair of 20× eyepieces, while the second comes with 30× eyepieces. (The 25×-to-75× zoom variant has seemingly been dropped from production at this point.) In both cases, focusing is done individually by twisting each eyepiece barrel. Although the binoculars do not come with a finderscope, a dovetail base is supplied that will accept any Vixen-manufactured finder, a must-have accessory in my opinion.

Each of these heavyweights is, in effect, a pair of 4.9-inch f/5 achromatic refractors mounted in tandem on a fork mount. The fork arms can be oriented vertically or may be tilted to offset the binoculars from the center of the tripod, making it much easier to view through the binocular eyepieces when the glasses are tilted at angles greater than 45°. At first, the fork mounting appears to be too lightweight to support the binoculars, but it proves adequate for the task. Despite some creaking when the binoculars are moved left or right, the alt-azimuth mount moves very smoothly. Panning back and forth or up and down is a real pleasure, especially along the Milky Way.

Star images are sharp edge to edge across the entire view, regardless of magnification. Coma and curvature of field, two common aberrations of binoculars, are absent. Imagine seeing in the same field the sparkling stellar jewels of the sword of Orion, with M42 set in the center, or the bespangled beauty of the Double Cluster in Perseus in an overflowing field of stars—truly awe-inspiring! Although coma and curvature of field are nowhere to be found, false color interferes when viewing bright objects. Celestial sights such as the Moon, Jupiter and its moons, and Saturn's beautiful rings are all sharp (yes, the rings are discernable at 30×, and can be suspected at 20×) but were marred by unnatural, purplish edges.

Zeiss (Carl). Long known as one of the world's premier sources for fine optics, Zeiss continues the tradition by offering the finest, and most expensive, binoculars of all. For instance, the Zeiss 8 × 56, 10 × 56, and 12 × 56 Victory B/GA T*P* AOS AK phase-coated roof prism binoculars include conventional center focus and fully phase-coated optics, but no tripod socket. Forget about coma, field curvature, or other aberrations that plague lesser binoculars. None of these problems are found here, only crisp stars scattered across a velvety dark field. Each features a wider-than-most field of view, a heavier-than-most weight, and the highest price of any similar-size binoculars mentioned here! Ahh, but the view is spectacular.

Zeiss's line of ClassiC binoculars are just that. The 7 × 50 B/GA T*I.F. use the classic Porro prism design, while the 8 × 56 B/GA T* P* ClassiCs feature roof prisms. Both feature Zeiss's exceptional T* fully multicoated optics, while the 8 × 56 roof prisms are also phase coated (denoted by the P* in the product name) for the best possible light transmission. Few ever complain about the optical quality of Zeiss optics, and you won't hear any complaints from me. These ClassiCs yield nothing less than perfect star images across their 7.4° field, just as we have come to expect from all Zeiss binoculars. Their 21-mm and 18-mm of eye relief, respectively, also make for very comfortable viewing, even for viewers who must wear thick-lensed eyeglasses. Just be aware that the barrels are coated with rubber armor that adds ruggedness to each pair but also a fair amount of weight. Both models are tripod adaptable.

Finally, Zeiss 20 × 60 BS/GA-T* image-stabilized binoculars offer a unique approach to the problem of quivering hands. Unlike models from Canon and Fujinon, the patented Zeiss stabilization system does not require batteries to power it. In fact, it contains no electronics whatsoever; instead, it's a clever system that relies on mechanics alone. Press a button and you hear a *clunk*, indicating that the stabilization mechanism is activated. The button releases tension on an internal *cardanic* spring, which is connected to a framework that holds the prism assembly in place. When the tension is released, the prisms can move in any and all directions, compensating for the observer's motions, while maintaining optical alignment. It's a very effective system, although the button must stay depressed for the stabilization to continue working. Thankfully, Zeiss includes a tripod socket for attaching the binoculars to a camera tripod, because this model is still heavy at more than 3.5 pounds. Optically, the images are very good, but keep in mind that these are only 60-mm binoculars. The same investment (or less) will also net a 80-mm Fujinon, a 100-mm Miyauchi, or a 125-mm Vixen binoculars, all of which are more astronomically suited.

Recommendations

At this point in the chapter, it should be painfully clear that there is much more to "simple" binoculars than you might have imagined at the start. And this isn't even the end of it. There are so many binoculars on the market today that this chapter could not possibly capture all of them. As a result, you will find even more capsule reviews in the *Star Ware* section of my Web site, www .philharrington.net. Be sure to stop by and take a look under "chapter 4 notes."

Of course, even after all of this, you still may be facing the same question that you faced at the start of this chapter: Which binoculars are right for me?

Table 4.1 **The Best of the Best**

Price Range	Handheld Binoculars (smaller than 10 × 70)	Giant Binoculars (10 × 70 and larger)
Under $50	• Celestron Upclose • Orion WorldView	• None
$50 to $100	• Bushnell Legacy • Nikon Action • Orion Scenix	• Barska X-Trail (15 × 70) • Celestron SkyMaster (15 × 70)
$101 to $200	• Orion UltraView • Pentax PCF WP II	• Oberwerk Giant • Orion Mini Giant
$201 to $300	• Celestron Ultima DX • Orion Vista	• Oberwerk 20 × 80 Deluxe II • Orion Little Giant II
$301 to $500	• Carton Adlerblick F592 • Fujinon MT SX	• Oberwerk 20 × 90 • Orion Mega View
$501 to $1,000	• Canon IS II 12 × 36 • Fujinon FMT-SX • Zeiss B/GA T*I.F. ClassiC	• Apogee RA-88-SA • Fujinon FMT-SX • Vixen HFT-A
Over $1,000	• Canon IS L IS WP 10 × 42 • Canon IS All Weather 18 × 50 • Zeiss Victory BT*P* AOS AK	• Fujinon ED-SX • Kowa Highlander Prominar • Miyauchi BJ100iBF

Which will deliver the best images for my budget? To try and sort through all the reviews, descriptions, and marketing hype, Table 4.1 attempts to answer those questions. The table divides the binoculars into several categories to aid you in picking that perfect pair.

I have always found the binocular universe to be a much more personal one than that seen through a telescope. Maybe it's because we are using both eyes rather than one, or perhaps it's simply because we are looking up, rather than down or to the side. In any case, no amateur astronomer should leave home without a pair of trusty binoculars at his or her side. Even if you own a telescope, you may find, as I have, that many of the most memorable views of the night sky come through those binoculars.

Let me close this chapter with one final point to keep in mind. You do not need an expensive pair of binoculars to enjoy our universe. Any pair of binoculars, from top-of-the-line Zeiss to a pair of K-Mart blue-light specials, can take their owners on many wonderful voyages. So, no matter what binoculars you end up with, use them to begin that journey tonight. And enjoy the trip!

5

Attention, Shoppers!

It is time to lay all the cards on the table. From the discussion in chapter 3, you should have a pretty good idea of what type of telescope you want. But we have only just begun! There is an entire universe of brands and models from which to choose. Which is the best one for you, and where can you buy it? These are not simple questions to answer.

Let's first consider where you should *not* buy a telescope. This is an easy one! Never buy a telescope from a department store, a consumer-club warehouse, a toy store, a hobby shop, an online auction, or any other mass-market retail outlet (yes, this includes those twenty-four-hour consumer television channels) that advertises a "640 × 60 telescope." First, what exactly does that mean anyway? These confusing numbers are specifying the telescope in a manner similar to a pair of binoculars; that is, magnification (640 power, or 640×) and the aperture of the primary optic (in this example, 60 millimeters, or 2.4 inches). Brands that use this sort of deceptive advertising include Baytronix, Bushnell, Fotar, Optrons, and Tasco. Several of these appear regularly on eBay and other auction Web sites, so be very skeptical when shopping around. These telescopes can be summed up in one word: garbage. In fact, the late George Lovi, a planetarium lecturer and astronomy author for years, is credited with describing these instruments as "CTTs," short for "Christmas Trash Telescopes."

> **The Golden Rule of Telescopes:** *Never buy a telescope that is marketed by its maximum magnification.*

Remember, you can make any telescope operate at any magnification just by changing the eyepiece. Telescopes that are sold under this ploy almost always suffer from mediocre optics, flimsy mounts, and poor eyepieces. They should be avoided!

So where should you buy a telescope? To help shed light on this all-consuming question, let us look at the current offerings from the more popular and reputable telescope manufacturers from around the world. All of these companies are a big cut above those department-store brands, and you can buy from them with confidence.

I have chosen to organize the reviews by telescope type, with manufacturers listed in alphabetical order. Department-store brands and models have been omitted, because they really fall more in the toy category than scientific instrument. It is disconcerting to see that some reputable companies mentioned in this chapter also sell such telescopes with their brand names emblazoned across the telescope tubes. Don't be fooled; these models are no better than the others being purposely left out of this review. Those small-aperture refractors and reflectors are far inferior to the better-made telescopes detailed here.

In the course of writing this book, I have looked through more than my fair share of telescopes, but it takes more than one person to put together an accurate overview of today's telescope marketplace. To compile this chapter, as well as the chapters to come on eyepieces and accessories, I have solicited the help and opinions of amateur astronomers everywhere. Surveys were distributed in print and online. I was quickly flooded with hundreds of interesting and enlightening replies, and I have incorporated many of those comments and opinions into this discussion. The Talking Telescopes online discussion group, which originated as an extension of this book, is to be singled out and commended as a great forum for the exchange of experiences and ideas. You'll find a link to the group from my Web site: www.philharrington.net.

Speaking of my Web site, there is such a glut of telescopes on sale today that trying to discuss them all in this book is impossible. Because I didn't want to sell these scopes short, I have put those discussions in the "*Star Ware* Extras" section of www.philharrington.net. All models to be found there are still listed in Appendix A in the back of this book, but they are annotated with a note directing readers to the Web site.

You can play a role in the fifth edition of *Star Ware* by completing the survey at the back of this book. An online version may also be found at www.philharrington.net.

The Mysteries of the Orient

One of the biggest changes to hit the amateur telescope market since the 1970s is the tidal wave of Asian-made telescopes. For years, most low-end department-store telescopes were constructed in China or Taiwan, but it wasn't until recently that companies began to target the upscale amateur-level telescope market. Now, most name-brand companies offer many instruments from those same exporters. Two names to look for throughout the discussions to come are Synta Optical Technology Corporation and Guan Sheng Optical Co., Ltd. Each manufacturer concentrates on achromatic refractors and Newtonian reflectors, although Synta is also producing apochromatic refractors and some components in Celestron's Schmidt-Cassegrain telescopes. Synta has had a long-standing relationship with both Orion Telescopes and Celestron International, and, in fact, bought the latter when that company fell on hard times in 2005. Meade Instruments also imports many of its smaller telescopes from Far Eastern sources, as well. Even Tele Vue Optics uses Asian sources for its optics, although the company checks each component very closely during final assembly in its New York plant.

Reports from owners around the world indicate that there is also some variability in optical quality among some imports. Part of this is likely due to a second tier of companies that are now copying Synta and Guan Sheng products to varying degrees of success. Therefore, it is imperative that you know the retailer's return policy before you purchase.

Refracting Telescopes

Although many small refractors continue to flood the telescope market, the discussion here is usually limited to only those with apertures of 3 inches and larger (yes, with a few notable exceptions). As a former owner of a few 2.4-inch (60-mm) refractors, I found that I quickly outgrew those instruments' capabilities and hungered for more. Why? At the risk of possibly offending some readers, let me be brutally honest: most 2.4-inch refractors* will produce marginal views of the Moon, the brighter planets, and possibly a few of the brighter deep-sky objects, but nothing else. They are usually supplied on shaky mountings that only add to an owner's frustration. If that is all that your budget will permit, then I strongly urge you to return to the previous chapter and purchase a good pair of binoculars instead.

The refractors included here are further divided into two categories to make comparisons a little easier. The first section discusses achromatic refractors, while the second addresses apochromatic refractors.

Achromatic Refractors

You may recall from chapter 3 that an achromatic refractor uses a two-element objective lens to minimize chromatic aberration. Some false color, called *residual chromatic aberration*, will always remain, however, unless the objective's focal length is very long. In general, the longer the focal length, the less residual chromatic aberration.

Celestron International. In the last few decades, Celestron, best known for its Schmidt-Cassegrain telescopes, has branched out into other designs. In the

---

*There are some exceptions to this, such as models from Stellarvue, Takahashi, Tele Vue, and William Optics. Still, their small apertures greatly limit their effectiveness as astronomical instruments.

CONSUMER CAVEAT: Minus Violet Filters

The Color Purple might have been a good movie, but it is not what you want to see around the edge of the Moon or surrounding bright stars and planets when viewing through an achromatic refractor. To help suppress residual chromatic aberration, add a minus violet filter into the mix, either by screwing it into the eyepiece or, better still, into the telescope side of your star diagonal. A minus violet filter helps to diminish the purplish fringe around bright objects. It will also darken the background sky, making for a more aesthetically pleasing view. On the down side, however, it will tint everything yellowish. It also has no positive effect on increasing a faint object's visibility—in fact, it should be removed when viewing faint objects, because the filtering effect could make them vanish.

Although a minus violet filter is not a miracle cure—it won't turn your $300 achromatic refractor into a $2,000 apochromat—it does offer an inexpensive way of suppressing a problem that plagues most achromatic refractors sold today.

refractor category, Celestron imports refractors from the Far East that range up to 6 inches in aperture.

The least-expensive Celestron refractors included here are the Firstscope 90 AZ and Firstscope 90 EQ. Both models share the same 3.5-inch f/11 optical tube assembly, which includes a fully multicoated objective lens assembly housed in a glossy black steel tube. Optical performance is surprisingly good for the aperture and price of these instruments. The alt-az version includes a 45° erect-image diagonal for upright terrestrial viewing, while the equatorial version includes a superior 90° star diagonal. Focusing is smooth, but even the steadiest hand will cause the view to jump because of the weak mountings that come with the instruments. The alt-azimuth version, actually the sturdier of the two, includes a pair of manual slow-motion controls for steplike tracking of sky objects. The equatorial version also includes manual slow-motion controls and can be outfitted with an optional clock drive, if desired. Setting circles adorn the equatorial mount, but like most instruments of this genre, they are little more than decorative.

For those interested in a super-portable terrestrial spotting scope that can do double duty as a low-power, rich-field astronomical instrument, consider the Celestron 102 Wide View. This 4-inch f/5 achromatic refractor measures only 21 inches long and weighs just 5 pounds. One would expect that an achromat with such a fast focal ratio would be plagued with residual chromatic aberration, and indeed it is prevalent around brighter objects. That said, most owners report that they like the images that the telescope produces very much. The biggest appeal is the wide field of view. Not surprisingly, it doesn't do well for viewing the planets, but then again, it's not designed with planets in mind. Jupiter, for instance, shows some of its belts, but appears to be surrounded by a purplish corona at 73×. But that aside, a telescope like the 102 Wide View makes the perfect grab-and-go second telescope for anyone who cannot afford (or who is unwilling to spend) the cost of an apochromat.

If you'd prefer to let the telescope do the pointing for you, then the 3.1-inch f/5 StarSeeker 80, 3.1-inch f/11 NexStar 80SLT, and 4-inch f/6.5 NexStar 102SLT are worth considering. Each model rides on a one-armed GoTo alt-azimuth mount. The StarSeeker is a fun, wide-field telescope that can seek those stars up to about 80×, when the telescope's fast focal ratio begins to falter from spherical aberration. The NexStar 80SLT can continue up to about 120× before suffering a similar fate, making it the better choice of the two for viewing the planets. The 102SLT, although a wide-field scope by nature, can also handle up to about 120×. In all three cases, image quality is very good for the respective aperture and price.

Those images are also easier to enjoy through these telescopes than they were through the older NexStar GT models that they replace, thanks to the much improved SkyAlign GoTo system. Inside, the SkyAlign's microprocessor contains data and finding instructions for more than 4,000 sky objects, including the Moon, planets, and a host of deep-sky objects. Once the time, date, and location are entered into the onboard computer, the user only has to aim the telescope at three bright stars. No need to tell the computer which star is which; it automatically calculates the angles measured between the objects and then compares them to the angles between known objects. After the telescope mount is initialized, use the hand controller to find any of the objects listed in the onboard database. The system works well, usually getting the selected object in the field of the low-power eyepiece. Some owners lament, however, that their telescopes worked fine at first but then went haywire about an hour later. Therein lies the biggest problem with all GoTo scopes: power drain. A set of eight AA batteries will be fully drained after only a night's worth of use. To save the expense of buying new batteries every clear night, connect to a 12-volt rechargeable battery instead. Also be aware that the 80SLT's long telescope tube may hit the tripod legs when aiming near the zenith, which will knock the GoTo system off its mark and force the user to reinitialize the onboard computer.

CONSUMER CAVEAT: Batteries

All GoTo telescopes are power-mad. They love to eat batteries. In most cases, it takes eight AA batteries to feed the onboard computers found in inexpensive GoTos, such as the two SLT refractors discussed here. In continuous use, the batteries may last, at most, only a couple hours. As they drain, the computer's aim will become erratic at best. Unless you have an endless supply of inexpensive batteries, it's best to purchase the optional AC adapter (if an AC outlet is available) or a 12-volt rechargeable battery (like an emergency jump-start battery). The latter is available from many auto parts and hardware stores.

Celestron's 4-inch f/9.8 C4-R appears to be nearly an exact replica of its former GP-C102, a product imported from Vixen Optical in Japan. Many owners agree that the C4-R works quite well, especially when considering its cost. One owner thought the optics were not quite up to the quality of the original Vixen's but certainly suitable for beginners on tight budgets. Most owners were less critical, typically saying that the telescope produced "sharp images and acceptable amounts of false color." False color is inevitable around such unforgiving objects as Venus and the limb of the Moon through an achromatic refractor of this sort, but detail in the clouds of Jupiter and the rings of Saturn are crisp, while double stars snap into focus. Deep-sky objects also show well, but the small aperture does prove limiting.

The C4-R's rack-and-pinion focuser, so critical to producing sharp views through a telescope, is reasonably smooth, although not as refined as in some instruments. Part of the problem is the sticky grease that is used as a lubricant. Why Synta insists on using this "glue-bricant" on all of its telescopes' focusers and mountings remains a mystery. Use a spray degreaser to remove the grease, and then relube using dry Teflon.

The CG-4 mounting that is supplied with the C4-R is another weak link. Although the CG-4's equatorial head is reasonably well made, the all-aluminum tripod wobbles under the telescope's weight. (Tip: extending the legs as little as possible will help to steady things.) Slow-motion controls in both right ascension and declination are smooth, however, with little binding noticed. Adjustment screws let the owner tune gear tension, should that ever become necessary.

The C4-R has two 6-inch f/8 big brothers: the C6-R and the C6-RGT. Now, I know that the purists in the crowd will immediately balk at that focal ratio. Recall the discussion in chapter 3 stating that achromatic refractors will suffer from false color due to residual chromatic aberration if their focal ratios are too fast—that is, too low an f/ number. According to the formula there, a 6-inch achromat should be no less than f/17.3 in order to eliminate this problem. That's more than 8.5 feet long! So, how good could an f/8 achromat be? Frankly, I had my doubts, too.

Many owner reports have been received and nearly all have given the C6-R high marks for optical quality, noting a surprising lack of false color on all but the brightest sky objects. Don't get me wrong, the color purple is very obvious around the Moon, Venus, Jupiter, and so on, but it is not a showstopper. A few owners also added that whatever false color existed, it appeared to lessen as magnification increased. Note, however, that even when residual chromatic aberration is not visible to the eye, it still affects image sharpness, since light at different wavelengths focuses at slightly different distances.

To Celestron's credit, it heard the complaints from owners of the old CR150-HD 6-inch refractor, who had nothing but bad things to say about the telescope's CG-5 equatorial mount and tripod. As a result, the C6-R pair comes on an improved Advanced Series CG-5 mount with far more substantial 2-inch tubular legs made of stainless steel. Measuring 32 inches long when collapsed and extendable to 56.5 inches, the new CG-5 legs are similar in design to those used with Celestron's more expensive NexStar Schmidt-Cassegrains.

The redesigned CG-5 mount's axes are also smoother and more stable. The GT version of the mount adds a NexStar computer-controlled GoTo system. Initializing the NexStar computer is straightforward and well-detailed in the owner's manual.

Meade Instruments. There is no denying that Meade's introduction of computer-driven introductory-level telescopes back in 2000 stirred up a lot of interest, both among new stargazers as well as seasoned amateurs. These scopes were the first of their kind, and while there were some bumps along the way in terms of reliability, Meade's Digital Series (DS) refractors and reflectors started a trend that continues to this day. The original DS instruments have long since been replaced by better models.

Today's DS-2000 product line includes the 3.1-inch f/8.8 DS-2080AT-LNT refractor that comes on a one-armed computer-controlled mounting that outwardly appears to be the twin of Celestron's small NexStar SLT models. Like the Celestron, the DS's drive motors and associated cables are all internal to the mounting, which eliminates the concern over dangling cables becoming entangling around the telescope during operation. There is now only one external cable, which runs from the telescope to the Autostar hand controller. The DS-2080AT-LNT's brain features Meade's patented Level-North Technology, which has trickled down from the company's more expensive instruments. The user needs only to turn on the power and the telescope automatically levels and finds north. Time and date are set at the factory, but the user must enter his or her location when prompted during the start-up procedure. Once set, the Autostar GoTo aiming control works well, provided the required eight AA-size batteries are fresh.

Meade also sells two larger achromats: the 5-inch f/9 AR-5 AT and the 6-inch f/8 AR-6 AT. Overall, the optical quality of these instruments is quite good, although as one should expect, views do suffer from residual chromatic aberration. A purplish halo can be seen around even moderately bright stars, and it is strikingly apparent around the Moon, Jupiter, and Venus. Turning to deep-sky objects, both refractors do well for their apertures. Image contrast is quite good on showpiece targets such as the Orion Nebula and the Hercules globular cluster M13, while close-set double stars are well resolved. Some false color disturbs the view of brighter doubles, such as Albireo.

Like the Celestron Advanced Series CG-5, the LXD75 German equatorial mount that comes with the AR-5 and the AR-6 is modeled after Vixen's venerable Great Polaris mount. Meade learned several lessons from the first incarnation of this mount, the LXD55, which had several design flaws. The LXD75 is sturdier thanks to internal steel ball bearings and a much improved tripod with stainless-steel tubular legs. Although far better than before, the LXD75 is still too light-duty to support the heavy 6-inch AR-6 steadily. With that problem and the slightly higher level of chromatic aberration, I would advise against the 6-inch telescope but will recommend the AR-5. It is a nice compromise between aperture and heft, and it is certainly a far better choice than some of the 4- and 4.7-inch achromatic refractors sold today that ride on inferior mountings.

Both AR refractors are motor driven and come with Meade's Autostar GoTo control. The Autostar's pointing accuracy is certainly good enough to get objects in the field of a low power eyepiece, as long as the mount's polar axis is at least approximately aligned with the celestial pole. If I had any complaint about the Autostar, it's how loud the motors are when slewing. There is a "quiet slew" feature on the drive that slows the slewing speed, thereby lessening the noise. Your neighbors will thank you.

One advantage that the LXD75 enjoys over the CG-5 is that it includes periodic error correction (PEC) control. The PEC control tweaks the tracking rate for periodic errors in the right ascension axis's drive, which are inherent to all worm-gear drives. Once the PEC circuit has been "trained" by the user, which must be done each time the LXD75 is used because it has no memory, it will minimize the need to correct for guiding inconsistencies. For visual observations, where keeping a target somewhere in the field is usually adequate, PEC is of little value. Its true benefit shines through, however, during long-exposure photography, when the slightest tracking deviation can ruin pinpoint images. PEC will not, however, correct for improper polar alignment. More about periodic error correction can be found on page 142.

Orion. Orion Telescopes and Binoculars of California imports several refractors from the Far East and sells them under the Orion name. Let's begin with the company's ShortTube 80, the very popular 3.1-inch f/5 achromatic refractor shown in Figure 5.1. Its images are surprisingly good, with some spectacular

Figure 5.1 *The ShortTube 80 from Orion Telescopes remains one of the most popular grab-and-go instruments sold today. The ST80, shown here, is mounted on a Bogen/Manfrotto 3421 gimbal-style camera mount. Photo by Dave Mitsky.*

wide-field views of Milky Way star clouds possible. But be prepared for strong false-color fringing evident around brighter stars because of the fast focal ratio. Potential purchasers should also note that the instrument is not suitable for high-powered planetary views. Although it is not an ideal first telescope for that reason, the ShortTube 80 is almost universally proclaimed by its owners as the perfect grab-and-go second (or third, or fourth) telescope. Most users warn, however, that the 45° star diagonal that comes with the telescope dims images noticeably. Replacing it with a 90° diagonal not only eliminates the problem but also makes sky viewing more comfortable. Given that it includes a built-in camera-tripod adapter plate, a detachable 6 × 26 "correct image" finderscope, and two Plössl eyepieces, the Orion ShortTube refractor is one of the best buys in small rich-field achromatic telescopes today.

Orion sells several versions of the ST-80. You can buy just the optical tube assembly if you wish, or purchase the scope on its Paragon photo tripod or EQ-1 German equatorial mount. The Paragon tripod is fine, but it is really designed for terrestrial viewing only. Even though the EQ-1 mount is the best choice of the bunch, it is still a little flimsy. Finally, for price-conscious consumers, note that the Sky-Watcher 804, reviewed elsewhere in this section, is identical to the ShortTube 80. Only the mount differs.

Recently, Orion introduced its 3.1-inch (80-mm) f/6 Express 80 short-tubed refractor. Identical in style and appearance to the William Optics Zenith-star 80 discussed later in this chapter (and Antares Sentinel), the Express 80 includes a 2-inch rotatable Crayford-style focuser, a retractable dew shield, a padded travel case, and a mounting bracket for attaching the telescope to a standard photographic tripod. The Express 80 does not come with a star diagonal or a finderscope (although it is predrilled to accept Orion's finderscope dovetail base), and it is sold alone or with the Orion EQ-2 equatorial mount. The EQ-2 is too weak to support the telescope adequately. Instead, buy the optical tube assembly à la carte and then purchase a Universal Astronomics UniStar Light alt-azimuth mount, or a Celestron CG-5 or an Orion SkyView Pro equatorial mount separately.

Images observed through the Express 80 are more contrasty than through the ShortTube 80, although some owners report problems with collimation and astigmatism. Assuming those problems are absent, the Express 80 proves a competent performer for wide-field viewing. Even up to 100×, the view is sharp when armed with premium eyepieces. Just don't be misled by Orion's use of the term *semi-apochromatic* when describing image quality. The Express 80 is an achromat, complete with the residual chromatic aberration that one would expect in an f/6 refractor.

If you prefer a more traditional refractor, Orion offers 3.1-inch (80-mm) f/11.4 and 3.5-inch (90-mm) f/10.1 refractors that are worth a look. The Explorer 80 and Explorer 90 include multicoated objective lenses that produce reasonably sharp results, with decent images seen during star testing. Some minor residual chromatic aberration is evident through the 90 around brighter objects, while it is all but absent in the 80. Mechanically, both Explorers come with focusers that are fairly smooth, although some owners have mentioned minor irregularities as they rack the units in and out. Each tele-

CONSUMER CAVEAT: "Semi-Apochromat"

Some achromatic refractors tout themselves as being "semi-apochromatic," inferring that they suppress false color better than other achromats. You should view most of those claims with a good amount of skepticism. By definition, a refractor that focuses two specific wavelengths, typically red and blue, to the same point is referred to as an *achromat*. The result is better correction of false color than a simple lens, but residual chromatic aberration is inevitable. The objective lens of an apochromatic refractor focuses three colors (red, blue, and yellow-green) to the same point, thus eliminating residual chromatic aberration. So, what then does a semi-apochromat do—focus 2.5 wavelengths? Although there is a technical definition, the term as used today is mostly marketing hyperbole.

scope rides atop Orion's AZ-3 alt-azimuth mounting, a reasonably sturdy platform that offers dual slow-motion controls on both axes. As you may have inferred already, I prefer these sorts of mounts over small equatorials, because they are almost always steadier. True, you can't install a clock drive and they don't have setting circles, but the slow-motion controls are usually smooth enough to let observers keep things pretty well centered.

The same optical tube assemblies are combined with a 6 × 30 finderscope, a 90° star diagonal, and the EQ-2 German equatorial mount to create the Orion AstroView 80 and AstroView 90. Although the optics are unchanged, the EQ-2 mount adds nothing but shakes and jitters. As is typical of many imports, the mounting itself is not as sturdy as it should be. As a result, you would do better to spend less and get the Explorer versions. Before taking the plunge with either style, however, keep in mind that while they show nice views of the Moon and naked-eye planets, the small apertures will dramatically limit deep-sky exploration. Instead, spend about the same amount of money on a 6-inch Dobsonian-style Newtonian reflector or save up some extra money and get a 4-inch refractor.

Orion's 3.9-inch AstroView 100EQ refractor is versatile enough for wide-field viewing of rich Milky Way star fields as well as for planet watching. Just avoid pushing the magnification too high. Optically, the Orion 100 is a strong performer, with very good image sharpness. Residual chromatic aberration is inevitable in a 4-inch f/6 achromat, but unless you spend your time gazing at stars like Vega and Sirius, chances are good that it won't bother you much. To diminish chromatic aberration when looking at the Moon, leave the lens cap on, but remove the cover over the central opening. That will change the telescope to a 2-inch f/12 refractor, effectively eradicating any hint of false color around the lunar limb (but at the cost of reducing the overall resolution). But even at full aperture, Jupiter and Saturn put on very nice shows, as does the Double-Double quadruple star system, a great test for an instrument like this.

Orion's AstroView 120, a 4.7-inch f/8.3 achromatic refractor, has attracted a good amount of attention since its release several years ago. Like so many

CONSUMER CAVEAT: Spherical Aberration in Imported Refractors

As mentioned in the discussion of the Orion AstroView 120 refractor, many imported instruments show varying degrees of spherical aberration, resulting in poorly focused, fuzzy images. I believe that variability is often due not to lack of quality control in optical manufacturing but rather in assembly at the factory. It turns out that the objective cells in many Asian refractors are often screwed down too tightly as they are assembled, which warps the lenses. If you find this to be the case, try loosening the screws along the rim of the cell that hold it together just slightly, no more than about a quarter turn. That should help relieve the pressure and lessen spherical aberration. But if it does not, never, ever disassemble the lens cell!

telescopes, the AstroView 120's Achilles' heel is its flimsy aluminum tripod legs. To steady things out, Orion sells the same telescope on its SkyView Pro equatorial mount. The SkyView Pro 120, seen earlier in Figure 3.2, is certainly a sturdier package than the AstroView version thanks to a brawnier mount and tubular steel tripod legs, but it will cost you about $100 more. Predictably, residual chromatic aberration spoils the view of brighter objects through both versions, but overall, the SkyView Pro 120 is a competent performer. Some owners have reported unacceptable levels of spherical aberration in their individual telescopes, although this problem seems to vary from unit to unit. I enjoyed views through a SkyView Pro 120 not long ago at the annual Astroblast Star Party near Franklin, Pennsylvania. This particular scope cleanly resolved the Hercules globular M13, while dark space was evident between the two tight pairs that make up the Double-Double quadruple star system. Spherical aberration, although evident in star testing, was minor. That particular telescope's biggest problem, however, was its mount's incredibly tight right-ascension axis. I had trouble moving it even with the axis lock completely loosened.

Many people who are looking for a second instrument to compliment their existing equipment are drawn to 3.1- or 4-inch f/5 refractors. Others who prefer to think a little bigger might consider the Orion AstroView 120ST. The 120ST (short tube) packages a 4.7-inch f/5 optical package into a sleek-looking black metal tube that includes all the features of the standard AstroView; only the focal ratio differs. False color and spherical aberration are ever-present, although if you keep magnification down below about 60×, these distractions are surprisingly tolerable. I do offer one caution, however: if you are looking for a transportable telescope to carry on board an airplane for instance, the AstroView 120ST is probably not for you. But for a relatively compact, lightweight instrument (total weight with the mount is 37 pounds) that gives some great low-power views, then this is your next instrument.

Sky-Watcher Telescopes. With its North American sales office in Richmond, British Columbia, the Pacific Telescope Company sells many Sky-Watcher–brand refractors manufactured in China by the Synta Optical Technology

Corporation. The least-expensive model included here is the 3.1-inch f/5 Sky-Watcher 804, which bears more than a passing resemblance to Orion's Short-Tube 80. The tube colors and basic packages vary somewhat but are otherwise identical.

In the 3.5-inch aperture bracket, the Sky-Watcher 909 is an f/10 instrument that comes on either an AZ3 alt-azimuth or EQ2 light-duty German equatorial mount. The alt-az mount strikes me as the more stable. Unfortunately, the AZ3 version comes with a 45° diagonal, which as you know by now, is not suitable for astronomical viewing.

At the opposite end of the aperture spectrum, we have the 6-inch f/5 Sky-Watcher 15075HEQ5 refractor. That's right, a 6-inch f/5 achromatic refractor. One look at the formula in chapter 3 should make you scratch your head and say, "What are they thinking?" An f/5 must be overflowing with false color. And you'd be right. Residual chromatic aberration is as expected around bright stars, the planets, and the Moon. Okay, that's conceded. But when magnification is kept to 50× or less, and the telescope is aimed toward suitably large deep-sky objects like the Pleiades or Double Cluster, the views can be breathtaking, just as they are through the 15075's smaller 4-inch f/5 and 4.7-inch f/5 short-tube cousins. The extra aperture only serves to bring out more detail in the Orion Nebula, Andromeda Galaxy, and the Lagoon/Trifid Nebulae pair.

The Sky-Watcher 15075 comes mounted on the HEQ-5 German equatorial mount, a smaller version of Synta's EQ-6 heavy-duty mount. Both axes move smoothly, although there is some gear backlash that causes a slight lag when the clock drive is first engaged. Although the HEQ-5 is easier to carry than the massive EQ-6, its spindlier tubular legs are also weaker.

Several other achromatic refractors sold by Sky-Watcher are reviewed elsewhere in this section but under different names. In the interest of space, and rather than simply reiterating their pros and cons here, Table 5.1 lists equivalences and points you toward those other discussions. I would offer just one word of warning: pay close attention to the supplied eyepieces and, especially, mountings. Often a telescope will appear less expensive when sold by one company versus a second, when in reality, the difference is due to a switch in standard equipment.

Stellarvue. Owned by amateur astronomer Vic Maris, Stellarvue has established itself as a supplier of some of the highest-quality achromatic refractors sold today. Like many companies, Stellarvue outsources its optics. Unlike most

Table 5.1. **Telescope Equivalencies**

Sky-Watcher	Celestron	Orion
1025	102 Wide View	—
1021	C4-R	—
1206	—	AstroView 120ST
1201	—	AstroView 120
15012	C6-R	—

of the competition, however, Maris actually assembles and tests each instrument personally. This lets him control quality very closely, ensuring that each instrument that goes out works as it should.

Stellarvue sells several 3.1-inch (80-mm) short-tubed refractors, and the company is constantly tweaking its product lineup to improve those offers. All models are based on Stellarvue's original f/6 Nighthawk, which I dubbed in the third edition of this book the "king of the short tubes." The Stellarvue Nighthawk performed amazingly well for its size. Owners raved about the images, with one saying that "the quality is good, portability can't be beat, and the views are fantastic."

Since then, Stellarvue has expanded the Nighthawk family. First, we have the less-expensive, lighter-weight Nighthawk 2. The Nighthawk 2 comes with a red-dot unity finder, but a star diagonal and eyepieces are sold separately. The instrument also includes a 2-inch Crayford focuser. A few mechanical frills from the original Nighthawk have been cut to save some money, such as using a dust cap that presses on rather than threads on, but overall, the Nighthawk 2 is a very good scope. Although some minor false color exists around brighter objects, it is less than through the multitude of 3.1-inch f/5 short-tubed achromatic refractors sold today. Spherical aberration is also nearly nonexistent. As a result, although this is not necessarily an instrument for the planets, it holds its own surprisingly well when magnification is pushed. Indeed, the Stellarvue Nighthawk still reigns as the king of the short tubes. True, it costs more than many imported, mass-produced f/5 short-tube refractors, but the extra money is worth it.

The f/7 Nighthawk Next Generation also includes a rotating focuser, but Maris states that its improved design holds collimation better than the Nighthawk 2. I would still prefer a nonrotating focuser for absolutely perfect alignment at all times. In terms of performance, the Nighthawk Next Gen was introduced too close to press time to draw any conclusions just yet, but given the longer focal length, better color suppression would seem likely. Is it enough to unseat the original Nighthawk as the king of the short tubes? Time will tell. As reports are received and compiled, summary findings will be posted in this book's online extra material found at www.philharrington.net.

Stellarvue also sells a 3.1-inch f/9.4 refractor known as the SV80D, which, like the f/6, comes in several versions that differ only by accessories. Again, all are assembled and tested personally by Maris. The SV80D, seen in Figure 5.2,

Figure 5.2 *The Stellarvue SV80D includes an innovative split tube. Keep the short center section in place for traditional viewing with one eyepiece and a star diagonal or remove it when using a binocular viewer. Photo courtesy of Stellarvue Telescopes.*

comes with a 2-inch Crayford focuser and a red-dot LED unity finder, which are standard fare these days. But what sets the SV80D apart from the rest is the two-piece tube. By removing the threaded extension, the SV80D can be used with a binocular viewer (which are discussed in chapter 6) without needing a compensator lens to achieve focus. It's a simple fix, but one that no one else thought of.

The SV80D can be purchased either as an optical tube assembly alone or with a binocular viewer, 2-inch star diagonal, and a pair of matched eyepieces as the Nighthawk 80BV. Stellarvue's M1 alt-azimuth mount and M4 equatorial mount are also available optionally. Both are adequate for visual observing only, although if you are considering an equatorial mount, choose the Celestron CG-5 instead. It costs the same as the M4 but is substantially sturdier. If long guided exposures are your aim, then aim first at a better mount, such as the Losmandy GM-8.

The Stellarvue SV80D does a great job on brighter sky objects. There is some minor color fringing around objects like the Moon and Venus, but it really does not degrade the image noticeably. Indeed, Jupiter displays a sharp, color-free disk criss-crossed by several bands at 100×, while Cassini's Division in Saturn's rings can also be spotted easily. Most deep-sky objects appear faint and indistinct, however, because of the limited aperture. But even though its aperture is small, the SV80D is not an inexpensive telescope. If the quiz in chapter 3 pointed you its way, be sure to consider all your options before purchasing.

Vixen Company, Ltd. Vixen is a highly regarded manufacturer of refractors, reflectors, and telescopic accessories. In the category of achromatic refractors, Vixen offers several conventional and a few not-so-traditional instruments.

Although Vixen manufactures smaller telescopes, the smallest models included here are the 3.1-inch f/11.4 80M and the 3.1-inch f/5 80SS achromatic refractors. Like all Vixen instruments, the 80M's optical quality is excellent for its aperture. The Moon is exceptionally sharp through this telescope, with crisp crater rims, sharply defined mountains, and very dark shadows. Jupiter and Saturn are also treats through this telescope, as are many showpiece double stars like Albireo. Of course, the 80-mm aperture restricts the view to brighter sky objects, which some users will find very limiting.

The 80M's mechanical quality is just as good as its optics, with a smooth 2-inch rack-and-pinion focuser, glossy white tube, and attractive trim. The 80M also comes with a red-dot finder and an interesting flip-mirror star diagonal. Like the flip-mirror accessories described in chapter 7, the 80Ms can be changed instantly from right-angle to straight-through viewing simply by turning a small knob. This is especially handy for photography. Attach your camera to the straight-through port and place an eyepiece in the right-angle eyepiece holder. Once a shot is composed through the eyepiece, flip the mirror out of the way and shoot without having to worry about whether the target is still in view. Vixen packages the 80M with either its Porta alt-azimuth mount or Great Polaris-E equatorial mount. Both mounts are suitable for visual observation, while the latter is better for photography.

Although the 80SS is plagued by more chromatic aberration than the 80M because of its considerably faster focal ratio, it still has substantially less aberration than other 80-mm f/5 refractors. Spherical aberration is also much less evident than through cheaper Chinese-made f/5 clones. Indeed, one of the finest views I've ever had of the Pleiades was through an 80SS. True, the 80SS costs about double what those mass-market short tubes retail for, but if you are more interested in quality than cost, the Vixen 80SS is an excellent choice. Just save enough to buy a decent mounting, which is sold separately. Either Vixen's Porta alt-azimuth mount or the Great Polaris-E equatorial mount would be suitable, as would be a Manfrotto or a Gitzo photographic tripod discussed in chapter 7.

If you are looking for Vixen quality but with a little more aperture, the 4-inch f/9.8 Vixen 102M refractor comes mounted on either Vixen's highly praised Great Polaris-Economy (GP-E) German equatorial mount or sophisticated Sphinx GoTo equatorial mount, both on extendable aluminum tripods. As the owner of an older Celestron C102, which is essentially the same instrument previously imported into the United States by Celestron, I can attest to the optical excellence of this classic telescope with great enthusiasm. Stars show textbook images, with nicely defined Airy disks encircled by sharp diffraction rings. If there is any spherical aberration, it is all but undetectable. Color correction is also very good for a sub-f/10 achromat.

A telescope is only as good as its mounting, and in this case, the Great Polaris-E and Sphinx mounts are both well qualified to support the 102M. Dampening time when the telescope is rapped with the heel of the hand is only one to two seconds, rating an "excellent" in my book. So, once again, the question becomes: is the Vixen better than a less-expensive Chinese clone refractor? Yes, absolutely! Expect to pay several hundred dollars extra for that superiority, but in the long run, most people find it worthwhile.

Let's take a closer look at the Vixen Sphinx mount, since it is also used with several other Vixen instruments. Unlike your typical German equatorial mount equipped with GoTo computer control, such as Celestron's Advanced Series CG-5, Vixen's Sphinx is not just copying a trend; it's setting the trend for others to copy. The biggest plus to the Sphinx over any other GoTo mount sold today is its hand controller, called the Star Book. This is no mere plastic box with cryptic digital readouts scrolled across an LCD screen. The Star Book is, in effect, a digital planetarium coupled perfectly to its mount. The Star Book's screen mirrors the view of the sky, with its crosshairs showing the scope's aim. The right side of the screen lists the current time and date, the right ascension and declination where the telescope is pointing, the right ascension and declination of the cursor position, and other data. Move the cursor around using the toggle buttons on the Star Book, press GoTo, and the telescope obeys. In many ways, the Star Book can be likened to controlling the telescope using a laptop computer, except that the Star Book is much more compact as well as faster to respond. Users of the Sphinx mount marvel at its sturdy design, convenient setup, and flawless operation. Some comment that initializing the GoTo computer is not as intuitive as more conventional systems, but once you get used to it, it goes quickly.

William Optics. Since they first appeared on the world market a decade ago, William Optics (WO) apochromatic refractors have established a very good name for themselves. Capitalizing on the company's growing reputation, brothers William and David Yang from Taiwan have introduced several new telescopes, including the midpriced 3.1-inch f/6 Zenithstar 80 achromatic refractor. William Optics also sells a shortened version, appropriately called the Zenithstar 80 Short, which is designed expressly for use with binoviewers. As mentioned earlier, this is a growing market segment and one that will likely expand further in the years to come. Like the Vixen 80SS and Stellarvue Nighthawks, the WO Zenithstar 80's multicoated objective produces reasonably sharp images when restrained to about 80×, although chromatic aberration is ever present. A small amount of astigmatism also detracts from image sharpness. As I warned with the Orion Short-Tube 80, Celestron NexStar 80SLT, and other short-focus achromats, owners should not try to make the Zenithstar to do something it's not intended to do. It's not for planet-watching. If that is what you are looking for in a telescope, don't even consider this one. Overall, I would judge the Zenithstar a good performer but not up to the standard established by the Stellarvue Nighthawk.

Apochromatic Refractors

If you are looking for the best of the best, then this section is for you. Apochromatic refractors, with their multiple-element objective lenses made of exotic materials, are generally considered to be as close to an optically perfect telescope as one can get. Images are almost always tack sharp and high in contrast, offering spectacular views of the planets and deep-sky vistas alike. But this perfection comes with a steep price tag. Is one right for you?

CONSUMER CAVEAT: Small Is Still Small

There is no doubt that many apochromatic refractors epitomize optical perfection. Trying to decide among the long list of candidates detailed here is difficult even after you have read all there is to read, and have even looked through several side by side, as I have. Two things are clear, however. First, all of the name brand apochromats have something going for them. That should become clear as you read the superlatives to come. Second, remember that even the finest 3.1-inch (80-mm) apochromat out there is still an 80-mm telescope. It will not outperform a 6-inch reflector or an 8-inch Schmidt-Cassegrain telescope costing less, sometimes substantially less. I have no doubt that the images may be sharper, crisper, and more contrasty through the smaller apo than the others, but they will also be dimmer. Apochromatic refractors don't have a big red "S" painted on them; they are not Super Telescopes. They can't leap over tall buildings in a single bound nor can they outrun speeding locomotives. As Scotty often said to Captain Kirk on *Star Trek*, "I canna change the laws of physics."

Astro-Physics. Astro-Physics is a name immediately recognizable to the connoisseur of fine refractors on rock-steady mounts. Owners Roland and Marjorie Christen introduced their first high-performance instruments in the early 1980s and effectively revived what was then sagging interest in refractors among amateur astronomers. Now, two decades later, Astro-Physics refractors remain unsurpassed by any other apochromat sold today.

As this edition of *Star Ware* is published, Astro-Physics is redesigning its line of refractors. Only the 160EDF, an exceptional 6.3-inch f/75 instrument, remains at present. Therefore, rather than offer speculation here, I'll ask that you visit the chapter 5 supplemental material found in the *Star Ware* section of www.philharrington.net. Information, test reports, and owner comments will be posted and analyzed there as the information becomes available. But two things are likely, given Astro-Physics' past record of performance. First, any instrument wearing the Astro-Physics name will undoubtedly be the finest of its kind. However, because of the high demand and limited production, delivery will likely take years. Considering the nearly instant availability of fine apo refractors from Takahashi and Tele Vue, waiting that long for an Astro-Physics refractor is difficult to justify.

Celestron International. Celestron International sells three apochromatic refractors, both of which are manufactured by its parent company, Synta, in China. The least expensive of the bunch is the Onyx 80EDF, a compact 3.1-inch f/6.25 telescope that comes in a custom-padded aluminum carrying case that's perfect for airline travel. Features include a sliding dew shield, a rotatable Crayford-style focuser, a dovetail mounting block, lens caps, and a small side-mounted aiming tube. Eyepieces and a star diagonal are sold separately.

To reduce chromatic aberration, the two-element objective in the Onyx 80EDF combines a lens made of Fluoro-crown glass (technically specified as O'Hara FPL-53) and with another of high-density crown glass. FPL-53 is the most advanced extra-low dispersion glass available, which lets optical designers shrink focal ratio while maintaining color correction. FPL-53's optical properties are very similar to those of calcium fluorite, which is used in the most expensive apochromatic refractors. To enhance image contrast, the objective's elements are also vacuum-coated with Celestron's Starbright XLT coatings.

The strength of the Onyx 80EDF is its superb low-power, rich-field views. Scanning the Scorpius and Sagittarius regions, for instance, is a real treat. The Onyx is capable of magnificent views of such diverse targets as the open star clusters M6 and M7 in Scorpius and the Lagoon/Trifid Nebulae region in Sagittarius. And although the Onyx's strong suit is not as a planetary telescope, it does a respectable job with Jupiter and Saturn. My own tests showed that images stayed sharp even up to 250× by using a 5-mm Orion Stratus eyepiece and Tele Vue 2.5× Powermate. I could make out a very slight purplish halo of false color encircling Jupiter's limb as well as around the edge of the Moon. But overall, the Onyx is a competent performer.

The C80ED is also a 3.1-inch instrument but with a slower f/7.5 focal ratio. The mechanical quality of the C80ED betrays its low price. The metal tube, thread-on metal dew cap, and focuser housing are all nicely finished in black with the company's signature orange trim, while the 2-inch rack-and-pinion focuser works very smoothly with no backlash.

Images are nearly free of false color, save for some hints around exacting test objects, such as Sirius and the lunar limb. Overall, however, they are certainly as color-free as those seen through some far more expensive instruments. There is some evidence of spherical aberration, but that is often due to the problem of the manufacturer tightening down too hard on the objective cell prior to shipment. Review the comments on page 90 for a way to fix that problem, if detected.

The C80ED is available in three configurations. The spotting scope version can be attached to a sturdy photographic tripod and works just fine for quick views once the 45° diagonal that comes with the scope is swapped for a 90° unit. The second variation, the C80ED-R, is sold on the Advanced Series CG-5 equatorial mount, which is more than acceptable for the size and weight of this little refractor. Finally, the upscale C80ED-RGT throws in Celestron's NexStar computerized guidance system to create the least-expensive GoTo apo on the market today. All three versions work well and rate very highly for the size class.

Celestron's C100ED is a slightly larger, 3.9-inch f/9 version of the C80-ED. Like its kid brother, the C100 is available alone as a spotting scope or as the Advanced Series CG-5 with the NexStar GoTo system as the C100ED-RGT or without as the C100ED-R. All three share many of the pluses and minuses as the 80ED. Minor color fringing is inevitable around tough test objects, but the relatively slow focal ratio coupled to the two-element ED objective does well at squelching more residual chromatic aberration. Also on the plus side are the sleek black tube emblazoned with orange highlights, a 9×50 straight-through finderscope, and a 2-inch rack-and-pinion focuser. The focuser is not as smooth as the 2-inch Crayford focuser found on Orion's equivalent, the 100ED discussed next.

Orion Telescopes and Binoculars. Orion was the first major company to bring economy-priced apochromatic refractors to the Western world when it introduced the 80ED. The 80ED, a 3.1-inch f/7.5 refractor made by Synta, uses a two-element ED objective lens to help diminish residual chromatic aberration while boosting image contrast. This is a true apo for the masses. The optics are amazing for the price, with the little 80ED often handling upward to 200× before significant image degradation occurs. Diverse targets such as Jupiter, Saturn, the Moon, the Pleiades, and the Orion Nebula all show good detail when the 80ED is teamed with high-end eyepieces.

As to mechanical quality, the 80ED is competent if a bit uninspiring. The dark-gray tube, for instance, is adequate for the task, but it will not win any prizes for design innovation. I have also heard that some owners complain that the 2-inch Crayford focuser is too sloppy in use. To the contrary, I find it to be smooth and easy to use, and it is easier to control than the grease-glued rack-and-pinion focuser on Celestron's C80ED. Photography, however, may be another matter. Although fine focus is easy to achieve, it is difficult to hold because of a locking thumbscrew that is not sufficiently strong to hold the focuser in place.

Capitalizing on its success with the 80ED, Orion quickly followed with the 4-inch f/9 100ED and later, the 4.7-inch f/7.5 120ED. Like the 80ED, the 100ED and 120ED each include an FPL-53 ED lens in its two-element objective to diminish false color. Overall results in the 100ED are very good indeed. One discriminating owner from Toronto explained: "I did a lot of side-by-side comparisons with my other scopes in the 4- to 5-inch range, and the 100ED equaled or bettered them all, though it was close in some cases. Perhaps not quite as good as the Tele Vue TV102, but pretty darned close, even up to 250×." The 120ED, however, suffers from more residual chromatic aberration than its smaller kinfolk owing to its larger aperture and faster focal ratio. Although images are better through the 120ED than Orion's 120-mm achromatic refractor, I find that the 100ED strikes a better balance between aperture, focal ratio, and aberration suppression. Like the 80ED, the 100ED and 120ED each include a sleek, dark gray tube and a 2-inch Crayford focuser but is otherwise unadorned. Some owners have mentioned they find the focuser to be too weak and sloppy for precise viewing at high power. In some cases, they have actually replaced the original focuser with one from Starlight Instruments (Feathertouch) or MoonLite Focusers. Even though the focuser costs a good percentage of the scope's original cost, they are pleased with the result.

All three Orion apos can be purchased à la carte or teamed with either Orion's SkyView Pro or Sirius's EQ-G German equatorial mount. Both are more than adequate for supporting either telescope. The SkyView Pro, an adaptation of Synta's EQ-4 (Celestron CG-5) mount, includes dual-axis manual slow-motions, tubular tripod legs, and the option of adding digital setting circles or GoTo computer control. The Sirius mount, based on Synta's HEQ5, adds a heavier duty equatorial head, an illuminated polar scope, GoTo control with a built-in 13,000-object data library, periodic error correction, and the ability to accept an autoguider for precise tracking during long photographic exposures. Overall build quality is very good, with smooth motions in both right ascension and declination. Vibrations also dampen quickly, although the tubular steel legs are a little spindly when fully extended. It's best to keep them collapsed as much as possible, which could cause some consternation when viewing through the eyepiece of a refractor. Power to the Sirius's fully internalized DC stepper motors can be provided from a car lighter plug using the supplied power cable, from an optional rechargeable battery, or with an AC-to-DC adapter. All things considered, the packages that include the Sirius mount would make very good astrophotographic platforms for an intermediate amateur armed with the proper CCD camera. Just keep in mind that although the scopes are small enough to qualify for grab-and-go status, the mount will probably have to be carried out in two or more trips.

As we will see, the SkyView Pro and Sirius mounts are also sold with several other Orion telescopes, including an 8-inch Newtonian reflector and a Celestron-supplied Schmidt-Cassegrains. All are discussed later in this chapter.

Sky-Watcher Telescopes. In late 2005, Pacific Telescopes announced its new line of Pro-Series telescopes, including 3.1-inch f/7.5, 3.9-inch f/9, and 4.7-inch f/7.5 ED apochromatic refractors. Given the performance of its other ED-refractor incarnations, including Celestron's C80ED and C100ED as well as Orion's 80ED, 100ED, and 120ED previously discussed, we can expect these to perform very well indeed. Each model incorporates a two-element, air-spaced objective lens. The inside element is made from extra-low dispersion (ED) FPL-53 fluorite glass, designed to minimize false color and maximize contrast. Yes, there is still a minor amount of false color (residual secondary spectrum) visible around very bright objects, but considering the comparatively low prices of these versus, say, a top-line Tele Vue or Takahashi instrument, we can afford to be a little forgiving.

Each of the Pro-Series apos is mounted on a GoTo computerized version of the company's HEQ5 German equatorial mount. All models in the Pro-Series come with 2-inch Crayford focusers, 9×50 finderscopes, and tripods with tubular legs of stainless steel. Be guided by the comments about the Orion apos for specifics, but overall, the Sky-Watcher Pro Series telescopes deserve high marks for their optics and mechanics alike.

Stellarvue. No other company has expanded its line of apochromatic refractors in recent years as much as Stellarvue. Of the many Stellarvue refractors that carry the highly valued description, *apochromat,* the smallest detailed here are a pair of 3.1-inch refractors, the f/6 SV80S (as in "short"), and f/7.5 Nighthawk T (also known as the SV80L). Both of these little powerhouses use three-element objective lenses, with a fluorite element sandwiched between two glass lenses to correct for spherical and residual chromatic aberrations, coma, and other higher-order optical flaws. Both instruments perform flawlessly, with exceptionally sharp images that ooze in contrast. Add to that a 2-inch Feathertouch focuser from Starlight Instruments—arguably the best focuser made today—and it's clear that these scopes are winners.

Moving up a notch in aperture, Stellarvue's newest family member is the 3.5-inch f/7 SV90T. To suppress false color to virtually undetectable levels, the inner element of SV90T's triple-element objective is made of calcium fluorite. Although the telescope was not widely available at press time, given Stellarvue's reputation, we can expect it to be an exceptional instrument. With the list of components including a hard-shell carrying case and a 2-inch Feathertouch dual-speed focuser, it certainly has all the makings of a modern classic. As with its Nighthawk achromatics discussed earlier, Stellarvue designed a removable tube section that allows the SV90T to focus either with a conventional star diagonal or with a binocular viewer plugged into the focuser. With that section removed, the telescope measures only 16 inches long, easily within commercial airline carry-on baggage limits.

Stellarvue also has instruments in the 4- to 4.5-inch range. The SV102T is a 4-inch f/7.8 scope with a three-element objective. False color is very nearly eradicated entirely, making this a great choice for planetary viewing. The SV102T includes the same great dual-speed Feathertouch focuser as the deluxe versions of the SV80s. A removable extension tube threads on and off so that the SV102T can be used with a binocular viewer. With the 4-inch-long extension tube in place, images focus just fine through eyepieces using the supplied 2-inch star diagonal. Then, remove the extension tube, and images focus correctly with a binoviewer inserted into the system. (As discussed in chapter 6, binocular viewers have internal light paths measuring about 4 inches long. As a result, eyepieces must be moved in to compensate, a problem for most conventional telescopes.)

The SV102T is overshadowed, however, by the 4-inch f/6.4 SV4, probably the most highly praised of all Stellarvue apochromats. The heart of the SV4 is an air-spaced triplet objective made of a very high homogeneity Super ED glass. Full multicoatings are used on all six surfaces to enhance light transmission and contrast. The standard SV4 includes a single-speed 2-inch Crayford focuser, while the more-expensive deluxe version throws in a two-speed Feathertouch focuser. Both include a removable extension tube for binoviewers. The many owners whom I have heard from all have nothing but accolades for the SV4. And I can see why. A recent test of the telescope that I did for *Astronomy* magazine confirmed textbook images. Star testing revealed exceptional optical quality, which was verified with every object I viewed. Whether aimed toward the Moon, Jupiter, Saturn, or bright deep-sky objects, it is easy to see why the SV4 is one of the most sought-after and highly recommended 4-inch apochromats available today.

Like their achromatic counterparts, the SV80S and the SV80L can be supported by a wide variety of mountings, including Stellarvue's own M1 alt-azimuth mount and the M4 German equatorial mount. Although it rides on a wooden tripod, the M4 is not as sturdy as Celestron's current generation Advanced Series CG-5 mount. The larger SV4 and SV102T can also ride on the CG-5 or on Stellarvue's well-built M6 alt-azimuth mount.

Larger apochromats in the 5- to 6-inch range are also in the offing for Stellarvue to replace the models that were dropped from production in early 2006 due to objective lens availability. The 5-inch f/7 SV5S and f/9.4 SV5L, each featuring a three-element objective from LOMO in Russia, are the first of these to come along. Although they are not in full production at the time of this book's publication, performance information on both models will be found on www.philharrington.net as it becomes available.

Vic Maris deserves great credit for his involvement with his customers. Few manufacturers stand behind their products like Maris. In addition to star-testing each instrument before it ships, he is constantly speaking directly with consumers, whether by phone, e-mail, or on the Stellarvue Yahoo! Group (http://groups.yahoo.com/group/Stellarvue). That speaks volumes about the man and about the pride he takes in his telescopes.

Takahashi. Takahashi apochromatic refractors have well-deserved reputations for their excellent performance. In fact, to many amateurs they represent the pinnacle of the refractor world—they just don't come much better. One owner explained, "Many telescopes feel flimsy and thin, but Takahashi telescopes feel like they are honed from solid metal." The only drawback is that this superior quality does not come cheaply!

The Takahashi FCL-90, also known as the Sky-90 II, is a wonderful 3.5-inch f/5.6 grab-and-go apo for astronomy on the run. Measuring only 16 inches end to end with its lens shade retracted, the Sky-90 is small enough to tuck into a briefcase, yet with a quality that is big enough to show some spectacular views of the night sky. The heart of the instrument is a doublet objective with a fluorite element. This design suppresses spherical aberration to extremely low levels, although some false color can be seen around such unforgiving objects as Venus, Vega, and Sirius. To improve the view even more, add into the mix the Takahashi Extender Q. The Extender Q lengthens the Sky-90's 500-mm focal length to an effective focal length of 800 mm, improving the overall color correction along the way. Beyond the Extender Q, the Sky-90 will accept all accessories (finderscopes, camera adapters, and so forth) designed to fit Takahashi's series of FS refractors. Overall, the Sky-90 is an excellent wide-field imaging telescope. Its price is comparable to the Tele Vue TV-102, which also offers excellent optical performance. The larger TV-102 doesn't have quite the image contrast of the Sky-90, but with its added aperture, it will produce slightly brighter views.

Takahashi's 4-inch f/8 FS-102II, also known as the FS-102NSV (New Short Version), is built around a doublet objective lens, with one element made of fluorite and the other of low-dispersion flint for exceptional color correction and image clarity. What makes this objective distinctive is that unlike other manufacturers, which place the fluorite element on the inside, Takahashi places it as the outermost element. This unique positioning claims to produce even better results. Although some optical experts feel this may be little more than advertising hype, there is no denying that the view is absolutely amazing (as it always has been with Takahashi apochromats). Indeed, the optics are so sharp that owners routinely report using magnifications in excess of 100× per inch of aperture, well beyond the 60× per inch ceiling mentioned elsewhere in this book. Rules can be broken given absolutely perfect optics. Planetary detail is simply stunning. Stars don't just draw to a focus, they snap into focus with a certainty not seen in many other telescopes. Trying to focus them in extremely low temperatures (about 10°F and lower), however, can be difficult because the focuser becomes extremely stiff (nearly unusable, as one owner puts it). The tube's paint job is also not as durable as it should be, so care must be taken not to scratch it during use.

Takahashi is in the process of introducing a new line of Triplet Super Apochromatic refractors, appropriately called TSAs. The first to be introduced, the TSA-102S, is a 4-inch f/8 instrument that includes all the pluses of the FS-102II previously, but it has the promise of even better suppression of false color and spherical aberration because of the improved objective design. An oversized 2.7-inch focuser (with adapters for 1.25- and 2-inch eyepieces) eliminates any possibility of vignetting when shooting through the instrument, even with the optional field flattener or focal reducer in place. With a tube that collapses to only 23 inches long, the TSA-102S is ideal for airline travel.

The 4.2-inch f/5.2 Takahashi FSQ-106 is a modified Petzval optical design that builds its objective lens around four separate lens elements set in two groups. Two of those four elements—one in the forward group, the other belonging to a widely spaced pair toward the back of the tube—are made of fluorite for excellent image contrast and false-color elimination. Astrophotographers especially appreciate the FSQ-106's huge 4-inch focuser, which lets them use large-format cameras with little or no vignetting, a common problem with many other apochromats. The focuser can also be rotated a full 360° without shifting focus, making it much easier to look through a camera's viewfinder when aimed at odd angles. Although a couple reports of problems with the instrument have been received, specifically pointing toward the lens cell, most users agree that it is a superb telescope. As one owner said, "The FSQ-106 is spectacular. I have never, I mean never, seen better contrast. The sky was ink black, stars sharper than pinpoints right to the edge. It was amazing what a 4-inch can do on objects like M13 [the Hercules globular cluster]."

Have even more money to spend on a top-rated refractor? Then consider the TOA-130. The TOA-130 is a 5.5-inch f/7.1 three-element instrument that uses specially formulated optical materials to create what Takahashi refers to as an "Orthoscopic-Apochromat," a technical-sounding term that is probably more advertising hype than anything. But even with that, the result is the highest order of optical correction in terms of chromatic and spherical aberrations. At the same time, the oversized 2.7-inch focuser, sliding dew cap, and meticulous attention to detail all add up to a versatile instrument that is just as suitable for visual observing as it is for photographic imaging. If I had one caution to add about the TOA-130, apart from the high price, it would be its weight. The solid construction adds up to an impressive 22 pounds, which is nearly 6 pounds heavier than the FS-128 and 8 pounds heavier than the Tele Vue NP-127. Therefore, a mounting that might be perfectly suitable for either of those lighter instruments will quiver under the extra mass of the TOA-130.

Making a good thing even better, Takahashi has added a 5.9-inch f/7.3 ortho-apo that the company calls the TOA-150. All of the pros, and all of the cons, of the TOA-130 can be applied here, as well. Again, the weight of the tube means that a substantial mount will be needed to keep the TOA-150 aimed steadily toward a target. But for amateurs who are looking for apochromatic nirvana, the TOA-130 and TOA-150 are two of the finest refractors available on the amateur market today.

The two largest Takahashi refractors, the 7.9-inch f/10 FCT-200 and the 9.8-inch F/10 FCT-250, are designed around triplet objectives that sandwich a fluorite element between two lenses made from German optical glass to produce uncompromising optical performance. Both of these beauties are observatory-class telescopes that are as close to optical perfection as you are liable to find. Retailing for between $200,000 and $250,000, these might well be the ultimate "amateur" telescopes.

Although their telescopes are sold without mounts, Takahashi makes an impressive line of German equatorial mounts, which are listed later in this chapter.

Tele Vue, Inc. Tele Vue manufactures several outstanding short-focus refractors for the amateur. The views they provide are among the sharpest and clearest produced by any amateur telescope, with excellent aberration correction.

The two smallest Tele Vues discussed here are the 3-inch f/6.3 TV-76 and the 3.3-inch f/7 TV-85. Both have two-element objectives of ED glass. Optical performance is outstanding, with minimal false color detectable around objects like the Moon, Venus, and the bright star Vega. Views of deep-sky objects are consistently sharp and clear. Like all Tele Vue scopes, both models are a joy to use thanks to their optical and mechanical excellence. The 2-inch focuser, for instance, is buttery smooth. You will need to supply a tripod or other support, because one is not included with either telescope. Each instrument's light weight means that there are many possibilities for an appropriate mount.

Tele Vue also sells two 4-inch apochromatic refractors, the NP-101 (NP is short for Nagler-Petzval) and the TV-102. Introduced in 2001, Al Nagler's NP-101 (shown in Figure 5.3) is a 4-inch f/5.4 apochromatic refractor, similar to its predecessors: the TV-101, the Genesis-sdf, and the original Genesis. The NP-101 adds an important twist, however. Its objective consists of four lens elements set in two groups (a modified Petzval design, hence the product name): a two-element full aperture objective up front and a two-element subaperture doublet set far back into the tube.

The TV-102's 4-inch f/8.6 objective is a simpler two-element design that uses a special dispersion (SD) element to reduce chromatic aberration in comparison to a traditional fluorite-crown apochromatic two-element objective. Overall, the optical design is similar to what is found in the Celestron C100ED and the Orion 100ED discussed earlier in this chapter. Tele Vue's quality control, improved fit and finish, and superior mechanics eclipse the others, however.

A shortened variant of the TV-102, called the TV-102i, is designed to focus with Tele Vue's binocular viewer attached to the end of the instrument. Optional extension tubes are also available to let the TV-102i reach focus with a standard star diagonal in place of the binoviewer.

The NP-101 and the TV-102 both produce tack-sharp images of the Moon, planets, stars, and brighter deep-sky objects that are nearly devoid of the residual chromatic aberration that plagues achromatic refractors. The NP-101's four-element objective lens system offers exceptional false-color suppression,

Figure 5.3 *The Tele Vue NP-101, one of the finest instruments of its type and aperture class.*

with nothing but pure starlight seen around Vega, Sirius, and other bright stars that really test a telescope's state of apochromatism. The Nagler-Petzval multielement objective is also designed to produce pinpoint star images right across the full field of view, whether through wide-field eyepieces or on images. Star tests show no sign of spherical aberration, astigmatism, or any other flaw that could dampen image quality in the slightest. No doubt about it, Nagler has done it again. The NP-101 is an amazing optical accomplishment, setting the bar by which all other apochromatic refractors are judged.

This level of excellence does not detract from the TV-102, which is an exceptional performer in its own right. But in side-by-side tests conducted at the Riverside Astronomy Expo, I found that it is not in the same league as the NP-101. Despite its longer focal length, residual chromatic aberration—although minor—is still present. But given its lower price, the TV-102 is still a good buy. Because the telescope is usually sold à la carte, buyers need to purchase a finder, a star diagonal, the eyepieces, and a mount.

Tele Vue's flagship is the NP-127, big brother to the NP-101. The same praise I have for the "kid" can be applied to the brawnier 5-inch f/5.2 NP-127 as well. What a wonderful telescope! Images are as sharp and crisp as anything out there, while the short focal length, when teamed with premium hybrid eyepieces like Tele Vue's own Nagler and Panoptic lines, produces breathtaking views that live up to the manufacturer's "space walk" claim. Like the NP-101, the NP-127 comes with a 2-inch star diagonal, a sliding dew cap, tube rings for mounting the telescope, and a custom-fitted hard-shell carrying case. Add a finderscope and a mount, like the Losmandy G-11, and you're set for a lifetime of observing enjoyment.

TMB Optics. Cleveland's Thomas Back has teamed up with Markus Ludes of Saarbrücken, Germany, to create and manufacture some of the finest apochromatic refractors for sale today. Back designs the optics and then coordinates through Ludes to have the instruments constructed in Russia. Optical tube assemblies range from the modest 3.1-inch f/6 and f/7 TMB 80 to the impressive 10-inch f/9 TMB 254, and even larger, custom-made instruments. I can only imagine the views through those monsters!

Although I have never seen the 10-inch TMB refractor, I have had a chance to look through several smaller models and can only say that the images I've seen are absolutely textbook perfect. A 4-inch TMB refractor, for instance, withstood the rigors of the star test with a perfection that I have seen few other telescopes deliver before. Planets, double stars, and other sky objects all showed high degrees of sharpness and contrast. It was a truly amazing telescope, as are all in the TMB lineup.

The instruments' mechanics are every bit as impressive as their optics, with silky smooth focusers from Feathertouch, retractable dew caps, and other amenities that we have come to expect on the finest instruments. This quality does not come cheaply, mind you, but TMB Optics does offer yet another option for those considering a refractor from Astro-Physics, Tele Vue, or Takahashi.

Buying a telescope that contains a Back-designed objective lens is highly recommended, but that company, not TMB Optical, will be responsible for any warranty issues. A reader wrote to inform me that he purchased a telescope with a TMB objective from another, unnamed company. There was a problem with the lens, but because it wasn't in an approved TMB telescope, TMB Optical did not honor the warranty. The lesson learned here is that you should know what you're buying before you buy it!

You will also find the name "Thomas Back" when looking at some of the products offered by Burgess Optical and William Optics, among others. Back's reputation as a designer of excellent lenses keeps him in high demand.

Vixen Company, Ltd. Vixen, a highly regarded name in astronomical telescopes, both in its home country of Japan as well as throughout the world, offers two lines of apochromatic telescopes. Four models make up the Vixen ED line: the ED80Sf (3.1-inch f/7.5), the ED81S (3.1-inch f/7.7), the ED103S (4-inch f/7.7), and the ED115S (4.5-inch f/7.7). Each features two-element, multicoated objective lenses of extra-low-dispersion glass combined with another element of flint glass. Images are clear, crisp, and surprisingly free of spurious aberrations through all three of these fine instruments. Color correction is very good for an ED apochromatic objective, although most experts feel that these, as a breed, fall slightly short of apos that use fluorite elements. As always, the Vixen mechanics are top notch, with smooth focusers and nicely appointed tube assemblies. The finders that come with each are rather small, and should be replaced with larger units as time and funds permit. One note about the ED115S: like the smaller ED103S, the 115S is sold with several Vixen mounts. The cheapest version comes on the Great Polaris-E (GP-E) mount, which is not sturdy enough to hold this scope's weight sufficiently. Spend the extra money and get the Vixen Great Polaris-D2 mount as a minimum.

William Optics. William Optics hit the market at about the same time the third edition of *Star Ware* was published in 2002. At the time, brothers William and David Yang from Taiwan offered two achromatic refractors, one apochromat, and several larger, customized instruments. The Yangs experienced some early quality concerns with their best-selling Megrez instruments, but they have since gotten that situation under control for the most part. I still receive e-mails from owners who mention continuing problems with pinched optics, an internal tube baffle coming loose, and missed shipping dates, but all also add that the company's customer service is quick to address the problems.

Two telescopes from William Optics carry the name Megrez, after the faintest star in the Big Dipper asterism. Many consumers find the use of the same name appended by a descriptive suffix to be very confusing, possibly leading someone to believe they are buying one product when, in fact, they are buying another.

Measuring 3.1 inches in aperture, the Megrez 80 II ED uses an ED element mixed into a three-lens f/7 objective to bring this scope close to a state of ideal apochromatism. Examining the Megrez 80 II ED at a recent Northeast Astronomy Forum, the hobby's preeminent trade show held annually in New York, I found images to be very good from within that indoor setting. The stark ceiling lighting in the show area, usually crowned with chromatic aberration through traditional achromats, showed minimal color through the ED. The new two-speed focuser is a welcome addition to a telescope that is already a fine mid-range instrument.

Although the Megrez II 80 ED is fine for visual observations, the Megrez II Super Apo, relying on a super-low-dispersion three-element objective designed by Thomas Back to achieve even better image correction, is better for photography. Owners of this f/6 instrument report little or no false color and sharp images all the way to 250× under steady seeing conditions. That translates to crisp views of Jupiter's belts, Saturn's rings, as well as some tough double stars. The compact tube of the Megrez II Super Apo easily fits into a carry-on bag, making it the perfect travel companion. It's an exceptional buy for the price.

The Yangs also make the 4.3-inch f/6.5 Fluoro-Star 110. Looking like a pumped-up Megrez, the Fluoro-Star's white tube and gold trim give the instrument a rich, refined look. Inside, the fluorite three-element objective, designed and manufactured in the United States by Telescope Engineering Company (TEC), brings light into sharp focus. (TEC makes its own line of exceptional telescopes, which are discussed in the online extension of this book at www.philharrington.net.) The Fluoro-Star 110 was redesigned in the summer of 2005 and now features a high-quality Crayford-style focuser with a calibrated drawtube that makes repeat focusing easy. Overall, the Fluoro-Star is the most-sophisticated, best-performing instrument ever to carry the William Optics name.

William and David Yang continue to evolve their products while addressing the needs of their customers. Look for new products to come out of the Yang's Taiwanese and Chinese production lines as time goes on. For instance, the new WO ZenithStar 105 ED refractor has only recently entered full production. With nice touches, like a dual-speed 2-inch Crayford focuser, a triplet air-spaced ED objective, and a retractable dewshield, this scope shows great promise as a lower-cost alternative to some more famous names in the apo marketplace.

Although the ZenithStar and Megrez telescopes can't compete with high-end models from Stellarvue, Takahashi, or Tele Vue, their lower prices fill a market niche that the others do not. And the Fluoro-Star's remarkable level of excellence shows great promise that should make those more-expensive names a little nervous. Stay tuned, it's going to be a wild ride.

Exotic Refractors

This final category of refractors is the exclusive domain of one company, which in half a dozen years, has changed the course of our hobby in a very special way. Although just about every amateur astronomer has observed the Sun through solar filters or by projecting its image onto a sheet of paper, few of us have had a chance to study it in detail using special filters that block all wavelengths of sunlight save for hydrogen-alpha. Unfortunately, that's where the action is! Instead of seeing a few sunspots scattered across an otherwise featureless disk, a hydrogen-alpha solar scope shows a seething Sun ripe with prominences, filaments, and other dynamic features.

Until recently, if you wanted to see those amazing solar features, you needed to outfit your telescope with an expensive set of filters that could cost well over the price of the telescope itself. Even then, the filters needed to be heated to maintain thermal equilibrium and to function properly, which meant that an external source of AC power was also required.

Then, within the past decade, Coronado Technology Group reinvented the hydrogen-alpha market by introducing several state-of-the-art filters and fully dedicated hydrogen-alpha refractors. The latter innovation is discussed here, while filters are found in chapter 7.

Let's start with the company's most-popular model, the Personal Solar Telescope (PST). This little 1.6-inch (40-mm) f/10 refractor, shown in Figure 5.4, brings the amazing world of hydrogen-alpha solar viewing to everyone. The quality construction, including an aluminum tube, fine-adjustment focuser, and integrated finderscope, all betray the PST's amazingly low price.

To aim the PST, scan around the Sun's location in the sky until its image is centered in the small Sol Ranger ground-glass window next to the eyepiece holder, and then look in the eyepiece. With the Sun in view, turn the attenuation ring to adjust the band-pass filter and, therefore, contrast. Turn it one way, and sunspots become evident. Reverse the twist, and sinuous filaments and prominences pop into view. Images are focused by turning a small thumb knob on the bottom of the internal prism housing. The eyepiece itself does not move

Figure 5.4 *The Coronado Personal Solar Telescope (PST), one of today's hottest astronomical products. The author's backyard Star Watcher Observatory, detailed in chapter 8, is seen in the background.*

when being focused; instead, focusing is done internally. A 20-mm Kellner eyepiece comes with the telescope, which although adequate for a quick glance, is low on image contrast and sharpness. My favorite views through the PST are with a 10-mm Tele Vue Radian eyepiece (40×). In addition to expanding the eyepiece selection, I would strongly recommend purchasing the custom storage case to keep the PST safe between observations.

There is a pair of 0.25-20 threaded inserts on the bottom of the black prism/filter/star diagonal housing for mounting the 3.25-pound PST onto a standard photographic tripod, a small German equatorial mount, or Coronado's own Malta tabletop alt-az mount. The latter costs $125 and is fine for quick views, but I prefer using a Manfrotto tripod equipped with a micro-adjustment tripod head to make it easier to keep up with the Sun's motion across the sky. I recommend against putting the PST on a small equatorial mount, however. Even though such a mount will make it easy to track the Sun, the telescope's nonrotatable 90° star diagonal will place the eyepiece at some pretty uncomfortable angles at times.

There are a few minor downsides, such as the rigid star diagonal or finding a way to cover your eyes to block sunlight from sneaking in around the edge of the eyepiece (I use a black cloth), but one cited most often by owners is that the PST is primarily a visual instrument. Although some fine photographs of the Sun through the PST can be found online, most users find that aspect of this model to be disappointing. For imaging, you would do best to pony up the extra money for a Coronado SolarMax 40 or larger instrument.

Also keep an eye on the objective lens' coatings. They often degraded on early PSTs, causing image contrast to suffer, although it did not pose a safety concern. To Coronado's credit, the company will replace the lens assembly at no additional cost.

The PST is also available in a Ca-K version, which isolates the calcium-k line of the Sun's spectrum rather than the hydrogen-alpha line. Turning the Sun blue rather than orange, the PST Ca-K is designed to show *plages,* which are active regions in the Sun's chromosphere above sunspots. It will also show the brightest prominences, but overall, images are very dim compared with the hydrogen-alpha PST. Stick with hydrogen-alpha if you'll be primarily viewing by eye.

Coronado also sells several costlier hydrogen-alpha solar scopes that are designed for more serious observations, both visual and photographic. The SolarMax 40 has the same optical specifications as the PST, but its superior construction and narrower band-pass filter (less than 0.7 angstrom versus the PST's, specified as less than 1.0 angstrom) means better resolution of solar granulation and other surface details. But at more than $1,000 costlier, the question becomes, are the images really that better? There is no question that, yes, images do show finer surface structure. But are they that much better? Perhaps they are to a dedicated solar observer, but for the rest of us, the PST is just fine. (It is also possible to add a second SolarMax 40 filter onto a PST using a small adapter plate. This combination lowers the band-pass to less than 0.5 angstrom. This so-called double stacking serves to enhance surface detail, although it also dims the image.)

Coronado's 2.4-inch f/6.6 SolarMax 60, 2.8-inch f/5.7 SolarMax 70, and 3.5-inch f/8.8 SolarMax 90 refractors take a good thing and make it even better. In case you think those fast focal ratios could result in excessive chromatic aberration, forget it. Remember, you're looking at the Sun at only one very precise wavelength. Chromatic aberration only results from mixing the full visual spectrum. The best view I've ever had of the hydrogen-alpha Sun was at a recent astronomy convention through a SolarMax 90 equipped with a binoviewer and matching eyepieces. Whether it was the detail in the filaments, granulation, or prominences, I found it difficult to pull away and let the next person in line take his turn. The SolarMax 70 is also available in a calcium-k version.

Reflecting Telescopes

Like refractors, reflecting telescopes can be further subdivided based on their optical design. By far, the most popular type of reflector on the market today is the one originally designed by Sir Isaac Newton, and so the first part of this section addresses Newtonian reflectors. The second portion of this section looks at Cassegrain reflectors, which are far more limited in number. Finally, I have created a third category for *exotic reflectors,* which includes a few unique instruments that are worth singling out.

Newtonian Reflectors

As mentioned in chapter 2, Isaac Newton was not the first person to design a telescope based around a concave mirror for focusing light, but his design remains the most popular by far. For many, a Newtonian represents the ideal first telescope, offering a large aperture at a comparatively modest price.

As we will see next, Newtonians come in many different shapes and sizes on a variety of mountings, but all seem to fall into three general categories. *Econo-Dobs,* ranging in aperture between 4.5 and 12 inches, are generally characterized by cardboard or metal tubes and Dobsonian-style altitude-azimuth mounts made from laminated particle board. Most come with one or two Plössl eyepieces and small finderscopes. The solid tubes can make hauling the equipment difficult at larger apertures, but also tend to keep the optics in collimation.

Contrast econo-Dobs with *primo-Dobs,* in apertures from 8 up to 28 inches and beyond. With low-riding mounts made of high-grade plywood, precision focusers, top-quality mirrors of Pyrex glass, and open-framed tubes, these reflectors represent the current state of the Dobsonian art. The primary mirror is typically mounted in a square wooden box, appropriately called the *mirror box,* while the secondary mirror, focuser, and finder are all mounted onto the upper cage assembly. In between, anywhere from four to eight poles hold everything together (actually, one telescope model uses 16 poles). The open-truss design greatly reduces the instrument's overall weight and disassembled size, making transport much easier than the same-size telescope with a solid tube. The primary mirror cell often consists of between 9 and 18 pads supporting the mirror's reverse side, while either a webbed sling or a teeter-totter post arrangement called a *wiffle tree* keeps the mirror from sliding around when the telescope is tilted skyward. Because of the extensive assembly required each time the telescope is set up, collimation must be checked and adjusted before every observing session. Also keep in mind that primo-Dobs are usually supplied without finders or eyepieces, which stands to reason since most people who buy them have already owned other telescopes and have probably accumulated a fair share of accessories along the way.

Be prepared to pay top dollar for these instruments and to wait for weeks, even months, for delivery since most primo-Dobs are not off-the-shelf items. Only after a deposit is placed, often nonrefundable, will work begin on your telescope. Depending on the number of orders in front of yours as well as the availability of parts, it may take up to several months for delivery. That's because most primo-Dob companies are one- or two-person shops, and in some cases may only be a part-time business. My best advice is to ask for an estimated delivery date before placing an order. But even then, as with any contracted job, whether it's for a telescope or an addition onto a house, expect that schedule to slip. Looking for instant gratification? Pick up an econo-Dob, most of which are ready for immediate delivery from the warehouse.

Finally, the third category of Newtonians comes on equatorial mounts. Actually, this group can be further divided into two subsections. All of the imported equatorial Newtonians, typically from China and Taiwan, belong to the first group. These usually come with one or two mediocre eyepieces and a 6×30 finderscope. Most also include one of the same mounts used with many of the achromatic refractors discussed earlier. Comparatively few Newtonians belong to the second subgroup, which features premium equatorial mounts. These instruments are far heavier, both physically as well as economically, than the others.

Some low-end equatorially mounted Newtonians are known as *Catadioptric-Newtonians* or *short-tube Newtonians*. Examples include Celestron's Firstscope 114 Short and NexStar 114SLT, Meade's DS-2130ATS, Orion Telescopes' Short-Tube 4.5, and the Sky-Watcher 1141 and 1301. All are designed to squeeze a relatively long focal length into a short tube. The design, actually called a Jones-Bird telescope after its inventors, uses a spherical primary mirror. An auxiliary corrector lens (akin to a built-in Barlow lens; see chapter 6) is usually set permanently into the focuser's tube to stretch the instrument's f/4-ish primary mirror to an effective focal length between f/7 and f/9 and to correct for spherical aberration.

A Jones-Bird modified Newtonian has good points and bad. First, I'll discuss the good. By using a shorter focal length mirror, the telescope's tube can be cut in half, making it easier to carry. The reduced weight also lessens the burden placed on the mounting.

Unfortunately, the disadvantages of today's inexpensive Catadioptric-Newtonians outnumber the advantages. The biggest problem is a dramatic increase in optical distortions, such as chromatic aberration (unseen in traditional reflectors) and spherical aberration. The former is introduced by the auxiliary corrector lens at the base of the focuser, which does not adequately compensate for the latter. The design also requires a larger secondary mirror than would be needed otherwise, thereby increasing the telescope's central obstruction and decreasing image contrast. That's because short-tubed reflectors actually use very fast primary mirrors, which also means that their optics must be collimated more precisely than a traditional Newtonian with the same effective focal ratio. Even with properly collimated optics, however, this ill-conceived design never focuses images as sharply as conventional Newtonians.

AstroSystems. This small company in LaSalle, Colorado, makes some big primo-Dobsonians. Truss-tubed TeleKits from AstroSystems are available either in kit form or fully assembled. Standard models range in aperture from 10 to 25 inches, although special orders are available up to 32 inches for those who have severe cases of aperture fever. Kits come with just about everything you need except for some basic tools, but they require intermediate woodworking skills to live up to their full potential.

Factory-assembled TeleKits include a combination of polyurethaned Baltic Birch plywood, aluminum truss tubes, and stainless-steel hardware. Each kit also includes several nice features that most other companies either charge extra for or don't offer at all. These include cooling fans built into the mirror cell, complete with rechargeable batteries and matching AC and DC chargers; large, spoked altitude bearings edged with Formica; a cloth shroud for sealing out stray light; and an especially useful QuickSwitch filter slide. This latter feature holds up to four 48-mm filters underneath the focuser for quick changes between filters without having to remove the eyepiece. Simply turn the thumb knob that's next to the focuser and a cable-and-pulley system moves a metal filter holder along a track, shifting from one filter to the next. A click-stop detent tells you when the filter is centered.

The upper cage assembly also includes Astrosystems' own 2-inch Phase 4 focuser, which works smoothly. Its unusual design uses a toothed timing belt drive to eliminate the backlash that is common to rack-and-pinion focusers as well as slippage, which can occur in Crayford-style focusers, especially at colder temperatures. Inside the upper cage, Astrosystems' secondary mirror holder uses thumbscrews that can be easily turned by hand to adjust the mirror's tilt. That eliminates the chance of dropping tools onto the primary mirror during adjustment. On the downside, however, the secondary holder uses four axial screws for adjustment, making it surprisingly difficult to set the mirror's angle precisely. Most others use three.

I have yet to hear from a TeleKit owner who is displeased with his or her telescope. Part of that reason has to be thanks to each telescope's optical quality. Buyers can select their telescopes' mirrors from Zambuto Optics, Pegasus Optics, or Discovery Telescopes. All three companies produce fine astronomical mirrors, although most amateurs agree that Zambuto's mirrors maintain the highest consistency in quality of any company in the industry. Every Zambuto-mirrored telescope I have ever used has been at least about $1/8$th-wave and produces sharp, contrasty images regardless of target. On good nights, the views of

Jupiter and Saturn can even rival those through the best refractors, while benefiting from the larger aperture for even better resolution.

Overall, Astrosystems' TeleKits get high marks for their mechanical design and execution. The design is clean and functional, assembly is simple, and fit and finish match that of the finest wood furniture. Whether purchased as a kit or already assembled, each TeleKit will give its owner a lifetime of viewing pleasure.

Celestron International. Long known for its Schmidt-Cassegrain instruments, Celestron also sells several reflectors. The Firstscope 114 EQ features a 4.5-inch f/8 spherical mirror mounted in a steel tube. This reflector comes with a 1.25-inch focuser, 10-mm and 20-mm Kellner eyepieces, and a StarPointer red-dot unity finder. Optically, the Firstscope 114 is surprisingly good, although its equatorial mount can be rather shaky. Focusing that 10-mm eyepiece can be especially taxing.

Now that Celestron is owned by Synta, it is not surprising that we are seeing many different combinations of the company's telescopes on various mounts. For instance, the Celestron StarSeeker 114 takes the optical tube assembly that Orion Telescopes has imported for several years as the StarBlast and mated it to the one-arm NexStar mount. The StarBlast is one of my favorite inexpensive telescopes sold today, as you will see in my review on page 125. Even though the StarSeeker 114 telescope works equally well, the GoTo mount negates one of the things I like most about the StarBlast: spontaneity. The StarBlast is at its best when swinging through the sky, not necessarily aimed at a particular target, but just sweeping randomly across the cosmos. The prescriptive nature of a GoTo mount means that the StarSeeker will miss those unexpected little discoveries, whether of an interesting asterism or just a chance meeting of a deep-sky object that the observer forgot about.

Celestron's NexStar 130SLT includes a 5.1-inch f/5 Newtonian optical tube assembly, which promises brighter views than the StarSeeker 114. Its fast focal ratio makes it ideal for wide-field deep-sky objects like the Pleiades and the Double Cluster, but not for the planets and close-up lunar views. The optics are fine for low power, wide-field views, but don't expect sharp images at much over 100×. Both models include Celestron's innovative SkyAlign computerized navigation system, which was detailed earlier in this chapter under the StarSeeker 80 refractor discussion.

Don't confuse the NexStar 130SLT with the cheaper StarSeeker 130. Although they are the same optically, the StarSeeker version comes with a flimsier aluminum tripod. The NexStar 130's tripod, with its tubular steel legs, will prove much sturdier. The extra $100 for the NexStar 130 is money well spent.

However, instead of any of these small GoTo scopes, I would recommend one of the Orion SkyView XT IntelliScopes, detailed later, which have the option of adding digital setting circles. Think of these as "PushTo" telescopes, as opposed to "GoTo." In this case, the observer pushes the telescope toward a preselected target until indicators on the small setting-circle control box's readout reaches zero. Although not as sexy, these are faster and quieter than

GoTo telescopes, draw much less battery power and are every bit as accurate. PushTos do not track the sky automatically, however, as GoTos do.

Moving up in size, we come to the recently redesigned, reintroduced Celestron StarHopper econo-Dobs. Available in four apertures from 6 to 12 inches, the Asian-made StarHopper steel tube assemblies include very good optics made from BK-7 glass. Each telescope's 2-inch rack-and-pinion focuser (1.25-inch on the 6-inch) works well, although as can be expected with nearly all telescopes of Far Eastern origin, there is a layer of grease-glue that can gum things up a little. The 10- and 12-inch models include a battery-powered cooling fan attached to the mirror cell to help speed up the mirror's acclimatization on a cool night.

Rather than rely on Teflon or nylon pads for their azimuth bearings as most other Dobsonian mounts do, StarHoppers ride on small roller bearings set into their bases. Although a good idea on paper, in practice they actually aren't as smooth as the simpler bearings they replace. The larger, heavier apertures are especially affected, since their increased mass actually causes the rollers to dent the mating surface, in effect creating speed bumps. At the same time, the side-mounted altitude bearings are quite small in diameter, which impacts their smoothness negatively. So, even though the *variable tension clutch system* is effective for adjusting balance, StarHopper motions are not as smooth as some other econo-Dobs.

Celestron's most advanced Newtonian reflectors include the 6-inch f/5 C6-N, C6-NHD, and C6-NGT; the 8-inch f/5 C-8N and C8-NGT; and the 10-inch f/4.7 C10-N and C10-NGT Advanced Series instruments. All of these Synta products come on German equatorial mounts, although the standard C6-N includes the light-duty CG-4 mount and weak, pressed aluminum legs. The rest come on the far sturdier Advanced Series CG-5 mount, which is quickly establishing itself as the best low-priced German equatorial mount sold today.

Except for the C6-N, each Advanced Series telescope can be purchased on a standard mount or one that is completely "GoTo-ized" (indicated by the "GT" suffix in the model name). The Advanced Series Computerized CG-5 includes a 40,000-object database and hand controller that clips onto a tripod-mounted bracket. Each instrument's computer and motors are powered with 12-volts DC for easy use in the field. Because of power requirements, Celestron does not include a battery holder, as some DC-powered telescopes do. Instead, the company supplies a 24-foot cord to plug into an automobile cigarette lighter or a rechargeable battery, which makes much more sense. A 110-volt AC adapter is sold separately.

In testing a C8-NGT for *Astronomy* magazine, I found that the scope had some of the finest optics I have ever seen in a Newtonian reflector. Star testing proved that the primary mirror was a nearly perfect paraboloid. Estimating its wavefront accuracy, I judged the mirror to be at least $^1/_{10}$th wave and possibly even better. That performance translated into some stunning views of sky objects. On a night when seeing conditions rated a 5 on a scale of 1 to 10 (inferring a fair amount of atmospheric turbulence), the C8-N delivered wonderful images of Jupiter and Saturn at 357× using a 7-mm Pentax eyepiece and a 2.5× Tele Vue Powermate.

Discovery Telescopes. Discovery began doing business back in 1991 under the name Pirate Instruments. After that short-lived retail venture, the company turned its efforts to the world of wholesale, supplying Dobsonian-style Newtonian reflectors to both Orion Telescopes and Celestron. In late 1998, owners Bill Larsen and Terry Ostahowski reentered the retail arena, this time as Discovery Telescopes. Discovery's Dobsonian-style telescopes are built right in its Oceanside, California, factory

Optical quality is a strong point behind all of Discovery's Dobsonian-style reflectors. The least expensive line, known as PDHQ reflectors, come in six apertures: 8-inch f/7, 10-inch f/6, 12.5-inch f/5, 15-inch f/4.2 or f/5, 17.5-inch f/4.1 or f/5, and a colossal 20-inch f/5. All optical components are made of Pyrex glass and housed in spiral-wound cardboard tubes painted black. Each instrument includes a 25-mm Plössl eyepiece and a Telrad one-power finder, as well as a nice 2-inch Crayford-style focuser. Enhanced aluminizing on both mirrors and nicely crafted rocker boxes made from birch plywood add to each instrument's beauty and functionality. Often, adding a finderscope or simply changing from a lighter to a heavier eyepiece will cause a poorly balanced telescope to shift in altitude, causing the target object to go whizzing out of the field of view. To help fix this problem, each PDHQ's balance point can be adjusted by sliding the mount's altitude bearings along a pair of metal tracks on either side of the tubes. Adding a finder? Slide the tube forward. Removing it later? Just slide the tube back. It's a simple but effective solution.

The only disadvantages to the PDHQ reflectors are the tubes' weight and girth. Recognizing that, all PDHQ models, save for the 8-inch, include a clever split-tube design that allows the front half to be separated from the back for easier carrying. Plug the two pieces back together by sliding four rods into matching guides and clamping the four latches closed. Collimation holds quite closely as long as the sections are snugged together with a similar amount of force each time. The weight of the 12-inch and larger PDHQs also cause the rocker box to flex laterally due to the thin wood used in construction. Owners report having to add brackets and other reinforcement to steady the view.

Discovery's Truss Design Dobsonian instruments range in size from 12.5-inch f/5 to 24-inch f/4.5 and are designed to go head-to-head with the likes of Starmaster and Obsession. A unique feature of its design is that the eight truss tubes that serve as the instrument's backbone come assembled in pairs. To put everything together, slip each pair's end brackets over threaded studs on the mirror box and upper cage assembly and screw them into place with large thumbscrews. Unlike the standard-bearing Obsession instruments (discussed later), which use captive hardware to hold the truss together, Discovery's design uses loose black thumbscrews, which can prove problematic. Other drawbacks include the secondary holder, which uses four adjustment screws instead of three, making it more difficult to collimate the scope.

Once the optics are adjusted correctly, owners rave about the great views, which star testing confirms are thanks to very good mirrors. Like the PDHQ telescope, all Truss Design reflectors feature optics made of Pyrex glass and coated enhanced aluminizing for 96% reflectivity. Not surprisingly, images, both of deep-sky objects as well as the planets, are also superb. The views I had of Jupiter and Saturn through a particular 12.5-inch Truss Design instrument that I tested for *Astronomy* magazine were especially impressive. Apochromatic refractors may be considered planetary telescopes, but there is no substitute for aperture on nights of steady viewing.

The only optical problem I detected with that 12.5-inch was a glow or a flare that seem to invade one edge of the view. No matter which way I looked through the eyepiece, tilting my head back and forth, the glow persisted. I finally discovered that the problem was because the upper cage assembly is too short. In other words, the focuser is so close to the front end of the telescope that light can shine over the edge of the assembly and right into the focuser. Placing a light shield fashioned from a piece of black poster board eliminated the problem, which was more pronounced from my suburban backyard than from a dark, rural setting.

Although Discovery Newtonians are very good, especially considering their price, delivery time can be lengthy depending on the model and production schedules. Promised dates are known to slip badly, which only emphasizes the need for open and honest dialogue between a company and its customers. I have heard from several Discovery customers who complain about the company's slow response time to inquiries, whether they were asking about delivery schedules or simply wanted product information. Unreturned phone calls, unanswered e-mails, and a general lack of responsiveness can lead a company down a slippery slope. Amateur astronomy is a small community where communication is absolutely critical. Hopefully, Discovery will take steps to improve its efforts before too much more time passes. As it stands now, this history makes it difficult for me to recommend them.

Edmund Scientific. Now a division of VWR Scientific Products, Edmund's share of the astronomical marketplace has dwindled over the last twenty-five years, with many of its once-popular products no longer sold. All that remains of this company, which was once the leading astronomical supplier, is a limited assortment of astronomical gear. Sad to see an icon pass.

The mainstay of Edmund's brand name astronomical products since the mid-1970s has been the 4.13-inch f/4.2 Astroscan 2001 rich-field Newtonian. The Astroscan is immediately recognizable by its unique design, which resembles a bowling ball with a cylinder growing out of one side! The primary mirror is held inside the 10-inch "ball," opposite the tube extension which supports the diagonal mirror and eyepiece holder. The telescope is supported in a three-point tabletop base that may also be attached to a camera tripod.

The primary mirror, advertised as 0.125 wave at the mirror surface, yields good star images when used with either the 15- or 28-mm Plössl eyepieces that come with the scope. As magnification increases, however, image quality degrades. This should come as no great shock, because small deep-dish Newtonians are really not suitable for high-power applications. Although the Astroscan does not come with a finder in the classic sense, it includes a peep sight that permits easy aiming of the telescope.

Guan Sheng Optical (GSO). GSO is a bit of a mystery company. Peruse the archives of either *Astronomy* or *Sky & Telescope* magazine, and you probably won't find it mentioned anywhere. Google it on the Internet, and you will come up with the company's Web site in Taiwan as well as a few European importers but not much else. Yet GSO makes some of the most highly regarded econo-Dob Newtonian reflectors in the world, like the 6-inch reflector seen previously in Figure 3.3. Trying to find them is the challenge.

That is because GSO doesn't sell its telescopes under its own name. Instead, the company plasters its telescopes with the names of its importers. Over the years, the company has worn brands like Hardin Optical, Zhumell, Teleskop Service (Germany), Sky Optic (France), Antares (Canada), and Astronz (New Zealand).

A rose is a rose, but what sets all GSO Dobs apart from the crowd is its exceptional optics. Most GSO Dobs that I have viewed through contained some of the best optics that I have seen in low-end telescopes. Although they may not be on par with the likes of Zambuto, Royce, Nova, and other custom mirror brands found in primo-Dobs, GSO optics show that it is possible to produce quality glass in a mass-production environment. Its chief competition, Synta, also makes very good mirrors, but GSO's show better consistency.

One small company that imports GSO Dobsonians into the United States is Anttler Optics. Anttler's 8-inch f/8, 10-inch f/5, and 12-inch f/5 Deep Space Observer telescopes (Guan Sheng's GS-680, GS-880, and GS-980, respectively) all share common traits, including steel tubes painted black, particle board bases covered with black laminate, 2-inch Crayford focusers, 8×50 finderscopes, and side-mounted eyepiece trays. Construction is on par with the SkyQuest XT Classic Dobsonians imported by Orion Telescopes described later in this chapter. Like the Orions, GSO Dobsonians rely on strong springs anchored between each plastic altitude bearing and the rocker box to keep the telescope from slipping down toward the horizon whenever a heavy eyepiece is used. An open primary mirror cell helps to adjust the optics passively, but Anttler also offers a battery-powered muffin fan behind the primary to cool them more aggressively.

Anttler also offers upgraded bases made of Baltic Birch, refigured mirrors with enhanced coatings, and other accoutrements to improve their GSO imports. So-called Anttler High-Performance Newtonian telescopes add a few hundred dollars to the price, but also correct many of the GSO telescopes' shortcomings. Motions are smoother, the secondary mirror holder is easier to keep adjusted, and the tube interior, lined with flat black flocking paper, rejects stray light better to improve image contrast.

Jim's Mobile, Incorporated (JMI). JMI produces some innovative large-aperture Newtonian reflectors and aftermarket accessories. Its 12.5-inch f/4.5 and 18-inch f/4.5 Next Generation Telescope (NGT, for short) Newtonian reflectors break the Dobsonian mold by combining an open-truss tube assembly with a split-ring equatorial mount (the same type of mount used with the 200-inch Hale reflector at Palomar Observatory). Image quality is consistently excellent. Of the models I have looked through, all have had great optics, with image resolution limited only by seeing conditions and aperture, not by mirror imperfections. Each primary mirror is secured in a fully adjustable cell that holds collimation very well even after the telescope has been disassembled and subsequently reassembled. The rotatable upper assembly lets users turn the finderscope/focuser to a more comfortable viewing angle when needed, although this may cause collimation to wander a bit.

Although JMI offers a wide variety of accessories for its telescopes, two are absolutely required, especially if you observe from light-polluted or damp environs. One is a black shroud that wraps around the telescope's skeletal structure to prevent light and wind from crossing the instrument's optical path. It also slows dewing of the optics, a problem common to all open-truss telescopes. The other necessity is a tube extension that sticks out beyond the telescope's front. The extension serves to slow dewing of the diagonal mirror, which lies very close to the end of the nose assembly, as well as to prevent stray light from scattering into the focuser from below and washing out the scene.

Three apertures make up JMI's New Technology Telescope, or NTT, line. The smallest, the NTT-12 is a 12-inch f/5 open-truss tube assembly mounted on a strange-looking "Dobsonian" mount. I put that in quotes because it is not your typical Dob mount. This unique design allows the telescope to be tilted, in effect turning a Dobsonian mount into an equatorial. (The equatorial wedge, needed to tilt the scope to match the observer's latitude angle, is sold separately.) To keep the telescope from tipping over, the mount's 16-inch-diameter azimuth (right ascension) bearing is captured inside three V-rollers. The 10-inch altitude (declination) gear is also captured to prevent it from slipping off the mount. The NTT-12 comes with Vixen's SkySensor 2000 PC GoTo controller for complete automation, which will work with the telescope set up in either alt-az or equatorial mode. Setting up the instrument in either configuration is much more cumbersome than with other 12-inchers on standard Dobsonian mounts because the triangular base, mirror "bucket," and mirror must be carried as a single assembly.

The other two NTTs, a 25-inch f/5 and 30-inch f/4.2, are both folded Newtonians on beefy, nontiltable altitude-azimuth mounts. By placing a second, optically flat mirror between the primary and diagonal mirrors, the instrument's overall length is cut dramatically. Now, instead of having to climb to towering heights to look through the eyepiece, the eyepiece is never more than six feet off the ground and so requires only a stool at most. Clever! The NTT also comes on a state-of-the-art computerized mounting.

The folded Newtonian design is not without its faults, however. The biggest problem is the huge flat secondary mirror at the front of the tube required to redirect the light back toward the diagonal. In the case of the NTT-25, it measures 9 inches across; larger than some primaries! This translates to a central obstruction equal to 36% of the primary's diameter (13% by area), which is huge for a Newtonian. The NTT-30's 11.5-inch flat secondary creates a 38% central blockage. Such a huge obstruction in the optical path reduces image clarity and contrast dramatically when compared with a conventional Newtonian.

Mag One Instruments. Looking for something different in a Newtonian reflector? Consider a Mag One telescope. Owner Peter Smitka has created four PortaBalls—an 8-inch f/5 or f/6, a 10-inch f/5, a 12.5-inch f/5, and a 14.5-inch f/4.3—which are some of the most unique telescopes to come along in years. Rather than copying conventional design, the PortaBall centers around a hollow sphere that houses the primary mirror. The fiberglass sphere, reinforced with a central flange, is hand-sanded and finished with a smooth coat of white paint.

Mirrors for PortaBalls come from Zambuto Optics, acclaimed by owners for their excellence. The spherical mounting completely encloses the back of the primary, which would make a conventional mirror cell impractical. Instead, each primary mirror is glued in its mounting with small blobs of silicone adhesive, a technique that proves troublesome when the mirror needs to be removed for cleaning or realuminizing. A pair of black-anodized metal rings and a short tube segment of a composite material make up the telescope's upper cage assembly. Six aluminum truss tubes join the upper assembly to the primary-mirror ball. Each PortaBall also comes with a Rigel Systems Quik-Finder unity finder as well as a unique Helical-Crayford focuser on the 8- and 10-inch Portaballs, or a JMI DX-1 (optional Starlight Instruments Feather-Touch focuser) on the larger models.

Several options are available for each PortaBall, including electrical packages that add a battery-operated fan to help cool the primary mirror to ambient temperature, as well as anti-dewing elements on the Quick-Finder and secondary mirror. Although many owners say that the Portaball's axis-less design is easier than conventional Dobsonians to track the sky by hand at high magnification, a specially designed equatorial platform is also available, if desired. Once the platform has been polar aligned and the telescope has been nestled into its waiting embrace, a Portaball will track the sky effortlessly like an equatorial mount.

Are there drawbacks to the PortaBall? Owners could think of only a few. One is that digital setting circles cannot be adapted to the unusual mounting design. Another is the issue of balance. Although Smitka custom builds each unit to maintain its aim with its owner's heaviest eyepieces, or no eyepiece, in the focuser, very heavy eyepieces, such as some from Tele Vue, Meade, and Pentax, may throw the balance off, especially when the telescope is aimed toward the horizon. Finally, the price of PortaBalls, which is higher than conventional Dobsonians, may turn off some people. But most who have purchased a PortaBall say they are well worth the extra cost.

MC Telescopes. Featuring two lines of nicely designed truss-tubed Newtonian reflectors on premium Dobsonian mounts, MC Telescopes have always attracted a lot of attention since they premiered a few years ago. All models share common design features with other primo-Dob manufacturers, such as Obsession, Starmaster, and others. For instance, each telescope includes an open, fully adjustable mirror cell made from welded steel tubing to promote rapid, even cooling of the primary mirror. A 12-volt fan mounted to the back of the cell helps to cool the tube, as well. The secondary mirror holders use thumbscrews for easy, no-tools adjustment. The secondary holder can be ordered with either three or four adjustment screws; by all means, get the three screws, which prove far easier to adjust accurately.

The company's Truss Dob line ranges in size from a compact 10-inch instrument to a respectable 20-inch model. Mirrors come from Royce, Nova, or Waite Research, depending on aperture and price point. All three choices are very good, although I would probably opt for Royce, if available. MC Truss Dobs use a traditional eight-pole truss design, each with ball-and-socket connectors at either end for simple assembly. A 2-inch Crayford focuser from MoonLite Focusers is standard, as is a Telrad unity finder and cloth light-block shroud. Overall construction quality is excellent. Wheelbarrow-style handles, which are necessary for transporting the larger apertures, are sold separately. Rather than screw them into the mounting base as most other telescopes of this design require, the MC handles are designed to slide under the carrying handles attached to the rocker box. A tight mechanical fit and small Velcro patches hold the handles in place. To take them off, push down on them gently.

MC Telescope's other line is called Double Truss, because each telescope is built around not eight but sixteen truss poles. Although based on the truss design, these two scopes do not have a solid mirror box. Instead, eight short truss tubes attach the mirror cell to the altitude-bearing axis, which looks like a wooden box but without a top or a bottom. Eight longer truss tubes attach the focuser assembly to the same altitude-bearing box. This unusual design cuts weight by about 20% compared with the MC Truss Dob versions, but at the same time, the scope will inevitably take longer to put together. The Double Truss scopes also cost about $700 more than the MC Truss Dobs.

Meade Instruments. Meade's LXD75 series of telescopes includes the 6-inch f/5 N-6 EC Newtonian reflector, which line for line, reads like Celestron's C6-NHD Newtonian, discussed earlier. Both include German equatorial mounts that are adaptable to computerized GoTo control and have plate glass mirrors, small finderscopes, and adequate eyepieces (one each). The 2-inch focuser that comes with the N-6 is barely adequate, however. Many owners of this model and the LXD75 Schmidt-Newtonians have changed to higher quality focusers, such as those from JMI or MoonLite, and are now much happier.

Optically, the N-6 is a good performer, although the short focal length makes collimation critical. Images are sharp, with best results between about 50× and 150×. Contrast, however, is compromised by the N6's thick spider vanes and its 2-inch central obstruction. A 33% obstruction (by diameter) is unusually large for a Newtonian. The C6-NHD uses much thinner spider vanes and has a smaller central obstruction for superior image contrast.

Note, however, that unlike other models in Meade's LXD75 series of telescopes, the N-6 mount does not include Autostar computer control. If you want a 6-inch Newtonian with GoTo aiming control, bypass the N-6 for Celestron's C6-NGT.

In late 2005, Meade unveiled a new series of LightBridge 8-, 10-, 12-, and 16-inch truss-tube Newtonians on Dobsonian mounts. The new LightBridge reflectors are manufactured by Guan Sheng Optical in Taiwan, known for its high-quality optics. Each scope comes with a 2-inch Crayford focuser, cylindri-cal metal mirror "boxes" and upper cage assemblies, and laminate-covered rocker boxes. The truss tubes are affixed in pairs and are assembled using large thumbscrews, making tools unnecessary for assembly (but a flathead screwdriver is still needed for collimation).

Each LightBridge model is sold in two variations, labeled *standard* and *deluxe*. The main difference between them revolves around the azimuth bear-ings in their bases. The standard version's base rides on hard plastic pads like most other econo-Dobs, while the deluxe version steps up with an improved roller bearing assembly sandwiched between the bottom of the rocker box and the ground board, like Celestron's StarHoppers. Each deluxe version also includes an advanced red dot unity finder with four interchangeable reticles, while the standard model uses a conventional red-dot finder. It is doubtful that owners could attach magnifying finderscopes onto any of the LightBridge tele-scopes because of balancing problems.

And therein lies one of the dilemmas with the LightBridge: balance. The weight of heavier 2-inch eyepieces will undoubtedly cause the telescope to drop down toward the ground, especially when it is aimed near the horizon. As this book is being written, Meade is developing a braking system to prevent this problem from happening. Meade is also planning to offer a cloth shroud to rap around the truss poles, which is necessary to keep stray light from scat-tering into the optical path and washing out the image.

Perhaps the biggest drawback to the LightBridge reflectors is, ironically, also one of the model's strongest selling points: the truss design. Nearly all large Dobsonian-based reflectors nowadays are built around the open-truss design, which lets us suburbanites transport our large telescopes in the backs of our small cars for travel to dark skies. The LightBridges are no exception to this convenience. But the fact that they are aimed toward beginning astronomers is troubling, because the telescope's collimation will have to be checked every time it is assembled. Although the primary mirror is center spotted for refer-ence, LightBridges are not sold with a collimation tool. In fact, Meade doesn't even offer one as an option.

So, are Meade LightBridge telescopes a good choice? If you are looking for a nicely designed visual telescope that can be easily disassembled and car-ried in the trunk of a small car, and you are secure enough in your ability to collimate a telescope each time it's used, and as long as you don't plan on using heavy eyepieces or observing in a light-polluted area, then yes, they are a good choice. The optics are very good in both versions, although I would recom-mend spending the extra money and getting a deluxe model for its better azimuth motion. If, however, any of those criteria leave you a little uncertain, you should look at a solid-tubed Newt/Dob instead.

NightSky Scopes. NightSky is a new company that has established itself as a reliable source for premium, truss-tubed Newtonian telescopes on Dobsonian mounts. Owner Jim Nadeau, amateur telescope maker turned manufacturer, offers six instruments ranging from a 12.5-inch f/5 to a 22-inch f/4.5. Unlike companies that farm out parts of their telescopes to cabinet makers and wood-workers, Nadeau builds each telescope personally using oak plywood. He is able to control the quality of each very precisely, and it shows.

Each NightSky Scope uses mirrors from Pegasus Optics, which is well known for producing some of the finest optics for amateur scopes today. Mirrors are coated with standard aluminizing (90% reflectivity) and then over-coated for protection. Other standard features include dual-speed Crayford focusers from MoonLite Focusers, custom-fit cloth light shrouds, Telrad finders, wheelbarrow handles for easier transport, and mirror box covers. Digital setting circles, an anti-dew electrical package, a focuser upgrade, and full GoTo control by StellarCAT are also available at extra cost.

One of the nicest standard features in each NightSky Scope is the multi-point primary mirror cell that holds the mirror firmly in place, yet allows it to "float" to prevent pinching and distortion. Nadeau's design, traditionally called a wiffle-tree support, surrounds the mirror with several flexible pads on short threaded rods that can be adjusted for precise centering. The NightSky cell holds collimation better than most sling-style cells used by other companies, because there is no chance of the sling stretching under the mirror's weight or swinging during transport.

Another unique feature of each NightSky Scope is the triangular truss assemblies used to hold the telescope together. Rather than use six or eight individual poles, Nadeau prefers the Starmaster approach of using four triangular pole assemblies. Each is made up of a pair of aluminum poles that have been TIG-welded to a base plate and then painted flat black. When assembling the telescope, each triangle is attached in three places—one to the upper cage/focuser and two to the mirror box. All bolts are held captive, so there is no chance of dropping them.

All NightSky Scope owners whom I have heard from have nothing but positive things to say about their telescopes. Images are consistently sharp and clear even at high power, while the mechanics are smooth and functional. Perhaps best of all is the price of NightSky Scopes, which can be substantially cheaper than the competition. When you add everything up—the optical quality, craftsmanship, and price—it's easy to see why NightSky Scopes is one of the stand-out new companies in this edition of *Star Ware,* and one to watch in future years.

Obsession Telescopes. These telescopes were the first large-aperture, alt-azimuth Newtonians sold to break the "light bucket" stigma. Back in the 1980s, the first commercial Dobsonians often had large but mediocre optics and poor quality tube and mount assemblies. Obsession founder Dave Kriege changed all that when he introduced the next evolution in the simple Dobsonian mount. His original Obsession 1 included many breakthrough features that are found in most high-end Dobsonians today, including truss-tube assemblies; large, semicircular altitude bearings; and open back, multipoint mirror cells. Indeed, this genre of Newtonian reflector should probably be called "Kriegerians" rather than Dobsonians to indicate their true origin.

Today's Obsession telescopes still feature sharp optics, a clever design, fine workmanship, and ease of assembly and use. Six models from 12.5 to 30 inches aperture are currently available. Mirrors are supplied by Galaxy Optics or Optical Mechanics Inc. (OMI). Both companies are well known for their optical quality as well as their history of standing behind their products should a customer ever require assistance. The secondary mirrors feature Brilliant-Diamond coatings, which is a nonaluminum dielectric coating that the manufacturer states has 99% reflectivity. Each primary comes with enhanced aluminizing (96% reflectivity).

The overall design of the Obsessions (see Figure 5.5) makes them some of the most user-friendly large-aperture reflectors around. Their open-truss tube design allows the scopes to break down for easy transport to and from dark-sky sites, although keep in mind we're still talking about big telescopes that are much more unwieldy than smaller instruments. Once at the site, the 12.5-, 15-, 18-, and even 20-inch Obsessions can be set up by one person in about ten to twenty minutes without any tools; the 25-inch scope requires two people, but it is still quick to assemble. First, insert the eight truss poles into their wooden blocks mounted on the outside of the mirror box to reduce overall girth, then place the upper cage assembly on top. Pass four tethered pins through mating holes and latch down their quick-release skewers. This no-tool design is especially appreciated by those of us who know what it is like to drive to a remote site, only to find out that the telescope cannot be set

Figure 5.5 *The Obsession 15, an outstanding 15-inch f/4.5 Newtonian reflector comes on a state-of-the-art Dobsonian-style alt-azimuth mounting.*

up without a certain tool inadvertently left at home. Even nicer is the fact that all hardware remains attached to the telescope, so it is impossible to misplace anything. The only way to lose Obsession hardware is to lose the telescope!

Many conveniences are included with Obsessions. For instance, a small 12-volt DC pancake fan is built into the primary mirror mount to help the optics reach thermal equilibrium with the outside air more quickly. Another advantage is that each instrument, except the 12.5-incher, comes with a pair of removable, metal wheelbarrow-style handles attached to a set of rubber wheels. The handles quickly attach to either side of the telescope's rocker box, letting the observer roll the scope around like a wheelbarrow. Of course, these handles do not solve the problem of getting the scope in and out of a car. If you own a van, a hatchback, or a station wagon with a low rear tailgate, the scope may need to be rolled out using makeshift ramps. You can also buy any Obsession reflector outfitted with Wildcard Innovation's Argo Navis Digital Telescope Computer (DTC) digital setting circles or the complete ServoCAT computerized aiming system preinstalled.

It should come as no surprise, given the apertures of these monsters, that their eyepieces tower high off the ground. Unless you are as tall as a basketball player, you will probably have to climb a stepladder to enjoy the view. A tall ladder is necessary for looking through the bigger Obsessions. For instance, when aimed at the zenith, the eyepiece of the 25-inch scope is a towering 10 feet off the ground.

True, Obsession telescopes are expensive, but after all, they are a lot of telescope—and are still the standard by which all other Dobsonian-style Newtonians are judged. For the observer who enjoys looking at the beauty and intrigue of the universe and has access to a dark sky, using Obsession telescopes can really become obsessive.

Orion Telescopes and Binoculars.

Orion offers several Newtonian reflectors bearing the Orion name. The most popular of all its lines are SkyQuest Dobsonian-mounted reflectors. SkyQuests come in two versions: XT Classics and XT IntelliScopes. Optically, they are the same, but it's what's under those optics that attracts the most attention.

All five XT Classics get glowing reviews from most owners. Each telescope is built around a parabolic mirror made of either plate glass (6- and 8-inch models) or BK-7 glass that is coated with standard aluminizing. Pyrex mirrors adjust to changes in temperature (such as occurs when taking a telescope outside from a warm house) more rapidly than either plate glass or BK-7, although the mirrors in the XTs seem to come around quite quickly (certainly within thirty to sixty minutes, depending on aperture).

Other mechanical attributes include rack-and-pinion focusing mounts that slide smoothly and evenly thanks to thin internal Teflon pads. The 4.5- and 6-inch models come with 1.25-inch focusers, while the larger models include 2-inch focusers (1.25-inch adapters are included). Each instrument's diagonal mirror is mounted on a four-vane spider mount that requires a small Phillips screwdriver to adjust. Several owners have mentioned some frustration at the initial collimation of their XT's, resulting in poor initial images. Improper adjustment of the diagonal was often the culprit.

Speaking of optics, the mirrors themselves are consistently quite good. Once everything is properly collimated, images have been acceptable in all that I have looked through personally. The XT4.5, with its spherical mirror, works best at magnifications below 120×, while the others, with their parabolic mirrors, can handle higher magnifications more readily.

All XT Classic telescopes include black steel tubes and Dobsonian-style mounts made of black laminate-covered particle board. The finish is nice, but those dark colors can make the telescope literally disappear on dark nights, save for the white Orion lettering. Motions in both altitude and azimuth, while not as silky smooth as on premium Dobsonians, are adequate. The CorrecTension Friction Optimization feature—pretty fancy language for a couple of springs—serves a dual purpose. The primary mission is to exert enough downward force to press the round altitude bearings into the side boards, increasing the friction between the two. The springs also keep the telescope's tube and ground board together, letting users carry the two together. That's handy with the 4.5- and 6-inch XTs, but some will find the XT8 too heavy to move as a single unit. The 58-pound XT10 and 81-pound XT12 are best moved in stages.

Augmenting the XT line are the 6- to 12-inch XT IntelliScopes, the first econo-Dobs set up at the factory to accept digital setting circles. Optically, they have the same specs as the XT Classics, although the XT10 and the XT12 have Pyrex mirrors rather than BK-7. All four models use adjustable knobs to regulate the tension on their vertical axes rather than springs as with the XT Classics. Either method is equally efficient.

The intelligence behind the IntelliScope includes a pair of 9,216-step digital encoders (one on the altitude axis, one on the azimuth axis) and a hand-held Computerized Object Locator. The azimuth encoder comes with the telescope, while the altitude encoder is included with the hand controller, which must be purchased separately. Let's be clear up front that these are not motorized GoTo telescopes. Apart from the 9-volt battery used to power the Computerized Object Locator, none of the IntelliScopes has any power requirements. Instead, you are the motor. After selecting one of the sky objects in the computer's memory, two directional guide arrows with numbers are displayed on the hand controller's illuminated LCD screen. The user pushes the telescope in the direction of the arrows until both numbers decrease to 0.0. If the telescope was initialized correctly, done by aiming at two preselected stars, then the target should be in the eyepiece field. The lack of motors also means that you'll have to keep the telescope aimed at the target manually (in other words, by nudging it).

Owners report good success using the IntelliScope, although many question the choice of colors for the Computerized Object Locator's illuminated buttons. The 12-button numerical key pad glows a dim red, which is perfect since it does not impact our eyes' night vision. The LCD readout, directional arrows, and "enter" buttons, however, shine a bright green, which will affect night vision. Why Orion chose green for some of the buttons remains a mystery. As one person noted, "I have to hold the computer against my body to dim its light while looking through the eyepiece." Also, in order to save battery life, the controller goes into power-save mode after 50 minutes of inactivity, requiring the user to reinitialize the computer before continuing.

Another popular telescope from Orion is its little 4.5-inch f/5 StarBlast, shown in Figure 5.6. When I reviewed the scope for *Astronomy* magazine (January 2004 issue), I was immediately struck by its cuteness factor. Orion is marketing the StarBlast as a kid's telescope, but in truth, anyone who enjoys panning across the Milky Way or viewing broad-field targets like the Pleiades and the Andromeda Galaxy will love the view. Outwardly, the green StarBlast looks like it's riding on half of a Dobsonian mount. Because the tube assembly is so lightweight, the designers rightfully felt that a full Dobsonian mount would just add weight but offer no additional benefit. They were absolutely correct. Like Orion's larger Dobsonian-mounted telescopes, the StarBlast mount is made from laminate-covered chipboard.

In its mounting, the StarBlast is only 25 inches tall and weighs only 13 pounds, which makes it a great grab-and-go telescope for impromptu observing sessions. Just be sure to bring along something to get the scope up off the ground, since it's much too short to use alone. I used a short, plastic patio table, but some might prefer a picnic table, a bar stool, or some other sturdy support. I know some owners who use an upside-down wastepaper basket or an empty 5-gallon paint can, but these will rock on uneven ground.

Orion also sells the StarBlast on a small equatorial mount for a little more money. Personally, I would stick with the original half-Dob version and place

Figure 5.6 *Orion Telescope's highly rated 4.5-inch StarBlast Newtonian reflector on a simple altitude-azimuth mount.*

the scope on a table or other support. Although the equatorial mount is up to supporting the scope thanks to the telescope's light weight, the half-Dob's motions are smoother and less apt to wobble during focusing.

Moving up a step in aperture, Orion offers two 5.1-inch Newtonians. The f/6.9 SpaceProbe 130 includes a 6 × 30 finderscope, as well as 10- and 25-mm, 1.25-inch format Kellner eyepieces. Its spherical primary mirror is acceptable at lower magnifications but begins to weaken beyond about 100×. At the same time, the benefit of the increased aperture over a 4.5-inch telescope is outweighed by the instrument's weak German equatorial mount. Rather than this instrument, consider the Orion XT4.5 or XT6 Dobsonians, or the company's SpaceProbe 130ST.

The SpaceProbe 130ST is an f/5 rich-field version of the regular Space-Probe 130. All mechanical parts are well made, with a smooth 1.25-inch focuser, nicely designed diagonal mirror mount with thin spider vanes, a spring-loaded 6 × 30 finderscope mount for quick adjustment, and an all-metal primary mirror cell. The 130ST is supplied with 10- and 25-mm Plössl eyepieces and the same light-duty equatorial mount as the longer model. In this case, however, the mounting is reasonably sturdy for most uses, because the telescope weighs so little. Owners consistently rave about the telescope's optical performance, especially on wide-field objects like the Orion Nebula and Pleiades. All of the latter's stars fit comfortably into the supplied 25-mm Plössl's field, which captures a generous 2° of sky. The 2× Barlow lens that is also included is nearly worthless, however, so don't use it to judge the scope's high-power prowess. The parabolic mirror is able to produce surprisingly good high-powered views of the planets and double stars but only with decent eyepieces.

The Orion AstroView 6 is a more robust version of the SpaceProbe 130ST. The 6-inch f/5 parabolic primary mirror improves image brightness by gathering 39% more light than a 5.1-inch mirror. Although that sounds impressive, in reality, the difference amounts to a gain of less than half a magnitude. Still, images will be more enjoyable through the AstroView 6 thanks to its stronger, sturdier mount. Vibrations, however, can still be troublesome, especially when trying to focus at magnifications in excess of 100×.

Looking for a larger equatorial Newtonian? Orion's SkyView Pro 8 EQ should be on your list of candidates. The 8-inch f/4.9 optical tube assembly is very good indeed, thanks to sharp optics and some nice frills, like a 2-inch rack-and-pinion focuser. My only wish is that it came with a better finder-scope; the supplied 6 × 30 finder is just too small to be useful. In the field, the SVP 8EQ produces images that are reasonably sharp as long as the optics are properly collimated. Coma softens the view toward the edges of the field, but that's to be expected with such a fast focal ratio. Although the SkyView Pro is a decent, mid-range mount, the SVP 8EQ's weight taxes the top limit of the mounting's capacity. In a head-to-head assessment, however, Celestron's C8-N eclipses the SVP 8EQ for its slightly longer focal length, which reduces coma to more manageable levels, as well as its sturdier tripod legs.

Orion also combines the same 8-inch f/4.9 telescope on its Sirius equatorial mount that creates a steadier but heavier instrument. A discussion of the

Sirius mount's merits can be found under the Orion 80ED and 100ED apochromatic refractor review earlier in this chapter (see page 98). But of all the telescopes that Orion combines with the Sirius mount, I believe the Sirius 8 is the best match. For one thing, the higher position of a Newtonian's eyepiece allows the user to keep the Sirius tripod's stainless-steel legs fully collapsed, greatly improving the mount's stability. Unless you're considering some serious through-the-telescope deep-sky photography, the Sirius 8 is more than enough telescope to satisfy most amateurs for years of viewing enjoyment. For those who prefer computer control, the Sirius 8 EQ-G throws in the optional GoTo hand controller in place of the non-GoTo controller that comes standard.

The 8-inch tube assembly is also available on the heavier duty Atlas EQ-G mount as the Atlas 8, while the 10-inch f/4.7 Atlas 10 EQ places an excellent 10-inch f/4.7 optical tube assembly on the same state-of-the-art base. The Atlas, a variant of Synta's EQ6 mount, has a black powder–coat finish and includes fully internalized motors for a very neat looking package. The dovetail assembly that holds the telescope to the mount can slide 6.5 inches along the saddle plate for precise balancing. The Atlas's robust tripod is built on three 2-inch-diameter stainless-steel legs. The two-part legs extend from 33 inches to 58 inches in length by using a quick-release clamp on each. Unless you need to raise the telescope to unusual heights, however, leave the legs fully collapsed for improved stability. A spreader, which includes a tray to hold three 1.25-inch eyepieces and three 2-inch eyepieces, screws into the underside of the mounting head and presses outward against the three legs to improve the mount's overall rigidity. The standard Atlas EQ mount includes a dual-axis motor control, an autoguider port, and a periodic error correction for fine-tuned tracking during astrophotography, while the EQ-G version adds GoTo computerized control. A GoTo upgrade kit can also be purchased later if desired. The kit comes with a new hand controller and involves replacing the mount's two internal drive motors as well as the circuit board.

The Atlas 10's primary is made of Pyrex for better thermal behavior in changing temperature conditions than the smaller Atlas (and Sirius) 8. The optical assembly's open-back mirror cell also helps to promote shorter cooldown times. Images are very crisp and star tests consistently show $1/4$-wave optics or better once the optics have cooled and collimation is confirmed (which should be checked before each use). Overall owner satisfaction is very high, although all owners are quick to caution about the telescope's weight. If you're looking for a lightweight instrument, the Atlas 10 probably isn't for you. The 54-pound Atlas mount (not including counterweights) takes some effort to carry, especially when you're tired and just want a quick peek. Those nights are best left to binoculars or a lighter telescope.

Sky-Watcher Telescopes. Based in British Columbia, Pacific Telescope Company teams with Synta Optical Technology Company to offer a variety of imported reflectors, as well as many refractors, which were previously listed.

The smallest Sky-Watcher Newtonian is the model 1145EQ1. The Sky-Watcher 1145 is a compact, f/4.4 rich-field telescope that makes a great grab-and-go instrument for those on-the-run observing sessions. Supplied with a small German equatorial mount and aluminum tripod, the 1145 is perfect for low-power viewing of broad star clouds and the like. The one-armed tabletop mount that comes with the f/4 StarBlast from Orion Telescopes is sturdier than the comparatively flimsy EQ1 mount supplied here, provided the StarBlast can be placed on a sturdy table or other support.

If you are looking for a general purpose 4.5-inch reflector, the longer focal length of the f/8 Sky-Watcher 1149EQ2 makes it a better choice than the 1145. To support the heavier tube, the 1149 is coupled to the slightly beefier EQ2 equatorial mount, which is satisfactory but still not noteworthy. The optics are fairly decent, even if the spherical primary mirror means that spherical aberration will begin to blur the view above 100×. Review chapter 2 for a discussion of spherical versus paraboloidal primary mirrors. Unfortunately, the 1149 only comes with a poor 5 × 24 finderscope.

Sky-Watcher also includes two 5.1-inch Newtonians that mirror the variety of its 4.5-inch models. Of these, the Sky-Watcher 13065PEQ2 is the most noteworthy. Featuring an f/5 paraboloidal primary, the 13065P is a nice, compact instrument for quick wide-field viewing with minimal aberrant intrusion, although it must be accurately collimated to function at its best. This scope is essentially a twin to Orion's SpaceProbe 130ST. Take a look at that review on page 126 for further thoughts.

There isn't much of a contest between the 13065P and the Sky-Watcher 1309EQ2. The 1309 is a conventional f/7 Newtonian that relies on a spherical mirror to focus images into the eyepiece. As a result, users can expect a realistic magnification limit of about 100× before spherical aberration degrades the view. There are better telescopes for the money.

Both of these instruments come mounted on the weak EQ2 German equatorial mount and an aluminum tripod. Both also come with a 6 × 30 finderscope. Although the finderscope is nothing special, the finder base is. As with all finders supplied by Synta to its various retailers, Celestron and Orion Telescopes among them, the finderscope ring mount uses two adjustment screws spaced 90° apart operating against a spring-loaded piston. It's a little thing, but brilliant both in its simplicity and user-friendliness. Whoever came up with this solution is a real innovator.

I have always had a soft spot for rich-field telescopes. They can produce amazing views of the night sky, especially star clusters and nebulae. That is why the 6-inch f/5 Sky-Watcher 15075PEQ3-2 has a special appeal. With its f/5 parabolic primary mirror, the 15075P has enough light-gathering prowess to show all of the Messier objects as well as many NGC members, including such challenging sights as the Rosette Nebula, North America Nebula, and California Nebula, given suitable viewing conditions. Like the 13065P mentioned previously, the 15075P includes nicely designed mechanics, including the EQ3-2 equatorial mount, which is sufficient for this tube assembly.

Also built around parabolic primary mirrors, the Sky-Watcher 2001PEQ5 and 2001PHEQ5 are a pair of impressive 8-inch f/5 Newtonian reflectors. Both feature 2-inch focusing mounts, 9 × 50 finderscopes, and well-designed primary and secondary mirror cells that allow for rapid optical cooling. The only difference between the two models is the mounting. The 2001P comes on the EQ5 German equatorial mount, while the other model is supplied with the upgraded HEQ5 mount. The HEQ5's tubular steel tripod is more than up to the task of supporting this scope, but the weaker aluminum legs of the EQ5 are not. Take that into consideration when debating between the two. Optically, the telescopes will deliver satisfactory performance for years of viewing pleasure.

The largest equatorially mounted reflector in the Sky-Watcher family is the 10-inch f/4.7 25012EQ6. Like Orion's Atlas 10, the 25012 features a steel tube, an open mirror cell for rapid cooling, a 9 × 50 straight-through finderscope, a polar-alignment scope built into the right ascension axis, and a dual-axis motor drive system. All these features come on top of the heavy-duty (and just plain heavy) EQ6 German equatorial mount and adjustable, tubular steel tripod. Take a moment to review the Atlas 10's summary found earlier in this section (see page 127) to read about some of this telescope's strengths and weakness. To summarize those thoughts here, the 25012 is an excellent performer optically, with bright, sharp images thanks to its very good Pyrex primary mirror. The mounting is fine for visual observing as well as some manually guided astrophotography, but it is neither equipped with a periodic error correction (PEC) circuit nor designed to accept an autoguider. As a result, guided exposures must be followed closely by eye using either a separate guidescope or an off-axis guider. Periodic error correction as well as complete GoTo control is available on the computerized EQ6 SkyScan mount, however. Speak with your dealer about upgrading from the standard EQ6 to the EQ6 SkyScan if these features are important to you.

Sky-Watcher Dobsonians come in three apertures: a 6-inch f/7.8, an 8-inch f/5.9, and a 10-inch f/4.7. The 10-inch includes a Pyrex mirror, which is less affected by sudden changes in temperature than the plate-glass mirrors that come with the 6- and 8-inch models. This means the mirror's performance will be less affected when moving from a warm house out into the cool night air. Glass-type aside, each aperture works very well once cooled to ambient temperature and properly collimated. Jupiter's Red Spot, Saturn's Cassini's Division, and a trove of deep-sky objects all show well through each instrument. Star testing will likely reveal some slight spherical aberration, but certainly within acceptable limits for an introductory telescope. One instrument I tested a while back, however, arrived with the secondary mirror seriously out of alignment. Thankfully, the thin-vaned spider mount is easily adjusted by using the instructions on the Sky-Watcher Web site.

Mechanically, Sky-Watcher Dobsonian mounts move a little stiffly in azimuth. I have found that by replacing the standard three plastic pads on the ground board with furniture slides (available at any competent hardware or home supply store), the problem is greatly diminished. The telescope also may prove to be a little tail heavy, but this problem is also easy to overcome by twisting the tension control handle that protrudes from the side of the Dobsonian mount.

Stargazer Steve. Many of us who are children of the 1960s and 1970s might have owned a 3-inch f/10 Newtonian reflector manufactured by Edmund

Scientific Corporation. It seemed to be the quintessential first telescope of the era. Although the Edmund 3-inch f/10 is no longer made, Steve "Stargazer Steve" Dodson, an award-winning amateur telescope maker from Sudbury, Ontario, has resurrected, enlarged, and improved the design to produce the 4.25-inch f/7.1 Sgr-4, an excellent first telescope for young astronomers. Images of the Moon, planets, and brighter sky objects are sharp and clear, thanks to its parabolic primary mirror. Many other 4.5-inch reflectors use primary mirrors with spherical curves, which, as mentioned in chapter 3, can lead to spherical aberration and fuzzy images.

The telescope tube, made from vinyl-coated cardboard and painted flat black on the inside to minimize stray-light reflections, is held on a simple altitude-azimuth mount of birch plywood and maple, finished with clear varnish. The mounting and the tripod are both lightweight yet sturdy and easily carried by young and old alike. Motions in both altitude and azimuth are smooth. A wooden knob adjusts the tension on the altitude axis, which is important when using eyepieces of different weights. The Sgr-4 comes with a 1.25-inch-diameter 17-mm Plössl eyepiece and a Rigel QuikFinder unity finder for aiming.

Stargazer Steve also offers several telescope kits for those who want the added enjoyment of actually making their own telescope. Everything is precut for easy assembly, so all you need to do is put the kit together by following the included videotaped instructions and finish the wood with paint or stain. If you start the project in the early afternoon, the telescope may even see "first light" that night.

The 4.25-inch f/10 Deluxe Planetary telescope kit includes the optics, cardboard tube, rack and pinion focuser, birch plywood mount, and Teflon bearings. A 10-mm Kellner eyepiece is supplied, as are two rings that the user sights through when aiming. Those rings are not suitable for aiming the instrument with any great precision, however, and should be replaced with either a lightweight 6 × 30 finder or a one-power red-dot finder, listed in chapter 7. Although images are sharp through the Deluxe Planetary telescope, most users find that the base is much too low to the ground for comfortable viewing. Instead, place the telescope and mount on a short stool or other support to raise the eyepiece to off-the-ground height. Although the Deluxe Planetary telescope shows wonderful close-ups of the Moon and planets, it is not as well suited for general viewing as Steve's second 4-inch kit, discussed next.

The 6-inch f/8 Deluxe Planetary telescope is outfitted similarly to its 4.25-inch namesake, but with some welcome improvements. For one, the f/8 primary mirror has a parabolic concave curve rather than a spherical curve like the smaller scope. Star tests reveal consistently excellent optical quality, producing sharp and clear images that are a delight to behold. They easily surpass many of the better-known econo-Dobs on the market and come highly recommended.

Steve's Truss Tube 6 kit, seen in Figure 5.7, is built around a small, open-tube optical assembly that is both functional and fashionable. Six 25-inch-long aluminum poles attach between the mirror rocker box at the telescope's base and the instrument's front focuser ring. The 2-inch low-profile helical focuser that comes with the telescope works smoothly, although it can take some time

Figure 5.7 *The 6-inch f/5 Stargazer Steve Truss Tube Kit 6 can be easily disassembled for compact storage, making it a perfect traveling companion by land, sea, or air.*

to adjust image sharpness when swapping eyepieces. Oversized altitude bearings let the telescope swing smoothly when moving up and down in the sky, while the Teflon-on-Formica azimuth bearing creates equally smooth left-to-right motions. A collapsible wooden tripod is also included to raise the telescope to a usable level. Everything will fit together perfectly, since Steve personally assembles each instrument during manufacture before disassembling for shipment.

Looking for a large-aperture travelscope to take on that next cross-country flight? The Truss Tube 10 kit just might be it. Like the 6-inch truss kit, the Truss Tube 10 includes everything needed to assemble the instrument quickly for same-night viewing.

Starmaster Telescopes. Starmaster continues to attract a lot of attention in the world of telescopes. Owned and operated by Rick Singmaster in Arcadia, Kansas, Starmasters come in a wide variety of sizes to suit nearly anyone looking for a well-made, Dobsonian-mounted Newtonian reflector. Currently, Starmaster has two lines of telescopes: Versa and Truss.

Starmaster's 8- and 11-inch Versa telescopes are beautifully crafted instruments that tempt consumers who want the best of the best. Rather than have their investment dollar go almost solely into large-aperture optics, Versa buyers prefer to sacrifice aperture for quality craftsmanship. But when you buy a Versa scope, you are buying only the optical tube assembly, not an entire instrument. You'll need to place a suitable mounting underneath. That's where the Versa's "versa"-tility comes into play. The tube assembly is designed to attach to either a Dobsonian-style rocker box using a pair of aluminum altitude rings or to an equatorial mount using a mounting plate. Both are sold separately by Starmaster.

As we have come to expect from all Starmaster telescopes, the Versa scopes deliver nearly flawless images. Whether examining fine planetary detail or probing the depths of distant nebulae, images are tack-sharp, rivaling those delivered by pricier apochromatic refractors. That's largely due to the telescopes' optics that come from Carl Zambuto Optics. Of course, Zambuto quality does not come cheap, which partially explains the high price of the Versa scopes versus other similarly sized reflectors. Is it worth spending $3,000 (the current price of the 8-inch Versa) on a telescope when you can buy the same aperture from a mass-market importer for about $450? Optically and mechanically, the $450 8-inch reflector will not be nearly as refined as the Versa, although it will likely deliver images that most users would find satisfactory. Only you can answer that question for yourself, but consider that the Zambuto-mirrored 10-inch Starmaster Hybrid Truss reflector actually costs less than the 8-inch Versa when you figure in the cost of the latter's rocker box. Also consider that putting the scope together from scratch can be a chore, because each of the poles (five for the 8, six for the 11) must be slid into matching collars, then screws are tightened using a Phillips screwdriver to hold the whole thing together. Of course, components can be left assembled between observing sessions, if desired.

For the intermediate-to-advanced amateur, Starmaster also offers several Truss telescopes, from a 14.5-inch f/4.3 to a monster 28-inch that comes in either f/3.7 or f/4.1. Each is based on the now-familiar eight-tube truss design riding on large, semicircular bearings and set upon a low-riding Dobsonian base. The 14.5- through 24-inch mirrors come from Carl Zambuto Optics, while the 28-inch mirrors are by Pegasus Optics. To prove their worth, each telescope comes with certification attesting to its optical quality. In fact, Singmaster personally tests each instrument before it is sent to the customer. As the late actor and commercial spokesman Orson Welles used to say in a television commercial about fine wine, no Starmaster telescope is sold "before its time."

Indeed, each and every Starmaster that I have viewed through has produced amazingly clear images. I can recall several memorable views I had through a 16-inch that were just breathtaking. No matter what we looked at, the image was about as good as I have ever seen it through that aperture. Globular clusters like M13 sparkled, while nebulae shone with an eerie glimmer usually seen only in photographs. And the planets Jupiter and Saturn also revealed a level of detail missed through most apochromatic refractors because of their limited apertures.

The 14.5-inch Starmaster Truss scope is referred to as a hybrid because of its unique assembly. The gist of the design is that the eight truss poles are preassembled in four trapezoidal groups that make setup and teardown go very quickly. If desired, the truss pole assemblies can be left attached to the upper cage assembly, and the entire unit lifted onto the mirror box, where it is attached with four thumb knobs. The truss tubes that come with the larger Starmasters come preassembled in pairs, which also makes assembly go quickly. Each truss model features furniture-quality oak-veneer plywood construction finished with a durable coating of clear polyurethane and rubber-coated aluminum truss tubes. Motions in both altitude and azimuth are smooth thanks to the Teflon-on-Formica bearing surfaces.

A unique standard feature of the larger Starmasters is a detachable mirror cell. By first removing the mirror (the heaviest component of a large-aperture reflector), carrying the telescope about becomes much easier. Once at the chosen observing site, the mirror cell is reattached to the telescope with little ill effect on collimation. A great idea! Detachable wheelbarrow-style transport handles are also available, but at an extra cost.

Each Starmaster Truss features oak-veneer plywood construction and comes with a 2-inch Crayford-style focuser, a Rigel QuikFinder unity finder, primary and secondary mirror covers, a truss pole carrying case, and a cloth light shroud. Extra-cost options include upgraded focusers and factory-installed digital setting circles or a full computer-driven GoTo mounting. The latter, called *Sky Tracker*, is amazing to watch. As with other GoTo telescopes, once initialized, just enter the object of interest into the hand controller, press the button, and the telescope slews automatically to its location with amazing accuracy. Yes, an expensive option, and as much as I frown on such things (to me, it's the journey, not the destination), I must admit that the Sky Tracker works very well.

Starsplitter Telescopes. This company, based in Thousand Oaks, California, offers a number of different Newtonian reflectors for amateur astronomers. All are mounted on Dobsonian mounts that, while solidly made, do not exhibit the level of excellence in finish as those manufactured by Obsession or Starmaster. Teflon pads riding on Ebony Star Formica let the telescopes move smoothly in both altitude and azimuth. In addition, each comes with a 2-inch low-profile focuser.

Starsplitter's Compact II line of telescopes is available in three apertures (10-inch f/6, 12.5-inch f/4.8 or f/6, and 15-inch f/4.5), while its Compact IV reflectors include 8-, 10-, and 12.5-inch models. All feature mirrors from Zambuto, Swayze, or Nova Optics. The Compact II's follow the more conventional eight-pole truss design utilized by many others, while the Compact IVs use four parallel tubes to bridge the gap between the mirror box and upper cage assembly. Some may question whether the four-tube arrangement is stable, or will it twist and flex in use. Although a rectangle is certainly not as stiff as a triangular truss, the Compact IVs should be fine unless they are heavily laden with accessories.

In both cases, each truss tube is attached to the mirror box by sliding it into a mounting block that is attached to the inside of the mirror box. Although the design works, it's not nearly as elegant as that used by Obsession, Starmaster, or for that matter, on the Starsplitter II line, reviewed next. The biggest problem is the potential for dropping something directly onto the primary mirror, since the mirror box's protective wooden cover must be removed before the truss tubes can be attached. Although the hardware is captive, I still get nervous leaning over a fully exposed primary while tightening or loosening thumbscrews.

Starsplitter also makes two lines of big telescopes: the Starsplitter II and the Starsplitter II Lite. The Starsplitter II series includes six instruments from a 15-inch f/4.5 to a 30-inch f/4.5, while the four Lites range from a 20-inch f/4 to a 28-inch f/4.5. Choose your telescope's mirrors from Nova, Galaxy, Pegasus, or Swayze Optics, all of which make decent optics. If I had to pick one, I would probably go with Galaxy Optics, followed closely by Nova. Starsplitter II telescopes are designed around open-truss tubes that allow for comparatively easy storage. All come with removable wheelbarrow handles for moving the instrument around prior to setup. The no-tool design uses captive hardware, making separate tools such as wrenches and screwdrivers unnecessary for setting up and tearing down the instrument. The truss tubes are held to both the mirror box and the secondary-mirror assembly with thumbscrews and mating wooden blocks.

From my research for the first edition of this book in 1994, I purchased an 18-inch Starsplitter II, reasoning that it was the largest telescope that could fit into my car at the time. As I reported in subsequent editions, everything continues to work quite well. The telescope takes about ten minutes to set up from scratch and another few minutes to collimate by turning one of three large knobs behind the mirror. The diagonal mirror, which is so critical to the telescope's overall performance, is held in a conventional spider mount, requiring a straight-blade screwdriver to loosen and tighten its pivot screws. This conventional system is not nearly as convenient as the innovative designs used by others, especially Obsession. The Galaxy optics in my 18-inch are very good, although their true quality did not come through until I checked and, subsequently, repositioned the diagonal mirror, which had been installed too far back in the upper cage assembly. Once I redrilled and relocated the mirror's spider mount (something that should have been caught during manufacturing), I am now seeing sights through this telescope with amazing clarity, contrast, and color.

So, the bottom line is that Starsplitter II telescopes are well-designed instruments, although their cabinetry is not quite as refined as Obsession or Starmaster instruments. They should also be tested before shipment, which mine clearly was not. The Starsplitter Compact II and IV lines, however, seem to cut a few too many corners, although, it must be recognized that they are also less expensive. Still, if I were in the market for a truss-style Dobsonian reflector but could not afford the cost of a Starmaster or Obsession, I would give AstroSystems TeleKits, Nightsky Scopes, and TScopes all very strong consideration.

TScopes. Founded by Ed Taychert in 2003, TScopes has also garnered a lot of high praise among veteran amateur astronomers who are always looking for that ultimate telescope. Fashioned in the Kriegerian style, with large altitude bearings sitting atop a low rocker box, TScopes come in three model lines ranging from 13 to 15 inches in aperture.

The company's original and least-expensive models, simply called TScopes, are available with either 13-inch f/4.6 or 14-inch f/4.7 primary mirrors. Sources for those mirrors include, in order of increasing price, Obsidian Optics, Raycraft Mirrors, Woden Optics, and Swayze Optics. Obsidian and Swayze are well-established companies used by other telescope manufacturers as well, while Raycraft and Woden are less common. The aluminized primaries all have 91% reflectivity, while each secondary mirror has 96% enhanced reflectivity. Extra-cost upgrades, including digital setting circles, GoTo computer control by ServoCAT, and improved focusers are also available, although wheelbarrow handles for lifting are not.

The TScopes FarStar and Type-3 Deluxe reflectors are especially interesting. The FarStar 13- through 15-inch scopes are some of the lightest weight instruments in their respective aperture classes. Setup and teardown are both very easy, but the real fun comes in their use. Motions are very smooth in both altitude and azimuth, with no binding or dead spots in any direction. FarStars are designed to be as light as possible without sacrificing stability by using an especially low-riding *space frame* rocker box that has three wedge-shaped sections cut out of each side to reduce weight. The FarStar mirror box is also designed with several triangular openings in the wood to lighten the load. I must admit that when I first saw the design, I was skeptical that it compromised instrument stability. But those fears proved unfounded.

The Type-3 Deluxe TScopes, usually referred to as T-3s, are similar in design and appearance to the original TScopes above, but use thicker wood, larger altitude bearings, and an all-metal primary mirror cell for improved stability and, naturally, slightly more weight. The T-3s also offer what Taychert calls his Peerless Optical Package, which includes an Obsidian Optics primary mirror stated to be $1/20$th-wave. I am a cynic by nature, so take these claims with a grain of salt. But the proof is in the pudding, as they say. After he looked through a T-3 Peerless TScope at a club observing session, a friend and experienced observer from New Hampshire related to me his impressions: "I'm rarely impressed by telescopes these days. I've pushed Dobsonians around with sticky motions, seen SCTs with horrific aberrations, and even looked through apochromatic refractors that seemed to be merely adequate. But a club member's 14-inch TScope [with the Peerless Optical Package] had the best Newtonian views I have ever seen. I was told the mirror was $1/20$th wave and had a Strehl Ratio of 0.996. While I have some skepticism about these numbers and know the complete optical system couldn't touch the mirror-only specs, the views were wonderful."

Vixen Company, Ltd. Japan-based Vixen offers several short-focus Newtonians reflectors. The smallest included here, the 5.3-inch f/5.3 R135S, is aimed toward the beginner market. Although it costs more than many Chinese and Taiwanese clones, and even though the aperture is somewhat limiting, the sturdy construction and high quality of the R135S's parabolic primary mirror makes this small Newtonian one that will let the owner hone his or her observation skills as well as experiment with astrophotography. The R135S is supplied with a red-dot unity finder but no mounting. For that, Vixen's own Great Polaris-Economy (GP-E) or Celestron's CG-5 equatorial mount would be ideal.

The R150S adds another 0.6 inch of aperture to create a fine 5.9-inch f/5 telescope. Both the R135 and the R150 come with a unique back-and-forth sliding focuser that moves fore and aft along a geared track rather than vertically in and out like conventional designs. Although the scheme works fine, it strikes me as more complicated than it needs to be. After all, a standard rack-and-pinion or Crayford focuser works fine as well and doesn't require the secondary mirror to be moved during focus.

Vixen's R200SS is a stubby 7.9-inch f/4 reflector that's perfect for wide-field viewing and terrific for transporting, but its fast focal ratio means that collimation must be dead-on. Owners praise the images but are quick to add that a coma corrector, such as a Tele Vue Paracorr, is needed for the best views. The R200SS includes a 2-inch rack-and-pinion focuser that also features a built-in adapter for attaching cameras directly to the telescope, a nice touch that eliminates the need for a separate camera mount (you will still need a T-mount for your camera—see chapter 7). Just make sure that you get the improved focuser that comes on the green-tubed R200s. The focusers used with older, white-tube R200s worked adequately but not as well as the newer version. Whether used for visual or photographic observations, the wide field of the R200 produces some spectacular views!

Each Vixen Newtonian is attached with a pair of rotatable rings to its excellent Great Polaris-E (GP-E) German equatorial mount. The GP-R200SS can also be purchased on the sturdier, costlier GP-D2 mount, a good upgrade for photography, or the state-of-the-art Sphinx GoTo mount. The GP-E and GP-D2 mounts are often supported by aluminum tripods, which are their only Achilles' heel. Better to get the optional wooden tripod. The Sphinx version comes on a sturdier tripod, although it also uses extruded aluminum legs.

Vixen products are not inexpensive, especially in this day and age of like-size clones from China and Taiwan that, at first glance, appear to be identical in design. Don't go by looks alone, however. Vixen quality exceeds those others by a wide margin.

Cassegrain Reflectors

Optical Guidance Systems. For the rich-and-famous, die-hard amateur astronomer, Optical Guidance Systems offers some impressive telescopes that are worth a look. The company's medium-to-large–aperture RC Ritchey-Chrétien Cassegrain reflectors rank among the finest instruments in their respective classes. Perhaps the most important benefit of the Ritchey-Chrétien design is its total freedom from coma. This, along with a lower f/ number than possible in a Classical Cassegrain, is especially important if the instrument will be used for astrophotography. The correspondingly larger central obstruction, however, will diminish contrast some during visual observations.

One of the strongest selling points of Optical Guidance System's telescopes is its superb optics, supplied by Star Instruments of Flagstaff, Arizona. To decrease overall weight, lightweight, conical mirrors are made from near zero-expansion ceramic material, then aluminized and silicon overcoated. Images are striking, with pinpoint stars across the field through its Ritchey-Chrétien instruments. The view is also quite impressive through the Classical Cassegrains, although not quite up to the same standard.

All OGS instruments, regardless of optical design, include oversized focusers from Astro-Physics, which are famous for their smooth, no-backlash movement. Options abound for all **OGS** Cassegrains. For instance, consumers can elect either aluminum or carbon fiber tubes. Carbon fiber has the distinct advantage of not retaining heat like aluminum does, which will keep focus steady even as the temperature changes as the evening wears on. Mountings, mounting rings, finderscopes and guidescopes, and other customized accessories are also available for all OGS Cassegrain instruments, but keep in mind that all except the 10-inch on the Takahashi NJP mount, and possibly the 12.5-inch on Software Bisque's Paramount-ME mount, are really designed for observatory installation, not portable use. All OGS equatorial mountings, whether fork or German-style, include the personal computer–based Telescope Control System, or PC-TCS, designed by Comsoft of Tucson, Arizona. PC-TCS includes all motors, controllers, and assorted paraphernalia needed to control telescope motion from an IBM-compatible computer.

RC Optical Systems. Owned and operated by Optical Systems, Inc. of Flagstaff, Arizona, RC Optical makes Ritchey-Chrétien–style Cassegrain reflectors that are just striking. Their optical and mechanical quality is top drawer all the way. Images are free of coma, a common problem with Classical Cassegrains. Like their chief competition, Optical Guidance Systems, RC Optical uses mirrors from Star Instruments, acknowledged as the leading supplier of Cassegrain optics. RC Optical places these fine optics in well-baffled tube assemblies made from carbon fiber, known for its strength, excellent thermal behavior, and light weight. RC Optical completes each telescope with an oversized focuser from Astro-Physics, electronic adjustment and focus of the secondary mirror, and built-in fans to promote active cooling, again just like OGS.

Complete RC Optics telescope tube assemblies include 10-inch f/9, 12.5-inch f/9, 16-inch f/8.4, and 20-inch f/8.1 models. None are for the faint of heart, however, because prices start at more than $13,000 for the 10-inch. For this, you get a nicely appointed optical tube assembly, but a mounting is extra. Again, like many in this category, these instruments are best installed in an observatory, although the 10- and 12.5-inchers could be made transportable with one of the heavy-duty mounts discussed later in this chapter.

Takahashi. This company manufactures several Mewlon Cassegrain reflectors based on the Dall-Kirkham version of the Cassegrain optical system. Recall from the brief discussion in chapter 2 that Dall-Kirkham Cassegrains use mirrors with simpler curves than either Classical or Ritchey-Chrétien Cassegrains, yet still produce excellent results. Takahashi Mewlon instruments currently come in four apertures: the 7.1-inch f/12 M-180, the 8.3-inch f/11.5 M-210, the 9.8-inch f/12 M-250, and the 11.8-inch f/11.9 M-300. All four are very sensitive to collimation, which is adjustable by turning screws on the secondary mirror mount. The primary mirror is not customer adjustable. Once the optics are properly set, images are tack (or is that "Tak?") sharp. Whether it's a view of a planet, a double star, or a distant star cluster, images through all Mewlons are always clear and crisp. A few Mewlons suffer from image shift when focusing, a by-product of the internal mechanism, but for the most part adjustment shows no backlash.

Takahashi also offers the 8.3-inch CN-212 convertible Newtonian/Cassegrain reflector. This two-way telescope is possible because both the Classical Cassegrain and Newtonian optical designs are based around parabolic primary mirrors. By swapping secondary mirrors, the CN-212 can change between an f/12.4 Classical Cassegrain and an f/3.9 Newtonian. A keyed secondary mirror mount helps to keep the secondary mirror in line, while the mating four-vane spider helps to maintain optical collimation after conversion. Looking for wide-field views? Go with the Newtonian. Want to switch to planetary observing? The Cassegrain is perfect. This is a fun instrument to use, because it gives you, in effect, the best of both worlds. Coma is evident, as we might expect, but overall, I find the images to be very good in either configuration.

Exotic Reflectors

Neither fitting into the traditional Newtonian nor Cassegrain families of reflectors, these unique instruments offer interesting possibilities for amateurs looking for something a little different to bring along to their club's next star party.

DGM Optics. Many people are under the misconception that mirror-based telescopes can never deliver the same tack-sharp views of the planets seen through high-end apochromatic refractors. Part of this fallacy is due to some less-than-tack-sharp optics, part is due to poorly collimated instruments, and part is due to large secondary mirrors that lower image contrast. While the first two issues are caused by either inferior optics or possibly operator error, the third is an intrinsic fault of many designs that need to have a secondary mirror right along the optical axis of the primary. But what if a reflector could be made without *any* central obstruction? Images should be comparable to apochromats, and indeed, possibly better because chromatic aberration would be absolutely nonexistent.

That is exactly the idea behind DGM's line of OA (short for off-axis) reflectors. The secret lies in its Pyrex primary mirrors. Although they look like conventional mirrors, looks can be deceiving. Slice a traditional primary mirror in half and you will find the center, or low point, of its concave curve exactly in the center. That is not the case with DGM's off-axis instruments. Instead, its concave curves are off-center, or more correctly, off-axis. To picture what a cross-section of a DGM mirror looks like, imagine taking a standard primary mirror and cutting out three or four small, round sections from near the rim, in much the same way that you might cut cookies out of a large sheet of dough. As you can probably imagine, one side of each small mirror would be higher than the opposite side.

Four models make up the OA arsenal, ranging in size from the OA-3.6 (3.6-inch f/11.1) to the OA-6.5 (6.5-inch f/10.4). All mirrors include enhanced (96%) aluminizing and overcoating for optimal brightness. The views through each really are quite amazing. You'd be hard-pressed to tell the difference between the sight of, say, Jupiter or Saturn through an OA scope and a premium apochromat of the same aperture. Images are clear and crisp, with no telltale "milkiness" or "mushiness" that often plague Newtonians. Given steady atmospheric conditions, magnifications can be pushed to better than 100× per inch of aperture before image quality begins to break down, something that even the finest apos have difficulty doing. But as I have long said, a high-quality, long-focus reflector is tough to beat. Eliminate the central obstruction, and you have a killer telescope. Best of all, you can get an OA telescope, mount and all, for about 30% (or less) of the cost for a similar apochromatic refractor. All you need to do is add a finderscope and eyepiece.

As much as I enjoy the optical excellence of the DGM scopes, I find their mounts to be a little wobbly, especially the lighter models. Doing my rap test on an OA-3.6ATS, images took an average of six seconds to settle back down, garnering only a fair rating. At the same time, I would also suggest considering a conventional Newtonian if you are thinking of a DGM OA instrument. For less money than the OA3.6-ATS, you can buy a nicely equipped 6-inch f/8 Newtonian on a German equatorial mount that, although it has a central obstruction, will deliver brighter views and better resolution.

Catadioptric Telescopes

As mentioned in chapter 3, there are four types of catadioptric telescopes popular among today's amateur astronomers. By far, the most common is the venerable Schmidt-Cassegrain design. Schmidt-Newtonians have never gained the following of their Cassegrain cousins. In fact, only one mainstream company, Meade Instruments, manufactures them.

Maksutov derivatives have multiplied greatly since the first edition of this book came out in 1994. Back then, if you wanted a Maksutov, you had one choice: Questar. Although the Questar still has a loyal legion of followers, many lower-priced alternatives, including both Maksutov-Cassegrains and Maksutov-Newtonians, are now readily available. Let's consider each segment one at a time.

Schmidt-Cassegrain Telescopes

Celestron International. Celestron is renowned as the first company to introduce the popular 8-inch Schmidt-Cassegrain telescope back in 1970. Although only one basic model—the Celestron 8—was sold back then, there are now many variations of the original design from which to choose. They range from bare-bones telescopes to extravagant, computerized instruments. Celestron reports, however, that all of its 8-inch Schmidt-Cassegrain telescopes share the same optical quality, regardless of model and price.

Images through Celestrons are usually good, sometimes excellent. Every now and then, a lemon sneaks through its quality control program, but I must say that I am impressed with the strides that the company (and Meade, discussed later) have made since the mid-1980s. Maintaining quality in a mass-produced instrument as sophisticated as a Schmidt-Cassegrain, while also trying to maintain an attractive price, is a difficult chore.

All Celestron Schmidt-Cassegrains can be purchased with or without Celestron's Starbright XLT coatings, which, according to the company, are designed to increase light transmission by approximately 16% over optics with its standard Starbright coatings. At its peak wavelength, the XLT coatings reflect 89% of the light entering the telescope into the eyepiece. In practice, that amounts to a visual gain of about 0.25 magnitude, a modest increase that can nevertheless mean the difference between spotting some faint fuzzies on the threshold or seeing nothing.

Nowadays, Celestron Schmidt-Cassegrains come in several apertures and on several different mountings. The unique NexStar 8SE is easily distinguished by its unique one-armed fork mount. (Actually, I think of it as more of a short, broad chopstick.) It looks like the telescope should bounce back and forth like a diver about to spring off a diving board. But looks can be deceiving, and the N8SE is proof of that. Although not as sturdy as a true, two-armed fork mount, the N8SE's mount and tubular-legged tripod prove sturdy enough for visual observations as well as some basic lunar and planetary photography using a lightweight webcam, Celestron's own NexImage, or Meade's Lunar Planetary Imager. The N8SE comes with a 1.25-inch star diagonal, a 25-mm Plössl eyepiece, and a StarPointer red-dot finder. I would prefer a real finderscope to the StarPointer, even on a GoTo scope, because they are much easier to confirm centering on alignment stars and are also handy in case the scope goes off target.

The N8SE's GoTo function seems to work quite accurately given fresh batteries. Precision is good enough to place each selected target somewhere in the field of a low-power eyepiece, even after a cross-sky slew. In all, the N8SE's onboard computer database has more than 40,000 objects in its memory. (Although technically correct, the advertising hyperbole fails to say that many are actually listed more than once under different designations, and more than 29,000 entries are single stars, which are probably of little interest.) All are accessible via a molded hand controller, which mounts neatly on the mount's arm when not in use. The N8SE can also be controlled via a computer's serial port by several different software programs.

Although most owners are satisfied with the N8SE, the telescope is not without a few idiosyncrasies. One problem is that the mounting's arm is not long enough for the telescope to pass underneath if a 2-inch star diagonal or a Barlow lens is inserted into the telescope's eyepiece holder. This could potentially lead to a collision if the telescope tries to swing near the zenith as it slews from object to object. Although a clutch on the altitude axis prevents any damage, once the clutch slips, the scope must be realigned with the sky before tracking and GoTo pointing can be resumed. Another complaint is that the NexStar goes through batteries—eight AAs—very quickly, a problem shared by other NexStar telescopes, as well. Be sure to pick up either Celestron's AC adapter or a strong rechargeable battery to supply power to the scope instead, because the scope is rendered practically unusable if the power goes out.

With the 8-inch model's success, Celestron expanded the Special Edition series to include the 5-inch f/10 NexStar 5 SE and the 6-inch f/10 NexStar 6 SE Schmidt-Cassegrains as well as a 4-inch Maksutov. The Celestron 5 has been around for years in one fashion or another and enjoys great popularity among amateurs looking for a compact scope that's great for traveling. The Celestron 6 comes with XLT enhanced coatings and has exceptional optics, as I found while testing one for *Astronomy* magazine. Image shift was negligible, while image quality was exceptional. In fact, I liked the scope so much that I bought it. What sold me was the view I had of the Double-Double multiple star in Lyra.

All four stars were perfectly clear at 305×, which is not bad for a 6-inch telescope under average sky conditions. I expect that the C6, which has an inch more aperture than the C5, will enjoy even greater popularity.

Photographers should also note that the N6SE and N8SE models have autoguider ports on their drive bases. This port directly interfaces with any of the Santa Barbara Instruments Group (SBIG) autoguiders on the market. The N5SE does not have an autoguider port.

Then, there are the two-armed 8-, 9.25-, and 11-inch f/10 Celestron Professional Computerized (CPC) models, which were developed to replace the older NexStar GPS models that were the subject of a lawsuit brought about by Celestron's chief competitor, Meade Instruments. Meade alleged that Celestron's "North and Level" computer routine, which was used to initialize the onboard computer, infringed upon its patented Autostar Level-North Technology system, and a judge agreed. To satisfy that ruling, Celestron decided rather than retrofit existing models with new technology, it would start from scratch both inside and out. The result is its SkyAlign alignment method. Once the current date, time, and location are determined from data received from GPS satellites, the user initializes the telescope by aiming it at three random stars (or planets or even the Moon). No need to tell the computer which star is which. The computer calculates the angles measured between the objects and then compares them to the angles between known objects.

The silver base of the CPC mount, previously seen in Figure 3.5, is larger and flatter than the old NexStar GPS models. To me, it looks like a flying saucer from a B-grade 1950s science fiction movie. Beauty is in the eye of the beholder, but as long as it works, that is what counts. And the CPC mount certainly does. For openers, Celestron finally listened to the cries of its customers and placed a centered alignment pin on the top of the tripod. Just set the telescope down on the pin and rotate the base around until it lines up with three captive screws attached to the underside of the tripod. Thread them together and the telescope is ready to go.

I am also glad to see that Celestron includes a decent 9 × 50 finderscope as well as locks on both axes on each CPC model. That way, even if power is drained during the night, each telescope can still be hand-powered like a non-GoTo instrument. They lack mechanical slow-motion controls, however, making tracking by hand a little inconvenient.

There are a few other downsides to the CPC scopes versus their NexStar GPS roots. Chief among them is the optical tube assembly's material. The NexStar line had tubes made from carbon fiber, which were not only lighter than Celestron's conventional metal tubes but would also acclimate to ambient temperature more efficiently. The CPCs returned to aluminum tubes, which take longer to settle down under the cooling night sky.

I also miss the NexStar GPS's hand controller mount, which clipped the unit into a bracket that was cleverly built into one of the fork arms. This proved especially convenient when transporting the telescope. Although the CPC's protruding plastic bracket is handy, both it and the hand controller must be detached and carried separately when transporting the telescope to keep from accidentally snapping off the bracket.

All CPC Schmidt-Cassegrains feature worm-gear clock drive systems that include an integrated periodic error correction (PEC) circuit for greater tracking accuracy when in polar mode. Theoretically, a worm-gear clock drive should track the stars perfectly if it is constructed and polar aligned accurately, but this is not the case in practice. No matter how well-machined a clock drive's gear system is or how well-aligned an equatorial mount is to the celestial pole, the drive mechanism is bound to experience slight tracking errors that are inherent in its very nature. These errors occur with precise regularity, usually keeping time with the rotation of the drive's worm. The PEC eliminates the need for the telescope user to correct continually for these periodic wobbles. After the observer initializes the PEC's memory circuit by switching to the record mode and guiding the telescope normally with the hand controller (typically a five- to ten-minute process), the circuit plays back the corrections to compensate automatically for any worm-gear periodicity. The NexStar PEC feature is permanent in that once the periodic error is compensated for, the instructions will be remembered for subsequent use. Although the permanent PEC circuit is appreciated by astrophotographers, it is only applicable in polar mode. That also requires old-fashioned polar alignment and the optional equatorial wedge, which must also be added for long exposures through the telescope to avoid rotation of the field during long exposures.

Of the Celestron Schmidt-Cassegrains that come on German equatorial mounts, the least expensive is the C8-S. Like the NexStar N8SE, the C8-S includes a 90° star diagonal and a 25-mm Plössl eyepiece, but features a small 6 × 30 finderscope rather than the StarPointer unity finder. Although that is a better choice for star-hopping, since the C8-S is not computerized, owners quickly discover they need to upgrade to a 50-mm finder to zero in on anything but the brightest objects. Best to buy a couple better-quality eyepieces at the same time to make the package complete.

The C8-S is carried on the improved Advanced Series CG-5 German equatorial mount, a Chinese clone of the Vixen Great Polaris. The Advanced Series CG-5 mount is fine for supporting the C8's optical tube assembly, whether for visual use, short exposures of the Moon and planets, or some wide-field, piggyback guided exposures. Much of that improvement over the previous generation comes from the mount's tubular tripod legs. These are a welcome change from the flimsy aluminum legs that plagued earlier CG-5s. Still, if you have your heart set on long exposures through the telescope, you would do better with one of the other C8 variants.

The C8-S's tube attaches to the mounting on a sliding dovetail plate, which allows users to balance the assembly precisely and easily, which is handy for when you upgrade to that 50-mm finderscope or a heavy 2-inch eyepiece. The CG-5 can accept a polar-alignment scope for fast, accurate alignment, but unless you are going to try your luck at guided photography, spend the money on a new eyepiece instead. You might, however, want to consider the optional DC-powered motor drive, which runs on four D-cell batteries.

For those who want a GoTo computer-controlled Celestron but don't have the money for one of the more expensive models, the C8-SGT (see Figure 5.8) is a good alternative. The SGT includes the NexStar hand controller, which fea-

Figure 5.8 *Celestron's C8-SGT couples its well-regarded 8-inch Schmidt-Cassegrain telescope with the Advanced Series CG-5 German equatorial mount.*

tures a two-line readout and a full-size key pad that I find easy to use even while wearing heavy gloves. There are jacks on the mounting for linking an autoguider to the CG-5GT drive system as well as for the optional CN 16 GPS accessory, which ties the onboard computer to the global positioning system for initializing the computer. Although GPS is handy if you don't know the exact location of your observing site, it is not a necessity by any means.

Celestron recently introduced the 6-inch f/10 C6 Schmidt-Cassegrain into the Advanced Series lineup. Like the C8-S, the C6 may be purchased with GoTo control, as the C6-SGT, or without, as the C6-S. Regardless of which you buy, both come on the Advanced Series CG-5 mount, which is exceptionally sturdy for this size instrument. When I tested the C6 for *Astronomy* magazine, I was immediately impressed with the optical quality of this compact instrument, which held its own even at over 400×. The view of the quadruple star Epsilon Lyrae alone lets me enthusiastically recommend the C6. I suspect that it will develop a loyal following like the old Celestron 5, which is now only available as a terrestrial spotting scope.

For amateurs who want a little more aperture than an 8-inch, Celestron offers its C9.25-S and C11-S telescopes on the Advanced Series mount, as well. Like the 8-inch, the larger tube assemblies attach to the equatorial mount's head by means of a tube-length dovetail bar. This freedom lets the user set telescope balance precisely—a big plus when adding cameras or other accessories.

Although the Advanced Series CG-5 mount is far sturdier than the precursor version, the extra weight of the larger telescopes push it to its carrying limit. The 20-pound C9.25 is acceptable for visual use, although accurate focusing at high magnifications can prove difficult because of vibrations. The

C11's extra 8 pounds tips the scales too much, however. Focusing, and even a moderate breeze, will vibrate the tube badly. Before you dismiss the idea of either of these telescopes, however, let's look at some economics. The complete C11-SGT GoTo scope with XLT coatings retails at $2,200. Odd, but the tube assembly alone may actually cost the same. In effect, you are getting a free mounting for a smaller telescope, which can always be sold on the used market for several hundred dollars if you have no use for it. Just a thought.

Finally, the CGE German equatorial mounts are well matched with the 8-, 9.25-, and 11-inch tube assemblies, as well as Celestron's flagship, the Celestron 14. Each comes on the sturdy, U.S.-made CGE German equatorial mounting. If this is to be a portable instrument that is carried into the field regularly, please keep in mind that although sturdily made, the mount and tripod together weight 100 pounds, with the heaviest single component weighing 30 pounds.

Each CGE telescope and mount slide together using a dovetail bar that is similar to that on the Advanced Series models. Setup takes between ten and fifteen minutes for the 8-inch, but upward to twenty minutes or more for the larger, heavier instruments. Those models take an effort to put together in the dark, especially the 14-inch, which is probably best kept in a permanent observatory. Hoisting that 45-pound tube assembly into place is not for the faint of heart!

Overall, owners are very pleased with their instruments. The CGE mount, which is based on Celestron's older, non-GoTo CI700 mount, is substantial enough to hold each telescope steadily. Each CGE scope is outfitted with the NexStar computer control system, which like the Advanced Series GT mounts, require rough polar alignment to work correctly. GPS sync is also available optionally by plugging in the CN 16 accessory kit.

As far as optical performance goes, Celestron Schmidt-Cassegrains perform quite well for the design. Because of the comparatively large central obstruction from their secondary mirrors, Schmidt-Cassegrains as a breed generally lack the high degree of image contrast seen in refractors and many Newtonians. It is also important to note up front that both Celestron and Meade Schmidt-Cassegrain telescopes suffer from something called *mirror shift*. To focus the image, both manufacturers have chosen to move the primary mirror back and forth rather than the eyepiece, which is more common with other types of telescopes. Unfortunately, as the mirror slides in its track, it tends to shift, causing images to jump. The current telescopes produced by both companies have much less mirror shift than earlier models, but it is still evident to some degree.

Taking all that into account, images through Celestron Schmidt-Cassegrains are usually very sharp and clear. Assuming that their optics are properly collimated and acclimated to the cool night air, each aperture delivers pleasing views of the Moon, planets, and deep-sky objects alike. And without a doubt, the best of the group is the 9.25-inch tube assembly. What sets the 9.25 apart lies in the optical design of the instrument. Although other SCTs use primary mirrors with focal ratios around f/2, the 9.25 uses an f/2.5 primary, which seems to make the difference. The design also benefits from a flatter focal plane with less coma and off-axis astigmatism.

Meade Instruments. Meade opened its doors in 1972 as a mail-order supplier of small, imported refractors. Meade's first homegrown instruments were 6- and 8-inch Newtonian reflectors, but the company was to make its mark with the introduction of the Model 2080 8-inch Schmidt-Cassegrain telescope in 1980. Since then, Meade has grown to become the world's largest manufacturer of telescopes for the serious amateur.

Meade's mainstay product has always been its line of 8-inch Schmidt-Cassegrain telescopes. Today it offers two variations: the SC-8 and the LX90GPS, in order of increasing price and sophistication. Each includes the same f/10 optical tube assemblies and ultra-high transmission coatings (UHTC), which increase light throughput by about 20% according to the manufacturer. As mentioned when discussing Celestron's optional XLT coatings, that amounts to an increase of about 0.25 magnitude over standard coatings.

Meade's light transmission and image contrast are comparable to Celestron's, with most owners rating the overall optical quality of their instruments as good to excellent. Some of the newest Meade SCTs that I have looked through have really impressed me as being quite sharp (as have several Celestrons). Star tests reveal good consistency from one to the next, with only slight deviations in images from one side of focus to the other. In-focus images, which are really what count, are clear and surprisingly contrasty, given the central obstruction. No, they still don't rival apochromatic refractors or even optimized Newtonians on the planets, but overall owners are quite satisfied. Meade's optical prowess has come a long way since its earliest 2080s.

Least expensive of the current instruments is the LXD75 SC-8, one of several telescopes coupled to Meade's LXD75 German equatorial mounts. For thoughts on the mount itself, review the comments found on page 86 of this chapter. The SC-8 is adequate for simple photography using a lightweight webcam or Meade's own Lunar Planetary Imager (LPI) or Deep-Sky Imager (DSI), but it is primarily designed as a visual instrument for those just getting into backyard astronomy. Although Meade does not offer a GPS add-on upgrade for the LXD75 like Celestron's Advanced Series CG-5 mount, they are otherwise pretty much comparable. (A well-designed aftermarket GPS interface unit for the LXD75 mount as well as several other Meade GoTo instruments called the StarGPS is available from PixSoft Inc.) Owners comment that pointing accuracy of the built-in Autostar GoTo system is very good, provided the mount's right ascension axis is aimed toward the general region of the celestial pole, and as long as the batteries are fully charged.

The 8-, 10-, and 12-inch fork-mounted LX90GPS scopes have several nice features, including a global-positioning receiver to determine date, time, and location automatically, a heavy-duty field tripod, an 8×50 finderscope, a 1.25-inch star diagonal, and a 26-mm Super Plössl eyepiece. The mounting's metal arms are cloaked with plastic covers that conceal Meade's Autostar computerized GoTo system. Although eight C-size batteries power the LX90GPS's

Autostar, purchasing an adapter to run off either a rechargeable battery or a standard 115-volt AC home outlet is strongly recommended. The LX90GPS's object database includes all objects from the Messier, NGC, and IC lists; all major members of our solar system; many asteroids, comets, Earth-orbiting satellites; as well as 200 blanks for users to add other objects that are not listed. The Autostar software is also upgradeable directly from the Internet. The LX90's GoTo motors can slew the instrument at up to 6.5° per second.

In actual use, the LX90's aiming ability proves to be very accurate, although some new owners complain when initializing the two-star alignment. After powering the telescope and inputting site information, the telescope automatically levels itself and finds true north. Then, it's on to the first alignment star, which is typically out of the finder's field of view—sometimes way out. The user must slew the telescope to it using the hand controller's directional buttons. After the user presses the enter button on the key pad, the telescope moves to the second alignment star. Again, it will probably not be in the field of view, but it should be a little closer. After centering it and pressing enter a second time, the telescope's GoTo should be accurate enough to put selected targets near the center of the 26-mm eyepiece's field.

Although this is an electronic telescope, it can still be used manually after the batteries drain. There are manual locks on each axis so the telescope can be held in place once aimed. Neither axis has a mechanical slow-motion control, however, so tracking the sky by hand can be a jerky experience. The LX90 is supplied with both a red-dot finder and an 8 × 50 finderscope, so aiming by eye can be done easily.

Although the 8-inch is quite sturdy, the 10-inch is at the top end of the mount's capacity. That leaves the 12-inch LX90, which may be too much for the fork and, especially, the drive base. Some LX90 owners also express minor complaints about the focusing being too stiff and a little uneven but overall seem very happy with their purchase. Keep in mind that since this is actually an alt-azimuth telescope, long-exposure astrophotography is only possible with an equatorial wedge, an add-on accessory available from Meade.

Shortly before this book went to press, Meade announced that it was canceling its line of LX200 Schmidt-Cassegrain telescopes in favor of the new line of LX200R Ritchey-Chrétien hybrid instruments. You will find them discussed later in this chapter.

Shifting optical designs for the moment, Meade is the only major manufacturer today to offer Schmidt-Newtonian telescopes. This continues a tradition that began in the 1980s with its Modular Telescope System (MTS) fork-mounted instruments. Today's Schmidt-Newtonian lineup, seen in Figure 5.9, includes the 6-inch f/5 SN-6, the 8-inch f/4 SN-8, and the 10-inch f/4 SN-10. All are carryovers from its former LXD55 model series, which premiered in 2001. Many owners complimented Meade on the optical quality of those telescopes, but complained about the poor mechanical quality.

Meade listened to those complaints and took steps to fix most of the problems but not all. Each of the Schmidt-Newtonians now comes with Meade's improved LXD75 mounting and Autostar computerized GoTo system. Although outwardly it appears the same as the old LXD55, internal ball bearings, tubular

Figure 5.9 *Meade's SN6 and SN8 LXD-75 telescopes are the only Schmidt-Newtonian telescopes sold on the market today.*

steel tripod legs, and other upgrades improve the LXD75's stability. The 10-inch strains the mount beyond its limit, however, and proves to be very shaky when focusing or used in even a light wind. The 6- and 8-inch models are much more stable. The LXD75 mount is powered by 8 D-cell batteries held in a separate power pack, by an external 12-volt battery (an auto cigarette lighter plug is available), or by plugging it into a 115-volt AC outlet (again, a cord is required).

Optical quality remains the strong suit of Meade's Schmidt-Newtonians. Once the optics are properly collimated and acclimated to the ambient outdoor temperature (both critical to the design), stars really sparkle, with reasonably sharp images nearly to the edge of the field. But as good as the optics may be, many owners are still disappointed with these telescopes. Most focus in on the all-aluminum 2-inch focuser, which is, as one owner described it, "practically unusable." Fine focus is critical in fast telescopes like these three Schmidt-Newtonians, and the supplied unit just doesn't cut it.

Orion Telescopes and Binoculars. Orion threw its hat into the Schmidt-Cassegrain arena in 2005 when the company introduced several New Orion Exclusive models ranging in size from 8 to 11 inches in aperture. In reality, these are not new but rather established optical instruments manufactured by Celestron that are being coupled to various mountings in Orion's collection from Synta in China. Orion has maintained close business ties with Synta for more than a decade, while Synta purchased Celestron in 2005. So, bringing all parties together in this manner was really just a matter of time.

Orion's Schmidt-Cassegrain telescopes include 8-inch f/10, 9.25-inch f/10, and 11-inch f/10 models, all with Celestron's XLT enhanced coatings. Head back to page 139 to read reviews of the Celestron tube assembly's optical and mechanical qualities, but to reiterate, expect the 9.25-inch to be the best of the bunch optically for reasons previously cited.

Let's concentrate here on the mountings that Orion has chosen for each instrument, although they have also been discussed previously. Orion has paired Celestron's 8- and 9.25-inch Schmidt-Cassegrain tube assemblies to create the SkyView Pro 8 XLT and SkyView Pro 9.25 XLT telescopes. Although the castings vary somewhat, this mount is comparable to Celestron's Advanced Series CG-5, making these two telescopes equivalent to the Celestron C8-S and C9.25-S, respectively. Both mounts are claimed to have a carrying capacity of 20 pounds, more than either tube assembly, although I find this rating to be a little optimistic. In practice, the 8-inch is fine, but the 9.25, like the C9.25-S, is pushing its luck.

Each model is also available on Orion's heavier duty Sirius EQ-G equatorial mount, an adaptation of Synta's HEQ-5 mount. This is a much better platform for each of these fine instruments, especially the 9.25. The only downside points back to its 1.75-inch-diameter tripod legs, which are the same as those on the SkyView Pro mount. They are going to wobble under the weight of either scope, especially the 9.25, if extended. Fully collapsed, they are acceptable, although then the observer will have to be seated to view comfortably. The Sirius mount's features, including GoTo computer control, an illuminated polar scope, periodic error correction, and autoguider capability, make it an excellent choice for amateurs interested in dabbling in advanced astrophotography.

Orion also sells two versions of the Celestron 11 on its Atlas German equatorial mount. Again, refer to the discussion of the C-11 under the Celestron banner earlier in this section. The Atlas mount is far sturdier than the CG-5 that Celestron uses with its C11-S, so in a battle between the two, Orion's version is the clear winner. Of course, the Orion Atlas 11's significantly higher price probably gave that away. Overall, the Atlas is fine for supporting the 28-pound, 11-inch telescope for visual observations as well as some simple astrophotography as well. But as mentioned earlier in this chapter, the standard Atlas mount does not have computer control, digital setting circles, periodic error correction, or the ability to be controlled by an autoguider.

Today's discriminating astrophotographer wants all of these conveniences, but to find them, he or she will need to spend several hundred dollars more on the Atlas 11 EQ-G. Although the most expensive of Orion's SCTs, it is also the most interesting from a technological perspective. The Atlas EQ-G mount, a slightly modified version of Synta's EQ6-G, features full GoTo control, periodic error correction, and autoguider capability. The Atlas is the sturdiest mount ever made by Synta; however, it is missing a few conveniences that others mounts have. For instance, while Meade's LX200 and Celestron's CPC periodic error correction systems are permanent, the Atlas-G's is not. Instead, the PEC circuitry must be trained each time it is used, which is not as convenient. The Atlas mount, although solid, is also cumbersome. When you consider having to make several trips to carry out the 28-pound tube assembly, the 54-pound mount and tripod, the three 11-pound counterweights, as well as a rechargeable 12-volt battery and other miscellaneous items, the Atlas 11 EQ-G (and non-GoTo Atlas 11, which weighs the same) might lose a little of its luster. By comparison, the tube and mount of the Celestron CPC-11 weigh 46 pounds and the tripod, another 19 pounds. The fully assembled Meade 10-inch LX90-LNT's tube, mount, and tripod weigh 53 pounds. Both would require an equatorial wedge for guided photography, but that would probably add less than 15 pounds and could be carried with the tripod.

For those not bothered by some of those issues, the Atlas 11 or Atlas 11 EQ-G could well be an excellent choice. The Celestron 11 is a well-proven design, so it would boil down to more of a question between this mount and the CPC fork. If you choose the Atlas, you should also purchase Orion's Atlas Mount Extension, which raises the telescope another 8.4 inches. The extension will bring the eyepiece to a more convenient height without having to extend the tripod legs.

Maksutov Telescopes

Celestron International. Celestron's NexStar 4 Special Edition teams a 4-inch f/13 Maksutov-Cassegrain and a one-armed alt-azimuth computerized mount. The mount is more than capable of supporting the lightweight optical tube assembly, but many owners comment on the so-so optics. Image quality is soft, especially at magnifications exceeding 150×. Even though the Meade ETX105 is a little more expensive than the N4SE, its optics produce sharper results. The N4SE's ergonomics, however, are superior to the ETX. The Celestron C130-M teams a 5-inch f/15 Maksutov-Cassegrain tube assembly with the Advanced Series CG-5 German equatorial mount. The same scope, sold as the C130-MGT, adds the NexStar computerized system for those interested in GoTo control.

Optically and mechanically, the C130-M is lukewarm at best. Most units seem to produce reasonably sharp images, although they are not as crisp as those through the Meade ETX125 or the Orion StarMax 127 Maksutovs. Although Jupiter shows two or more belts and Saturn's rings easily display Cassini's Division, neither is as clear as through, say, a 4-inch achromatic refractor. Internal baffling, so important to a Cassegrain-style instrument like the C130-M, appears to be adequate, as there is no evidence of flaring or image washout.

Both C130-Ms come with 32-mm Plössl eyepieces and 10 × 50 finderscopes, which are far more useful than the tiny 8× finderscope that comes with Meade's ETX, described later. Focusing, accomplished by moving the primary back and forth inside the tube, is smooth and surprisingly free of image shift. Like the ETX, a flip mirror is built into the telescope directly below the permanent 90° star diagonal. The mirror serves to steer light either toward the diagonal or straight out the back of the instrument, where a camera body can be attached for photography.

Intes Micro. Based in Moscow, Intes Micro makes well-regarded Maksutov telescopes. Its Maksutov-Newtonian lineup includes several models, including the 5-inch f/6 MN56, 5.5-inch f/5.5 MN55, 6-inch f/5 MN65, 6-inch f/6 MN66, 7-inch f/6 MN76, and 8-inch f/6 MN86. All are world-class performers. Small central obstructions and no distracting diffraction spikes create views that are comparable in nearly every way to similarly sized apochromatic refractors. Star clusters sparkle, nebulae glow softly, and planets show some amazing detail through these instruments. Given good seeing conditions, optical quality is sharp enough to let you go crazy, upping the magnification limit to 80× or more per inch of aperture. But what won't look the same as an apochromat is your credit card bill. The MN56, for instance, retails below $1,000, less than half of some 4-inch apochromats. You can buy quite a few eyepieces and other accessories with that savings!

Owners point out a few quirks, however. For one, the finderscopes are quite poor, with terrible eye relief and ill focusing. Get a new one from the list in chapter 7. The Crayford focusers also lack adequate travel to bring some eyepieces into focus. An extension tube will quickly solve the problem, however. Finally, cooldown time is slow.

The MN55 and the MN65 are billed as photo-visual instruments, with faster f-ratios for shorter exposures and wider fields. Their focusers offer extra travel for prime-focus imaging, although I wish those focusers were dual-speed units, which would allow for easier critical adjustment. But that aside, through-the-telescope photographic results are impressive provided the instruments are placed on suitable mountings, such as the Vixen Great Polaris-D2 or the Losmandy G-11.

On the Maksutov-Cassegrain front, Intes Micro lists numerous Alter models that range in size from the compact 5-inch Alter M503 to the observatory-class 16-inch Alter 1608. Once again, owners sing their praises, especially of image sharpness that, as one person put it, "whooped" his Schmidt-Cassegrain. Like an SCT, however, do not expect high levels of image contrast, since central obstruction from the secondary mirror is in the realm of 30% (measured by diameter). That is bound to soften things up a bit. Another similarity to Schmidt-Cassegrains is the method of focusing, which moves the primary mirror back and forth along a track inside the tube. Although image shift is negligible through Intes Micro scopes, owners comment on play (or *lag*) when reversing direction. Again, comments about excellent images after the telescope has stabilized are common.

Pricing may be a stumbling block to these instruments' popularity. Eight-inch SCTs from either Meade or Celestron cost less than the 6-inch Alter 603, and while the Alter tends to produce sharper images, the 8-inch SCT will show more in terms of magnitude penetration and resolution, assuming competent optics on both sides, of course.

Meade Instruments. The little Meade ETX-90, one of the most popular telescopes to be introduced in the 1990s, is a 3.5-inch f/13.8 Maksutov-Cassegrain designed for maximum portability while also delivering outstanding images. It certainly succeeds on both counts, and does so at a terrific price. Images are absolutely textbook, with good contrast and clear diffraction rings. Focusing is precise with no mirror shift detected, giving some wonderful views of brighter sky objects, such as the Moon and the planets. To quote one owner, "The Meade ETX delivers optical performance well beyond its price class." All the models I have seen also exemplify fine optics, a great triumph in low-cost, mass-production optical fabrication techniques. The optics are housed in a deep-purple metal tube that is smooth and nicely finished.

The ETX-90 is currently available in two versions: the computer-controlled AT Premier Edition and as an unmounted terrestrial spotting scope. Both feature the same optical tube assembly, complete with a built-in 90° star diagonal and flip mirror that swings out of the way when the ETX is coupled to a camera, a 26-mm Meade Super Plössl eyepiece, a nice screw-on dust cap, as well as a 45° erecting prism diagonal for upright terrestrial views. Unfortunately, the tiny 8 × 21 finderscope that comes with the ETX spotting scope will likely prove unusable for most, simply because it is mounted so close to the tube. I find it difficult, if not impossible, to look through as the telescope raises in altitude, causing my nose to scrunch up against the eyepiece. It might be best to replace the finder with one of the smaller one-power aiming devices described in chapter 7.

The ETX Premium Edition (often abbreviated ETX-PE) comes mounted on a miniaturized, computer-driven fork equatorial mount made mostly from molded plastic. The DC-powered Autostar computer runs on eight common AA-size batteries. Some respondents commented on the amount of backlash in the drive gears, but once it works itself out, the drive tracks well, keeping objects in view for half an hour or more. Once the user inputs his or her location (which could even be just a zip code), the telescope levels itself to magnetic north using its internal compass, calculates where true north is, then slews to the first of two alignment stars, like other GoTo scopes. Rather than a small, frustrating finderscope, the Premium Edition includes a red-dot Smart Finder designed to make it easier to align the telescope with the stars targeted during this process. It's still tough to get your head in the right spot because of the telescope's small size, but this method is certainly better than the 8 × 21 finder.

Despite my enthusiasm for the ETX, there are some drawbacks. Optically, it's great, but mechanically, it leaves something to be desired. I already mentioned the poor finderscope on the Spotting Scope and AT versions, but another difficulty shared by all three is focusing. Focusing is smooth, but the small, aluminum knob is difficult to grasp when looking through the eyepiece. Things are just too close together. The plastic fork mounting is also poorly engineered in my opinion, even though the Autostar works fine.

So, here we have it: the ETX-90 is a very good telescope, containing outstanding optics, but in a so-so package. Because of its small size and, especially, its mediocre finderscope, I would not recommend it as a first telescope for a beginner. But for someone who may already own a 6- or 8-inch Dobsonian and wants a nice planetary grab-and-go instrument, the ETX is perfect.

Given the high praise from all corners that the ETX-90 has garnered, Meade introduced the 4.1-inch f/14 ETX-105PE and the 5-inch f/15 ETX-125PE. Like the 90, the larger ETXs earn high scores for their consistently fine optics. One owner explained that his ETX-125 continuously resolves tightly spaced double stars with ease, stars that his larger Schmidt-Cassegrain didn't even show as elongated. Saturn's Cassini Division is a common sight through all. Star testing demonstrates that, like the ETX-90, both the 105 and the 125 come with optics that are very close to ideal. Internal baffling effectively blocks any flaring when viewing bright objects. Only some small amounts of coma can be detected with some long focal-length eyepieces. None of the ETXs includes a tripod, but Meade recommends its 884 Deluxe Field Tripod, which proves a sturdy support for any of the models.

Orion Telescopes and Binoculars. Orion sells several Maksutov-Cassegrain instruments that are proving to be very popular among amateurs. In the summer of 2001, Orion introduced the 3.5-inch f/13.9 StarMax 90, the 4-inch f/12.7 StarMax 102, and the 5-inch f/12.1 StarMax 127 Maksutov-Cassegrain instruments, all imported from China. More recently, Orion added the 5.9-inch f/12 SkyView Pro 150 Maksutov. Their attractive appearance coupled with remarkably low prices continues to attract wide attention. The three smaller apertures are also sold without mounts under the product line name Apex.

Like many imported refractors, each Orion Maksutov's focuser is threaded to accept a standard camera T-mount for astrophotography. Unlike the ETX, the star diagonal is not built in, so users need to remove it and attach their camera before attempting photography. Although not quite as convenient, this system is certainly serviceable.

The StarMax optics are very good in terms of overall quality, although not quite as sharp as those of the ETX. A minimal amount of spherical aberration is evident at high power and during star testing, but this does little to detract from image quality. Contrast is quite good, especially considering the 30% central obstruction created by the secondary mirror.

The StarMax 90 comes with a small, low-set 6 × 20 finderscope, which may prove difficult to look through because of how close it is located to the telescope tube. The 102 and 127 models feature 6 × 26 finders on taller mounts that make them easier to use, while the 150 is sold finderless. Small mounting blocks on the bottoms of the StarMax 90 and 102 can be used to attach each to a standard 0.25-20 tripod head, while the 127 and the 150 come with tube-length dovetail rails to attach the telescopes to Orion's AstroView and SkyView Pro mounts, as well as others like the Celestron CG-5.

Speaking of mounts, the StarMax EQs are paired with four different German equatorial mounts. The 90 rides atop Orion's lightest mount, the EQ-1, while the 102 is paired with the EQ-2. The 127 is matched to either Orion's AstroView mount or its heavier duty SkyView Pro. This latter combination is better at soothing the jitters for high-power viewing and through-the-telescope short exposures of the Moon and planets, and it also offers the option for PushTo aiming. Overall, the StarMax Maksutovs are superior to Meade's ETX series mechanically but not quite up to its level optically.

Exotic Catadioptrics

Meade Instruments. When Meade introduced the RCX400 series of telescopes in 2005, the company once again created quite a stir in the world of amateur astronomy. Internet discussion groups were abuzz with the prospect of a Ritchey-Chrétien Cassegrain telescope selling at prices that were well below those of premium R-C systems. Could Meade do what the others do for substantially less money?

Not really. As mentioned in chapter 3, Meade's RCX telescopes are not true Ritchey-Chrétien reflectors. Instead of using hyperbolic mirrors, the RCX telescopes use a spherical primary mirror (like a Schmidt-Cassegrain) and a specially designed corrector plate that collectively attenuate the light in such a way that, by the time it reaches the hyperbolic secondary mirror, it is effectively attenuated as it would be in a pure Ritchey-Chrétien (which does not require a corrector plate). The RCX design is actually very clever, but it is not a Ritchey-Chrétien.

Meade's RCX400 line includes four apertures: a 10-inch, 12-inch, 14-inch (see Figure 5.10), and 16-inch, all of which run at f/8. The shorter focal lengths (as compared to SCTs) mean shorter exposures for astrophotographers, the primary audience for this sort of telescope. Knowing that, Meade includes several innovative design features that astrophotographers will appreciate. Each telescope, for instance, features a carbon-fiber tube, famous for its strength, light weight, and, especially, its thermal stability. That means despite dropping temperatures as a night wears on, the telescope's focus will not be as adversely affected as in a metal-tubed instrument.

Focusing is also a joy through each RCX400. Rather than move the primary mirror back and forth to achieve focus, Meade decided to hold it in place and slide the corrector plate/secondary mirror assembly in and out instead.

Figure 5.10 *Meade's RCX400 is a state-of-the-art virtual Ritchey-Chrétien telescope thanks to the front corrector plate's influence on entering starlight before it strikes the spherical primary mirror and hyperbolic secondary mirror.*

Image shift is greatly minimized as a result. Focusing is by electronic means only, which makes a fresh set of eight C-size batteries a necessity; better yet, get the optional power cord and use a rechargeable 12-volt battery instead. Focusing can be done at four different speeds and can be set with nine user-defined focuser positions that make it easy to return to the precise focus point for several different eyepieces night after night. The same mechanism used to move the corrector assembly during focus is also used to adjust instrument collimation. Some early owners complained of trouble getting both to work correctly, but more recent reports cite few difficulties. Unfortunately, neither focusing nor collimation can be done without power, making users dependent on their batteries.

Astrophotographers, who require absolutely rock-steady support, will appreciate the RCX fork mounts. The RCX mounts come equipped with the Autostar II GoTo system, which works well once it has been initialized and set. Owners also sing the praises of the new RCX400 tripod, which is both beefier and easier to transport than the giant field tripod often used to support Meade's largest Schmidt-Cassegrains. One improvement is that the three legs can be removed for transport and storage but easily locked back into the tripod's base using quick-release levers. Of course, the legs are loose when carried, though wrapping them with a bungee cord should keep them from flopping around.

That brings up the topic of transporting the RCX400. Unless you are prepared to be accompanied by a helper always, RCX400 telescopes are best housed in an observatory, with the possible exception of the 10-inch. All four RCX400 models include pairs of handles on their fork arms, but their poor placement offers little help when horsing the telescopes onto their tripods.

Optically, Meade has done what its advertising promises. It has created a telescope optical design that will produce sharper images than Schmidt-Cassegrain instruments. Schmidt-Cassegrains are often plagued with spherical aberration, blurring the views of fine planetary details and tight double stars. This effect is certainly reduced in the RCX instruments. Image contrast is on a par with SCTs despite the slightly larger central obstructions in the RCX foursome. Part of this is likely thanks to the ultra-high transmission coatings that come standard with each. In side-by-side comparisons, the sharper images are apparent, but are not so obvious that an SCT owner is going to put his or her telescope up for sale the next morning. The improvement, visually anyway, is subtle. Rabid astrophotographers are the ones who will benefit most from the RCX instruments, and they are the most likely to invest the extra money to get the latest, most advanced compound telescope on the market.

Almost immediately after the bell tolled the start of 2006, Meade rolled out five LX200R faux Ritchey-Chrétien telescopes from 8 to 16 inches in aperture. These telescopes combine the philosophy of the RCX400 line with the proven mechanics of Meade's well-established LX200 Schmidt-Cassegrain telescopes, which these replace. All LX200R telescopes also rely on a corrector plate to tweak the light striking the primary in a way that it will ultimately mimic a true Ritchey-Chrétien, just like the RCX400s, by the time it reaches the secondary mirror. Unlike the f/8 RCXs, however, each LX200R is an f/10 instrument.

Each LX200R telescope takes the company's Smart Mount Technology to the next level. Ranging in size from a humble 8-inch to a monstrous 16-inch, all LX200Rs are mounted on computer-controlled fork mounts that automatically align themselves using data from global positioning satellites. After the telescope is set up and initialized, just select a target from its built-in listing of more than 147,000 objects (a big number, although that includes more than 1,800 individual lunar features and 98,000 stars), press the Go To button, and the LX200R will automatically slew to it, albeit noisily, at rates up to 8° per second. The LX200R requires a 12-volt DC power source, either from eight internally held C-size batteries, a car battery (cables are sold separately), or from a wall outlet with an optional 115-volt AC adapter. As with other alt-az GoTo fork mounts, the LX200R does not permit long-exposure astrophotography because of field rotation as the telescope tracks the sky. To compensate for that, you will need either an equatorial wedge or Meade's #1220 field derotator. The latter compensates for field rotation automatically by turning the camera in time with the field of view.

The LX200R's improved Autostar II control and Smart Drive make initialization quite simple. Most users will probably use the telescope's automatic alignment method. Lock both axes, turn on the power, and wait until the hand controller reads "automatic alignment." Press Enter and the telescope goes through the initial process of finding magnetic north using its built-in magnetic compass, while also acquiring time, date, and location data from orbiting GPS satellites. This may take several minutes depending on the availability of the GPS signals, but once set, the telescope will move to the first automatically selected alignment star. Center the star in the eyepiece (it might be far, so use the finderscope as needed) using the controller's arrows and press Enter. Repeat the procedure for a second alignment star and you're done. LX200Rs also offer other, more traditional ways of initializing the GoTo computer, if desired. All produce amazingly accurate results, consistently placing targets in the field of a low-power eyepiece time and time again. Not only is the Autostar II more accurate than earlier versions, its menus are also much more convenient to use—no more scrolling through different screens to find important system features.

The Smart Drive's periodic error correction (PEC) circuit lets the user compensate for minor periodic worm gear inaccuracies. The Smart Drive's PEC remembers the steps needed to compensate for the inaccuracies forever once they are input manually by the user. The Celestron CPC SCTs also have so-called *permanent* PEC.

Other nice touches include the LX200R's focus lock and electronic microfocuser, both of which prove extremely useful especially for photography. While coarse focusing is done by turning a knob on the back of the telescope, the LX200R also includes a motorized tube that moves the eyepiece in and out behind the telescope. The electronic microfocuser can be set at four different speeds but is meant only to augment the manual focuser. On its own, the microfocuser only has about 0.5 inch of travel, which is not enough to act as the only focuser but is more than enough to tweak the focus precisely without causing the image to shift.

Although the introduction of the LX200R occurred much too close to this book's publication date to include a detailed discussion here, I suspect that Meade may have a real winning combination here if the optics prove as good as those in the RCX line. Whither the Schmidt-Cassegrain? No, not necessarily. But among advanced astrophotographers seeking to rid themselves of SCT bloat, these instruments will undoubtedly prove popular.

Then, there are the 16-inch and 20-inch RCX Max scopes, which combine the virtual Ritchey-Chrétien optical tube assemblies with Meade's monster Max mount. These are too new to have been evaluated here, but you'll find more information in the *Star Ware* section of the Web site www.philharrington.net.

Vixen Company, Ltd. Vixen calls its 8-inch f/9 VC200L optical system VISAC, which stands for "Vixen Sixth-Order Aspherical Catadioptric." Unlike conventional Cassegrains, which only use primary and secondary mirrors to focus light into their eyepieces, the VC200L relies on a compound-curve, parabolic primary mirror; a convex secondary mirror; as well as a three-element corrector/flattener lens built into the base of the focuser. The secondary is mounted in the front of the open tube in an adjustable, four-vane spider mount, similar to those used in traditional Cassegrain reflectors.

Owners comment that the VC200L yields flat, distortion-free images from edge to edge thanks to the corrector lens, while spherical aberration is minimized because of the aspheric primary mirror. The result is pinpoint stars from edge to edge, an important consideration for through-the-telescope photography. The only common problem that owners have noted is that, visually, the VC200L suffers from low image contrast due to its large central obstruction—39% by diameter. As a result, this is not the best telescope for planetary observations.

Vixen also offers a variant on its VISAC design called the VMC200L. Here, the separate set of corrector lenses is redesigned and relocated directly in front of the secondary mirror, which lets Vixen use a spherical primary that is simpler, and therefore less expensive, to make. The difference puts about $400 back into the consumer's pocket. From a visual point of view, it's unlikely that anyone will notice much difference in images through the VC200L and the VMC200L when operating at the same power. For about the same price, however, you can get the Celestron 9.25, which will enjoy both brighter images as well as more contrasty views than either of the Vixen catadioptrics.

The VMC line also includes two larger models, the 10.2-inch f/11.6 VMC260L and the 13-inch f/13.1 VMC330L. Both are sold as optical tube assemblies only. Due to their weight, both will require robust mounts to support them steadily, which raises an already high price tag even more. Once again, larger conventional Schmidt-Cassegrain telescopes are available for less money. Although images will be brighter, SCTs will not correct for spherical aberration as well as Vixen's VISAC.

Mounting Concern

When the Beach Boys sang "Good Vibrations," they certainly were not singing about telescope mounts! One of the most common complaints of telescope owners remains dissatisfaction with their mountings. In an effort to remedy this situation, many retrofit their instruments with substantially larger, sturdier support systems. Table 5.2 lists most of the higher quality mounts that are available today. Some are suitable for small, portable telescopes, while others are designed to heft the largest instruments described in this chapter.

Some further explanation of the table is needed. First, the mounts are listed in alphabetical order, then according to how much telescope they can carry steadily. Note that the carrying capacity figures are based on the manufacturers' estimates, not actual trials. For the smaller mounts, these are usually good indications of a mount's usefulness for visual observations. If you are interested in doing astrophotography, I recommend that you reduce these figures from 25% to 30%. The column heading "Type" should be self-explanatory. Here, "GEM" indicates a German equatorial mount. Be sure to also pay close attention to the type of support. Tripods and piers are, in general, indicative of portability, while permanent piers are for observatory-bound instruments. In many cases, the supports are not included in the base price of the mount. In that case, they are listed as "sold separately," for optional. Other accessories, such as polar-alignment scopes, digital setting circles (DSC), GoTo computer control, and clock drives are also noted as "opt" if they are available optionally at extra cost. Finally, the mounts have been queued within several price ranges: A: $500 and less; B: $501 to $1,000; C: $1,001 to $2,000; D: $2,001 to $5,000; and E: $5,001 and up.

From this table, it's clear that there are a lot of telescope mounts from which to choose. Before making a choice, consider this advice. First—and you should be well aware of this by now from reading the earlier telescope reviews—stamped aluminum tripod legs are weak. They tend to twist and vibrate more than other designs, transmitting those motions to the mounting and, in turn, the telescope. Tubular steel legs, such as those supplied on Celestron's improved Advanced Series CG-5 and Orion's SkyView Pro, Sirius, and Atlas mounts, are far more stable.

If your mount has a stamped aluminum tripod that gives you the shakes, you can modify it to help improve stability. One simple fix seems obvious, but it is often ignored: make sure everything is tight. Often, as telescopes are transported and swung to and fro, screws can loosen. Connections that are even slightly loose can cause problems. Places to watch: the altitude adjustment and the mount-leg connections. Another trick is to fill the tripod legs with an inert material. Some have suggested sand, but foam insulation from an aerosol can is a better choice. The foam sprays out of the can as a liquid, but quickly expands and solidifies, filling the voids in the tripod legs. Best of all, weight gain is negligible. Placing a set of the vibration pads sold by Celestron, Orion, or Meade (see chapter 7, or make your own in chapter 8) under each leg also helps, as does suspending a weight, such as a plastic gallon container of water, from directly under the mount.

Table 5.2. *Telescope Mounts*

Manufacturer/Model	Type	Carrying Capacity	Type of Support	Slow-motion Control	Clock Drive	Setting Circles	Polar Scope	DSC or GoTo?	Price Range
Astro-Physics Mach1GTO	GEM	45	Sold separately	Yes	Yes	Yes	Opt	GoTo	E
Astro-Physics 900GTO	GEM	70	Sold separately	Yes	Yes	Yes	Opt	GoTo	E
Astro-Physics 1200GTO	GEM	140	Sold separately	Yes	Yes	Yes	Opt	GoTo	E
Celestron Advanced Series CG-5	GEM	25	Tubular steel tripod	Yes	Opt	Yes	Opt	GoTo opt	A
Celestron CGE	GEM	65	Tubular alum tripod	Yes	Yes	Yes	Yes	GoTo	D
DiscMounts DM-4	Altaz	15	Sold separately	No	No	No	No	DSC opt	B
DiscMounts DM-6	Altaz	40	Sold separately	No	No	No	No	DSC opt	C
FAR Labs Dyna Cradle	Altaz	11	Sold separately	No	No	No	No	No	A
Helix Hercules Single Arm	Altaz	13	Sold separately	No	No	No	No	DSC opt	A

Model	Mount	Capacity	Tripod					DSC	
Helix Hercules Fork (3 models)	Alt-az	8" = ~20 10" = ~30 12" = ~35	Sold separately	No	No	No	No	opt	A
Losmandy GM-8	GEM	30	Square alum tripod	Yes	Yes	Yes	Opt	Opt	C
Losmandy G-11	GEM	60	Tubular alum tripod	Yes	Yes	Yes	Opt	Opt	D
Losmandy HGM Titan	GEM	100	Tubular alum tripod	Yes	Yes	Yes	Opt	Opt	E
Meade LXD75	GEM	25	Tubular steel tripod	Yes	Yes	Yes	Yes	Opt	B
Meade Max	GEM	250	Structural beam tripod or pier	Yes	Yes	Yes	Yes	GoTo	E
Mountain Instruments MI-250 GoTo	GEM	65	Sold separately	Yes	Yes	No	Opt	GoTo	E
Mountain Instruments MI-500	GEM or Fork Eq	120 (fork) 180 (GEM)	Sold separately	Yes	Yes	No	Opt	GoTo	E
Mountain Instruments MI-750	GEM or Fork Eq	200 (fork) 280 (GEM)	Sold separately	Yes	Yes	No	Opt	GoTo	E

(continued)

Table 5.2. (continued)

Manufacturer/ Model	Type	Carrying Capacity	Type of Support	Slow-motion Control	Clock Drive	Setting Circles	Polar Scope	DSC or GoTo?	Price Range
Mountain Instruments MI-1000	GEM or Fork Eq	300 (fork) 420 (GEM)	Sold separately	Yes	Yes	No	Opt	GoTo	E
Optical Guidance Systems HP-75	GEM	130	Sold separately	Yes	Yes	Yes	No	GoTo	E
Optical Guidance Systems OGS-75	GEM or Fork Eq	130	Sold separately	Yes	Yes	Yes	No	GoTo	E
Optical Guidance Systems OGS-100	GEM or Fork Eq	200	Sold separately	Yes	Yes	Yes	No	GoTo	E
Optical Guidance Systems OGS-140	GEM or Fork Eq	300	Permanent pier	Yes	Yes	Yes	No	GoTo	E
Optical Guidance Systems OGS-190	GEM or Fork Eq	450	Permanent pier	Yes	Yes	Yes	No	GoTo	E

Orion Telescopes AZ-3	Altaz	6	Stamped alum tripod	Yes	No	No	No	No	A
Orion Telescopes EQ-1	GEM	6	Stamped alum tripod	Yes	Opt	Yes	No	No	A
Orion Telescopes EQ-2	GEM	9	Stamped alum tripod	Yes	Opt	Yes	No	No	A
Orion Telescopes AstroView	GEM	13	Stamped alum tripod	Yes	Opt	Yes	Yes	No	A
Orion Telescopes SkyView Pro	GEM	20	Tubular steel tripod	Yes	Opt	Yes	Yes	DSC or GoTo opt	A
Orion Telescopes Sirius/Synta HEQ-5	GEM	30	Tubular steel tripod	Yes	Opt	Yes	Yes	GoTo	B
Orion Telescopes Atlas/Synta EQ-6	GEM	40	Tubular steel tripod	Yes	Yes	Yes	Yes	GoTo opt	C
Parallax Series 125	GEM	35	Wood tripod or pier	Yes	Yes	Yes	Opt	GoTo opt	D
Parallax HD150C	GEM	100	Pier	Yes	Yes	Yes	Opt	GoTo	E

(continued)

Table 5.2. (continued)

Manufacturer/ Model	Type	Carrying Capacity	Type of Support	Slow-motion Control	Clock Drive	Setting Circles	Polar Scope	DSC or GoTo?	Price Range
Parallax HD200	GEM	120	Pier	Yes	Yes	Yes	Opt	No	E
Parallax HD200C	GEM	140	Pier	Yes	Yes	Yes	Opt	GoTo	E
Parallax HD300C	GEM	300	Permanent pier	Yes	Yes	Yes	Yes	GoTo	E
Software Bisque Paramount ME	GEM	150	Sold separately	Yes	Yes	Yes	Yes	Yes	E
Stellarvue M1	Alt-az	9	Aluminum tripod	No	No	No	No	No	A
Stellarvue M4	GEM	15	Wood tripod	Yes	Opt	Yes	Yes	No	B
Stellarvue M6	Alt-az	30	Wood tripod	No	No	No	No	DSC opt	B
Takahashi Teegul	Alt-az	15	Tripod	N/S	No	No	No	No	B
Takahashi P2Z	GEM	15	Wood tripod	Yes	Yes	Yes	Yes	No	C
Takahashi EM-11	GEM	20	Sold separately	Yes	Yes	Yes	Yes	Opt	D
Takahashi EM-200	GEM	35	Sold separately	Yes	Yes	Yes	Yes	GoTo opt	D

Takahashi NJP Temma II	GEM	65	Sold separately	Yes	Yes	Yes	Yes	GoTo	E
Takahashi EM-400 Temma II	GEM	85	Sold separately	Yes	Yes	Yes	Yes	GoTo	E
Takahashi EM-500 Temma II	GEM	90	Sold separately	Yes	Yes	Yes	Yes	GoTo	E
Takahashi EM-2500	GEM	155	Pier	Yes	Yes	Yes	Yes	GoTo	E
Takahashi EM-3500	GEM	350	Permanent pier	Yes	Yes	Yes	Yes	GoTo	E
Tele Vue Panoramic	Alt-az	8	Wooden tripod	No	No	No	No	DSC opt	B
Tele Vue Gibraltar	Alt-az	13	Wooden tripod	No	No	No	No	DSC opt	B
Universal Astronomics MicroStar	Alt-az	12	Sold separately	No	No	No	No	DSC opt	A
Universal Astronomics MacroStar	Alt-az	25	Sold separately	No	No	No	No	DSC opt	A
Universal Astronomics UniStar Light	Alt-az	15	Sold separately	No	No	No	No	DSC opt	A

(continued)

Table 5.2. *(continued)*

Manufacturer/ Model	Type	Carrying Capacity	Type of Support	Slow-motion Control	Clock Drive	Setting Circles	Polar Scope	DSC or GoTo?	Price Range
Universal Astronomics UniStar	Alt-az	30	Sold separately	No	No	No	No	DSC opt	A
Vixen Porta	Alt-az	5	Stamped alum tripod	Yes	No	No	No	No	A
Vixen Icarus D	Alt-az	9	Wooden tripod	Yes	No	No	No	No	A
Vixen Great Polaris-E	GEM	15	Stamped alum tripod	Yes	Opt	No	No	Opt	B
Vixen Great Polaris	GEM	15	Stamped alum tripod	Yes	Opt	Yes	Yes	Opt	B
Vixen Great Polaris-D2	GEM	22	Stamped alum tripod OR pier	Yes	Opt	Yes	Yes	Opt	C
Vixen Sphinx	GEM	22	Stamped alum tripod	Yes	Yes	Yes	Yes	GoTo	D
Vixen Atlux	GEM	48	Tubular alum tripod OR pier	Yes	Yes	Yes	Yes	GoTo	E
William Optics EZ Touch	Alt-az	~20	Wooden tripod	No	No	No	No	DSC opt	B

Even with these improvements, wood is still much better at dampening vibrations than metal. Unfortunately, few companies seem to supply wooden tripods anymore. That being the case, consider purchasing a set of aftermarket legs. NatureWatch of Charlottetown, Prince Edward Island, Canada, sells nicely crafted, extendable tripod legs made from oak that are designed specifically as replacements. Purchasers say that the difference is amazing, and well worth the price.

Looking at the mounts listed in Table 5.2, several stand out as excellent values for their price and purpose. In the below-$500 category, the Celestron Advanced Series CG-5 German equatorial mount rises above the other equatorial mounts in its price class. The beefy steel legs and improved internal bearings help greatly, while the optional NexStar GoTo computer system, discussed earlier in this chapter, adds greatly to the system's convenience. Meade's nearly identical LXD75 mount also gets high marks, and adds the benefit of periodic error correction, which proves very useful for through-the-telescope photography.

As far as alt-azimuth mounts go, the Vixen Porta is exceptional for small, lightweight refractors like the Tele Vue TV-85 or the Stellarvue SV80. Slow-motion controls are remarkably smooth, while the mount's compact design makes it perfect for air travel. I also find the UniStar series from Universal Astronomics and Hercules models from Helix Manufacturing to be extremely steady. Their basic, robust designs make them suitable for supporting telescopes that are far larger and heavier than appears possible at first glance. The only drawback in both cases, a lack of slow-motion controls, is compensated by its smooth motions and nicely designed handle that make tracking the sky about as effortless as a manual mount can be. All are sold without tripods, although both companies also sell surveyor-style tripods that work very well. Their mounts also can be outfitted with digital setting circles for an extra cost.

Few companies address the needs of amateur astronomers confined to wheelchairs. Trying to nudge a chair under the eyepiece of a refractor or a catadioptric telescope, or worse, up to a Newtonian's eyepiece is a nearly impossible challenge. To help handicapped stargazers enjoy the view a little more conveniently, FAR Laboratories offers a right-angle pier called the Dyna Pier. The Dyna Pier's 19-inch-long horizontal arm offsets the telescope away from the pier's 31-inch-high vertical support post, allowing the wheelchair easy access underneath the telescope, whether it is a refractor or a Cassegrain-style catadioptric telescope. Simply place the telescope, mount and all, onto the pier using one of FAR Lab's customized mounting plates. FAR also sells a small alt-azimuth mount called the Dyna Cradle for a lightweight refractor, like the Tele Vue TV-85 or Orion ShortTube 80. The Dyna Pier itself is strong enough to support up to a fork-mounted 8-inch Schmidt-Cassegrain telescope. Although its standard model is designed to attach to a permanently mounted *floor plate*, custom portable packages are also available.

Between $500 and $1,000, Vixen's Great Polaris-E (*E* for "economy") mount is the best buy in equatorial mounts. Vixen's GP-D2 and Sphinx mounts are also outstanding products for heavier instruments but are more expensive. Orion's Atlas and computerized Atlas EQ-G mounts are also very good, although they are really suitable only for diehard amateurs because of their weight. Casual stargazers will likely find them too heavy to lug around just for some quick, laid-back observing. The Atlas mounts, adaptations of the EQ-6 German equatorial mount from China's Synta Optical Technology Corporation, have DC-powered, dual-axis stepper motors; a hand controller; and an illuminated polar-alignment scope. The Atlas EQ-G adds GoTo control, an autoguider interface, and periodic error correction.

Staying in the same price range, the DM-4 alt-azimuth mount is a wonderfully constructed unit from DiscMounts, Inc. Skillfully constructed of a mix of machined and anodized aluminum, stainless steel, and a pair of 4-inch-diameter ball-bearing races, the DM-4 is capable of supporting up to 4-inch refractors and 8-inch Schmidt-Cassegrains without any fuss. For larger instruments, the pricier DM-6 (the "6" signifying 6-inch bearings and corresponding larger construction) can handle scopes up to 40 pounds, given a suitably robust tripod. Like the Helix and Universal Astronomics models discussed earlier, neither the DM-4 nor the DM-6 has, nor do they need, slow-motion controls. Observers have no problem keeping up with the sky simply by nudging the telescope by hand. Both can be outfitted with Sky Commander digital setting circles if desired.

Doubling the investment, the Losmandy GM-8 and G-11 are both very well executed, with intelligently designed features. (Note that once you figure in the cost of dual-axis drive motors, the Vixen GP-D2 will end up costing about the same as the GM-8, which comes with drive motors as standard features.) Celestron's CGE also offers another usable alternative in this elite group. And finally, you can't go wrong with a mounting from Astro-Physics, Takahashi, or the Mountain Instruments MI-250, although they are all expensive. The Takahashi NJP Temma II is especially impressive considering its portability, although its Temma II GoTo control requires a personal computer to use. Motions are smooth with no gearing backlash evident. The Astro-Physics Mach1GTO, though just gearing up for production as this book is published, also looks impressive. Given Astro-Physics' reputation, the Mach1GTO will undoubtedly be widely sought after by serious amateurs, especially those who are involved in advanced astrophotography.

For permanent observatory installations, mounts from Optical Guidance Systems, Parallax Instruments, and Software Bisque are all magnificent. If it's the right size for your telescope, Software Bisque's Paramount ME is my favorite among this elite group, although that is certainly not meant to disparage the others in this stratospheric price range, all of which are fine products in their own right. The Paramount ME is characterized by a total absence of gearing backlash, small periodic error, and extremely smooth, solid motions in the right ascension and declination. And since Software Bisque is a pioneer in telescope computer control, the software that comes with the Paramount is state-of-the-art.

All this talk about automatic tracking and digital control may have left Dobsonian owners questioning their telescopes' simplicity. One of the biggest drawbacks to Dobsonian telescopes is that they do not track the stars automatically. Instead, observers have to nudge them to keep up with the sky. One way around this issue is to put the telescope, mounting and all, onto an equatorial tracking table (see Figure 5.11). Equatorial tables, or platforms as they are also known, are designed to pivot along with the sky just like an equatorial mount, bringing anything placed on top of them along for the ride. A DC-powered motor attached to a long tangent screw mounted to the table pushes the unit in time with the sky once the table has been polar aligned.

Polar aligning the table to match your observing site is just as important as it is with any equatorial mount. Most tables can be tweaked 5° or so to match your latitude, but if a large change is expected, also expect to buy a second platform to match that location. In other words, a platform designed to work from Rochester, New York, won't be much help from the Texas Star Party. By design, they can only track the sky for a finite period, usually between thirty and ninety minutes, before the table needs to be reset on its base. Their pros and cons are discussed more fully in chapter 8, where you will also find plans for making your own table from scratch.

The most sophisticated equatorial platforms on the market come from, coincidentally, Equatorial Platforms. Owner Tom Osypowski manufactures

Figure 5.11 *An equatorial tracking table can transform a Dobsonian-mounted telescope into a motor-driven instrument that tracks the sky automatically. Photo courtesy of Round Table Platforms.*

several models constructed from either welded aluminum or finely crafted wood for different latitudes, telescope apertures, and weights. All use 12-volt motors powered either by internal or external batteries. For people interested in guided photography through their telescopes, Equatorial Platforms offers models equipped with dual-axis drive motors. Owners report excellent success with their platforms, both for visual as well as photographic observations.

Johnsonian Engineering manufactures two battery-powered equatorial tracking platforms that are marketed through Jim's Mobile (JMI) in North America. Their least expensive platform, the Nightrider 1, is designed for 6- to 13-inch Dobsonian-mounted reflectors weighing up to 90 pounds. The Nightrider 1 stands about 7 inches tall and is made of laminate-covered particle board (melamine, if you will). For larger instruments, Johnsonian states that its heavier duty NightRider 2 can handle up to 180-pound 16-inch Dobsonians. A series of grooves is machined into the base that lets the table track the sky correctly from anywhere between latitudes 0° and 55° (NightRider 1) or 20° and 55° (NightRider 2), either north or south of the equator. Set the table's rollers into the groove that corresponds to your latitude setting, and it's ready to go. Both Johnsonian platforms can accept an optional hand controller, which is a necessity for astrophotography.

The models from Equatorial Platforms and Johnsonian Engineering are all very well made, but might be a little rich for your blood and budget. Round Table Platforms sells two models at more down-to-earth prices to accommodate different instrument weights. The smaller platform is designed for telescopes up to about 75 pounds, while the larger platform can support instruments as heavy as 100 pounds. Both add about 5.25 inches to the height of the telescope. Owners of both models report very good success, especially when viewing the planets at high power. Neither platform offers a hand controller; instead, power is controlled from a small box mounted under the platform. As a result, Round Table Platforms are not suitable for guided photography through the telescope; however, they will work for short wide-field guided exposures.

Finally, TL Systems sells the EzCBP Cylindrical Bearing Equatorial Platform, an equatorial table kit. All parts except the wood for the platform itself are supplied, although you also need a few power tools as well as a good knowledge of carpentry. Mixed reports from owners—some very positive and some very negative—have been received, so caution would be advised here.

An alternative to equatorial tables comes from Tech2000. Its Dob Driver II is a dual-axis drive system run by a sophisticated microcomputer that lets any altitude-azimuth mount (although designed primarily for Dobsonians) track the sky automatically. Two 12-volt DC motors drive the mounting's axes by means of stainless-steel wheels and/or timing pulleys and belts, depending on the design of your telescope. Once set up and engaged, the Dob Driver will automatically step the telescope up or down, left or right, to stay centered on the selected field. The user can then choose from four operation modes—track, pan, guide, and seek—from the small hand controller that is supplied. But take heed because the Dob Driver II's installation requires some surgery on your telescope mount, including cutting and drilling. Instructions are provided, of course, but some people might still find it a little too challenging. Also be aware that the Dobsonian mount's altitude and azimuth axes must also be perfectly perpendicular for the Dob Driver II to track correctly. Depending on how accurately the mount was assembled, further modifications may be needed.

A more advanced and elegant system comes from StellarCAT. You have already seen the name ServoCAT mentioned with several of the primo-Dobs discussed earlier in this chapter. Obsession, NightSky Scopes, TScopes, and others offer this computerized tracking system from StellarCAT as an option when purchasing one of their telescopes. But what if you already own a Dobsonian? Good news! StellarCAT sells retrofit kits that will let owners install the system themselves. The ServoCAT Junior is designed specifically for 8- to 14-inch telescopes, while larger, heavier instruments need the muscle of the standard ServoCAT system. Either comes as a complete installation kit or as individual parts for those tinkerers who prefer to use their own gears, and so forth. StellarCAT states that installation typically takes four to five hours complete from start to finish, but for those who would rather leave the carpentry to someone else, send StellarCAT your telescope's rocker box and they will install the system for you. Owners report success when using either as a stand-alone tracking system or when coupled with either the Sky Commander or the Argo Navis digital setting circle systems for accurate GoTo operation.

The Scorecard

With so many telescopes and so many companies from which to choose, how can the consumer possibly keep track of everything? Admittedly, it can be difficult, but hopefully Appendix A will help a little. It lists all the telescopes mentioned previously, sorted by price range. Placement within each range is based on "street" prices, not necessarily the manufacturer's suggested retail prices. Frequently, MSRPs are artificially inflated, perhaps in an effort to make consumers believe that they are getting deals.

There are many things to look for when telescope shopping. If you are thinking about buying binoculars or a refractor, make certain that all the optics are at least fully coated with a thin layer of magnesium fluoride (abbreviated MgFl) to help reduce lens flare and to increase contrast. As mentioned in chapter 4, fully multicoated optics are even better. For reflectors and catadioptrics, check to see if the mirrors have enhanced aluminum coatings to increase reflectivity. Find out if the telescope comes with more than one eyepiece. Is a finderscope supplied? If so, how big is it? Although a 6×30 finderscope might be fine to start, most observers prefer at least an 8×50 finderscope; anything smaller than a 30-mm finderscope is worthless. If the telescope does not come with a finder, one must be purchased separately before the instrument can be used to its fullest potential.

Next, take a long, hard look at the mounting. Does it appear substantial enough to support the telescope securely, or does it look too small for the task? Do what I call the rap test. Hit the side of the telescope tube lightly with the heel of your hand while peering through the eyepiece at a target. If the vibrations disappear in less than 3 seconds, the mounting has excellent damping properties; 3 to 5 seconds rates a good; 5 to 10 seconds is only fair; while greater than 10 seconds is poor. Remember everything that you and I have gone over in this chapter up to now, and, above all, be discriminating.

Without a doubt, the best way of getting to meet many kinds of telescopes personally is to join a local astronomical society or attend a regional star party (you'll find a list in the *Star Ware* section of www.philharrington.net). Chances are good that at least one person already owns the telescope that you are considering and will happily share personal experiences, both good and bad. Plan on attending a club observing session or "star party" as it may be called. Here, members bring along their telescopes and set them up side by side to share with one another the excitement of sky watching. To find the club nearest you, contact a local museum or planetarium to find out if there is one in your area. If you have access to the Internet, visit www.philharrington.net for links to directories of astronomy clubs around the world.

Alternatively, if there is no astronomy club near you, but if you have access to the Internet, considering joining one or more astronomy discussion groups. I'll extend a personal invitation to visit and join my "Talking Telescopes" online discussion group, which boasts thousands of members from around the world. Again, information can be found at www.philharrington.net.

If you attend a club observing session or star party, be sure to look through every instrument there. Bypass none, even if you are not interested in that particular telescope. When you find one that you are considering, speak to its owner. If the telescope is good, he or she will brag just like a proud parent. If it is poor, he or she will be equally anxious to steer you away from making the same mistake. Listen to the wisdom of the owner and compare his or her comments with the advice given in this chapter.

Next, ask permission to take the telescope for a test drive, so that you may judge for yourself its hits and misses. Begin by examining the mechanical integrity of the mounting. Tap (gingerly, please) the mounting. Does it vibrate? Do the vibrations dampen out quickly or do they continue to reverberate? Try the same test by rapping the mounting and tripod or pedestal. How rapidly does the telescope settle down?

Working your way up, check the mechanical components of the telescope itself. Does the eyepiece focusing knob(s) move smoothly across the entire length of travel? If you are looking at a telescope with a rack-and-pinion or Crayford-style focusing mount, does the eyepiece tube stop when the knob is turned all the way, or does it separate and fall out? Is the side-mounted finder-scope easily accessible?

When you are satisfied that the telescope performs well mechanically, examine its optical quality. By this time, no doubt the owner has already shown you a few showpiece objects through the telescope, but now it is time to take a more critical look. One of the most telling ways to evaluate a telescope's optical quality is to perform the star test outlined in chapter 9. It will quickly reveal if the optics are good, bad, or indifferent.

How should you buy a telescope? Some manufacturers sell only factory-direct to the consumer, while others have networks of retailers and distribu-

CONSUMER CAVEAT: The Check Is in the Mail

Okay, I admit it. I hold a grudge. In fact, a problem that I had with a well-known manufacturer/importer of astronomical equipment led to this section of the book. I won't mention the company by name, but it is hoped that the company will be made conspicuous by its absence.

Happily, difficult experiences are the exception, not the rule. Don't be afraid to ask questions before ordering. Ask specifically about the company's shipping charges and return policy. Does it impose ridiculous restocking charges on returns? If you have problems with a mail order, visit this chapter's "online extras" in the *Star Ware* section of www.philharrington.net for advice on what to do and who to notify.

tors. When it comes time to purchase a telescope, shop around for the best deal, but do not base your choice on price alone; be sure to compare delivery times and shipping charges as well. Some of the more popular telescopes, such as those from Celestron, Tele Vue, and Meade, are available from dealer stock for immediate delivery. At the opposite end of the telescope spectrum are other companies whose delivery times can stretch out to weeks, months, or even more than a year! Consult Appendix C for a list of distributors, or contact the manufacturer for your nearest dealer.

Once you decide on a telescope model, it is best, if possible, to purchase the telescope in person. Not only will you save money in crating and shipping charges but you will also be able to inspect the telescope beforehand to make sure all is in order and as described.

Congratulations, It's a Telescope!

Whether your telescope is new or used, resist the urge to uncrate your baby immediately once you get it home. Instead, read its instruction manual from cover to cover. When you are done, read it again. Absorb all the information it has to offer. Do everything slowly and deliberately. Remember, the universe has been around for billions of years; it will still be there when you get your telescope together! If you have any questions, call the dealership where you bought the instrument. Although the staff may not know the answer, they should at least have the manufacturer's phone number (if not, check Appendix C). Some companies even have technical assistance lines set up for just such an emergency.

What's that, no manual? Or maybe the manual that came with the telescope is poorly written (sadly, many are!). In those cases, check the generic assembly instructions found in chapter 9. There are also links to online manuals on this book's Web site.

Finally, with everything together, don't be surprised if you look outside and it's cloudy! Coincidence? I'm not so sure about that. The curse of the new telescope seems strong enough to defy explanation. But take heart. Eventually, the weather gods will tire of their evil game and the clouds will part. When that happens, take your prize outside for "first light." Pick out something special to look at first (I always choose Saturn, if it's up) and enjoy the view! By following all the steps here as well as the other suggestions found throughout the chapters yet to come, you are about to embark on a fantastic voyage that will last a lifetime.

6

The "Eyes" Have It

Have you ever tried to look through a telescope without an eyepiece? It doesn't work very well, does it? Sure, you can stand back from the empty focusing mount and see an image at the telescope's focal plane. But without an eyepiece in place, the telescope's usefulness as an astronomical tool is greatly limited, to say the least.

Until recently, eyepieces (see Figure 6.1) were thought of as second-class citizens whose importance was minor compared with a telescope's prime optic. With few exceptions, many eyepieces of yore suffered from tunnel vision as well as an assortment of aberrations. The 1980s, however, saw a revolution in eyepiece design. In the place of their lackluster cousins stood advanced optical designs that brought resolution and image quality to new heights. With the possible exception of selecting the telescope itself, picking the proper eyepiece(s) is probably the most difficult choice facing today's amateur astronomers.

Although eyepieces (or *oculars*, as they are also known) are available in all different shapes and sizes, let's begin the discussion here with a few generalizations. Figure 6.2 shows a generic eyepiece with its components labeled. Let's look a little deeper. Regardless of the internal optical design, the lens element(s) closest to the observer's eye is always referred to as the *eye lens*, while the lens element(s) farthest from the observer's eye (that is, the one facing inward toward the telescope) is called the *field lens*. A field stop is usually mounted just beyond the field lens at the focus of the eyepiece, giving a sharp edge to the field of view as well as preventing peripheral star images of poor quality from being seen.

Although the eyepiece must be sized according to the diameter of the eyepiece optics, the barrel (the part that slips into the telescope's focusing mount) is always one of three diameters. Most amateur telescopes use 1.25-inch-diameter eyepieces, a standard that has been around for years. These are even

Figure 6.1 *A selection of eyepieces. Left to right in the back row: Orion 17-mm Stratus, Vixen 17-mm Lanthanum LVW, Tele Vue 22-mm Panoptic, Tele Vue 12-mm Nagler Type 4, and Tele Vue 10-mm Radian. Eyepieces in the front row include, from left to right, the 9-mm HD orthoscopic, a generic 10-mm Plössl, and a no-name 12-mm Kellner.*

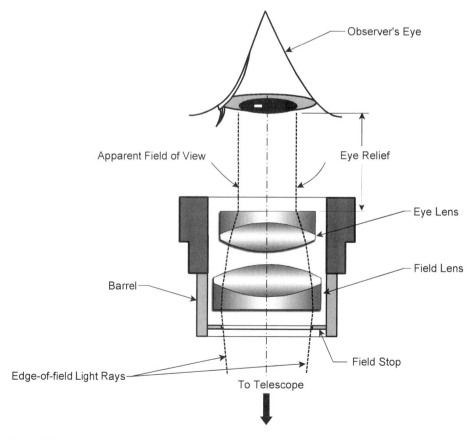

Figure 6.2 *A generic eyepiece showing internal components.*

supplied with most inexpensive, department-store telescopes nowadays. (Not long ago, most of these used inferior 0.965-inch-diameter oculars.) Finally, many of the telescopes discussed in chapter 5 also accept giant eyepieces with 2-inch barrels.

Before looking at specific eyepiece designs, let's first become fluent in terms that describe an eyepiece's characteristics and performance. There are surprisingly few. Perhaps most important of all is *magnification*—or maybe I should say lack of magnification. As previously outlined in chapter 1, magnification is equal to the focal length of the telescope divided by the focal length of the eyepiece. Therefore, the longer the telescope's focal length, the greater the magnification from a given eyepiece.

Wouldn't it be nice if there was one specific magnification value that would work well in every telescope for every object in the sky? Sadly, this is simply not the case. For certain targets, such as widely scattered star clusters or nebulae, lower magnifications are called for. To get a good look at the planets or smaller deep-sky objects—for instance, planetary nebulae and smaller galaxies—higher magnifications are required. If the magnification is too high for a given telescope, image integrity will be sacrificed.

Just how much magnification is too much and how much is just right? A good rule of thumb is to use only as much magnification as needed to see what you are interested in looking at. If you own a smaller telescope (that is, 8 inches or smaller in aperture), the oft-repeated rule of not exceeding 60× per inch of aperture is suggested. This means that an 8-inch telescope can operate at a maximum of 480×, but remember this value is not cast in stone. It all depends on your local atmospheric conditions and the instrument. Given excellent optics, you may be able to go as high as 90× or even 100× per inch on some nights, while on others, 30× per inch may cause the view to crumble.

On the other hand, larger telescopes (that is, instruments greater than 8 inches in aperture), especially those with fast focal ratios, even with excellent optics, can seldom meet or exceed the 60×-per-inch rule. Instead, they can handle a maximum of only 30× or 40× per inch.

The choice of the right magnification is one that must be based largely on past experience. If you are lacking that experience, do not get discouraged; it will come with time. For now, use Table 6.1 as a guide for selecting the maximum usable magnification for your telescope.

Another important consideration when selecting an eyepiece is the size of its *exit pupil*, which is the diameter of the beam of light leaving the eyepiece and traveling to the observer's eye, where it enters the pupil. You can see the exit pupil of a telescope or a binocular by aiming the instrument at a bright surface, such as a wall or the daytime sky (not the Sun!). Back away and look for the little disk of light that appears to float just inside the eye lens.

How can you find out the size of the exit pupils produced by your eyepieces in your telescope? Easily, using either of the following two formulas:

$$\text{Exit pupil} = D \div M$$

where:

D = the diameter of the telescope's objective lens or primary mirror in millimeters
M = magnification

or:

$$\text{Exit pupil} = F_e \div f$$

where:
F_e = the focal length of the eyepiece in millimeters
f   = the telescope's focal ratio (its f-number)

Knowing the diameter of the exit pupil is a must, because if it is too large or too small, the resulting image may prove unsatisfactory. Why? The pupil of the human eye dilates to about 7 mm when acclimated to dark conditions (although this varies from one person to the next and shrinks as you age). If an eyepiece's exit pupil exceeds 7 mm, the observer's eye will be incapable of taking in all of the light that the ocular has to offer. Many optical authorities would be quick to point out that an excessive exit pupil wastes light and resolution. This is not necessarily the case for the owners of refractors. For instance, if you own a 4-inch f/5 refractor and wish to use a 50-mm eyepiece with it, the resulting exit pupil from this combination is 10 mm. Although the exit pupil is technically too large, this telescope-eyepiece combination would no doubt provide a wonderful low-power, wide-field view of rich Milky Way star fields when used under dark skies.

The key phrase in that last sentence is *when used under dark skies*. If the same pairing was used under light-polluted suburban or urban sky conditions, the contrast between the target and the surrounding sky would suffer greatly.

Table 6.1 **Telescope Aperture vs. Maximum Magnification**

Telescope Aperture		Magnification	
Inches	Millimeters	Theoretical (60×/inch)	Practical
2.4	60	144	100
3.1	80	186	125
4	102	255	170
6	152	360	240
8	204	480	300
10	254	600	300
12	305	720	300
14	356	840	300
16	406	960	300
18	457	1080	300
20	508	1200	300
25	635	1500	300
30	762	1800	300

CONSUMER CAVEAT: Power Mania

Notice how Table 6.1 tops out at 300× for all telescopes beyond 8 inches. Experience shows (and there are those reading this now who I am sure will disagree vehemently) that little is gained by using more than 300× to view an object regardless of aperture. Only on those rare nights when the atmosphere is at its steadiest—that is, the stars do not appear to twinkle—can this value be bettered. If you observe from an area with extremely steady atmospheric conditions (southernmost Florida comes to mind), then go ahead and crank up the magnification until image quality drops off. But for the rest of us, keep it down to 300× or less.

This is due to the fact that the eyepiece is not only transmitting starlight but also sky glow (light pollution)—*too much* sky glow.

What about using a 50-mm eyepiece with a 4-inch f/5 reflector? Sorry, not a good idea. With conventional reflectors as well as catadioptric instruments, obstruction from the secondary mirror will create a noticeable black blob in the center of view when eyepieces yielding exit pupils greater than about 8 mm are used. With these telescopes, it is best to stick with eyepieces of shorter focal lengths.

On the other hand, if the exit pupil is too small, the image will be so highly magnified that the target may be nearly impossible to see and focus. Just as there is no single all-around best magnification for looking at everything, neither is there one exit pupil that is best for all objects under all sky conditions. It depends on what you are trying to look at. Table 6.2 summarizes (rather subjectively) my personal preferences.

Although magnification gives some feel for how large a swath of sky will fit within an eyepiece's view, it can only be precisely figured by adding another ingredient: the ocular's *apparent field of view*. Nowadays, most manufacturers proudly tout their eyepieces' huge apparent fields of view because they know that big numbers attract attention. Unfortunately, few take the time to explain what these impressive figures actually mean to the observer.

The apparent field of view refers to the eyepiece field's edge-to-edge angular diameter as seen by the observer's eye. Perhaps that statement will make more

Table 6.2 **Suggested Exit Pupils for Selected Sky Targets**

Target	Exit Pupil (mm)
Wide star fields under the best dark-sky conditions (e.g., large star clusters, diffuse nebulae, and galaxies)	6 to 7
Smaller deep-sky objects; complete lunar disk	4 to 6
Small, faint deep-sky objects (especially planetary nebulae and smaller galaxies); double stars, lunar detail, and planets on nights of poor seeing	2 to 4
Double stars, lunar detail, and planets on exceptional nights	0.5 to 2

sense after this example. Take a look at Figure 6.3, left image. Peering through a long, thin tube (such as an empty roll of paper towels), the observer sees a very narrow view of the world—an effect commonly known as *tunnel vision.* This perceived angle of coverage is known as the apparent field of view. To increase the apparent field of view in this example, simply cut off part of the cardboard tube. Slicing it in half (Figure 6.3, right image), for instance, will approximately double the apparent field, resulting in a more panoramic view of things.

In the world of eyepieces, the apparent field of view typically ranges from a cramped 25° to a cavernous 80° or so. Generally, it is best to select eyepieces with at least a 50° apparent field because of the exaggerated tunnel-vision effect through anything less. An apparent field in excess of 65° gives the illusion of staring out the porthole of an imaginary spaceship. The effect can be really quite impressive!

Naturally, eyepieces with the largest apparent fields of view do not come cheaply! Some, especially the long-focal-length models, are quite massive both in terms of weight and cost. Typically, they must be made from large-diameter lens elements and may be available in 2-inch barrels only. Some are so heavy that you may actually have to rebalance the telescope whenever they are used. More about this when specific eyepiece designs are discussed later in this chapter.

By knowing both the eyepiece's apparent field (typically specified by the manufacturer) and magnification, we can calculate just how much sky can

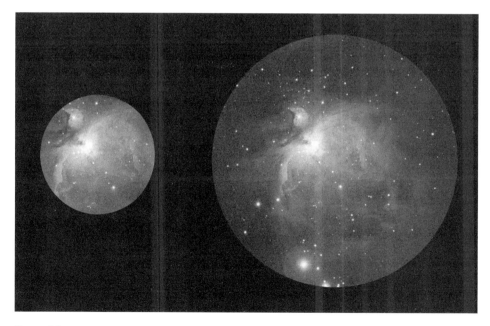

Figure 6.3 *Simulated views through a telescope with an 80-inch focal length using a 25-mm eyepiece with (left) a 40° apparent field of view and (right) an 80° field of view. Photograph of the Orion Nebula, M42, by Kevin Dixon.*

squeeze into the ocular at any one time. This is known as the *true* or *real field of view* (often abbreviated TFOV) and can be approximated* from the following formula:

$$\text{Real field} = F/M$$

where:
F = the apparent field of view (AFOV)
M = magnification

To illustrate this concept, let's look at an 8-inch f/10 telescope and a 25-mm eyepiece. This combination produces 80× and a 2.5-mm exit pupil. Suppose this particular eyepiece is advertised as having a 45° apparent field. Dividing 45 by 80 shows that this eyepiece produces a real field of approximately 0.56°, a little larger than the full Moon.

Another term that is frequently encountered in eyepiece literature but rarely defined is *parfocal*. This simply means that the telescope will require little or no refocusing when one eyepiece is switched for another in the same set. Without parfocal eyepieces, the observer may lose a faint object while refocusing should the telescope be moved accidentally during the change. Please keep in mind that even when an eyepiece is claimed to be parfocal, that does not mean that it is *universally* parfocal. Just because two eyepieces of the same optical design, say a 26-mm from Brand X and a 12-mm from Brand Y, each claim to be parfocal, they are most likely not parfocal with each other. At the same time, two eyepieces of different optical designs may be manufactured by the same company and declared parfocal, but they are not likely to be parfocal with each other.

Finally, a well-designed eyepiece will have good *eye relief*. Eye relief is the distance away from the eyepiece that the observer's eye must be to see the entire field of view. In general, the shorter an eyepiece's focal length, the shorter its eye relief. Of course, some eyepiece designs are worse than others. Less-expensive eyepieces may offer eye relief of only one-quarter times the ocular's focal length. This is much too close to view comfortably. On the other side, many modern designs maintain an eye relief of 20 mm regardless of the eyepiece's focal length, making observing more enjoyable, especially for those who must wear eyeglasses. Of course, there can also be too much of a good thing. Excessive eye relief can make it difficult for the observer to hold his or her head steadily while hovering above the eyepiece.

---

*Technically speaking, to calculate the real field of view accurately, you also need to take into account the diameter of the field stop. In practice, the formula here is accurate enough for most applications, but if the real field of view needs to be known to the nearest fraction of an arcminute, it is best to time how long a star located near the celestial equator (Mintaka in Orion's belt is a favorite) takes to cross the eyepiece field. Due to Earth's rotation, the star will appear to move at a rate of 15.04° per hour or 15.04 arc-minutes per minute of time.

Image Acrobatics

Since eyepieces can suffer from the same aberrations and optical faults as telescopes, it might be wise to list and define a few of their more common problems. Some have already been defined in earlier chapters, but now we will concentrate on their impact on eyepiece performance. (Of course, if any of the following conditions exist all of the time regardless of eyepiece used, then the problem likely lies with the telescope, not the eyepiece.)

If star images near the field's center are in focus when those near the edge are not, and vice versa, then the eyepiece suffers from *curvature of field*. The overall effect is an annoying unevenness across the entire field of view.

A more common flaw found in lesser eyepieces is *distortion*, which is most readily detectable when viewing either terrestrial sights or large, bright celestial objects such as the Moon or the Sun. This condition is usually characterized by a warping of the scene in a way that is similar to the effect seen through a fish-eye camera lens. Distortion is exacerbated in telescopes with relatively fast focal ratios, which you'll find noted throughout eyepiece reviews. Keep in mind that when an eyepiece is said to produce soft images across, say, the outer 20% of the field, that actually amounts to a significant portion of the view.

Chromatic aberration, previously defined in earlier chapters, is nearly extinct in today's eyepieces thanks to the use of one or more achromatic lenses. Still, some less-sophisticated eyepieces, often sold as standard equipment, suffer from this ailment. If an ocular transmits chromatic aberration, the problem will be immediately detectable as a series of colorful halos surrounding all of the brighter objects found toward the edge of the field of view; the center of view is usually color-free. This condition is also referred to as *lateral color*.

Spherical aberration has also been all but eliminated in most (but not necessarily all) eyepieces of modern design. If an eyepiece is free of spherical aberration, then a star should look the same on either side of its precise focus point. When spherical aberration is present, however, the star will change its appearance from when it is just inside of focus compared with when it is just outside of focus. This predicament is the result of uneven distribution of light rays at the eyepiece's focal point. Today, if spherical aberration is present, chances are it is being introduced by the telescope's prime optic (main mirror or objective lens) and not the eyepiece. At low powers (large exit pupils), it can also be introduced by the observer's own eye.

Just as with objective lenses and corrector plates, most eyepiece lenses are now coated with an extremely thin layer of magnesium fluoride. Coatings reduce *flare* and improve *light transmission,* two desirable characteristics for telescope oculars. A bare lens surface can scatter as much as 4% of the light striking it. By comparison, a single-coated lens reduces scattering to about 1.5% per surface. As already mentioned in chapter 3's discussion of binoculars, a lens coated with the proper thickness of magnesium fluoride exhibits a purplish hue when held at a narrow angle toward a light. Top-of-the-line eyepieces receive multiple antireflection coatings, reducing scatter to less than 0.5%. These show a greenish reflection when turned toward a light.

Eyepiece Evaluation

Galileo, Kepler, and Newton had it pretty easy when it came to selecting eyepieces. Look at their choices! Galileo used a single concave lens placed before the objective's focus. It produced an upright image, but the field of view was incredibly small and severely hampered by aberrations. Kepler improved on the idea by selecting a convex lens as his eyepiece and placing it behind the objective's prime focus. It gave a wider, albeit inverted view but still suffered from aberrations galore. Progress in eyepiece design was slow in the early years.

Let's take a look at just how far the art of eyepiece design has progressed while evaluating which ones are best suited for the telescope you chose earlier.

Huygens

The first compound eyepiece was concocted by Christiaan Huygens in the late 1660s (just as with a new telescope design, an eyepiece usually bears the name of its inventor). As can be seen in Figure 6.4a, Huygens eyepieces contain a pair of plano-convex elements. Typically, the field lens has a focal length three times that of the eye lens.

In the past, Huygens eyepieces were supplied as standard equipment with telescopes of f/15 and greater focal ratios. At longer focal lengths, these oculars can perform marginally well, although their apparent fields of view are narrow. In telescopes with lower focal ratios, however, image quality suffers from spherical and chromatic aberrations, image curvature, and overall lack of sharpness. Do you own a Huygens eyepiece? If so, it will probably have an *H* on the barrel; for instance, an *H25mm* indicates a Huygens eyepiece with a 25-mm focal length. You would do well to replace it.

Ramsden

Devised in 1782, this design was the brainchild of Jesse Ramsden, the son-in-law of John Dollond (you may recall him from Chapter 2 as the father of the achromatic refractor—small world, isn't it?). As with the Huygens, a Ramsden eyepiece (see Figure 6.4b) consists of two plano-convex lenses. Unlike that earlier ocular, however, the Ramsden elements have identical focal lengths and are flipped so that both convex surfaces face each other.

In most cases, the lenses are separated by about two-thirds to three-quarters of their common focal length. This is, at best, a compromise. Setting the elements closer together improves eye relief but dramatically increases aberrations. Going the other way will decrease the design's inherent faults, but eye relief quickly drops toward zero. Therefore, like the Huygens, it is probably best to remember the Ramsden for its historical significance and pass it by in favor of other designs.

Kellner

It took more than six decades of experimentation before an improvement to the Ramsden eyepiece was developed. In 1849, Carl Kellner introduced the

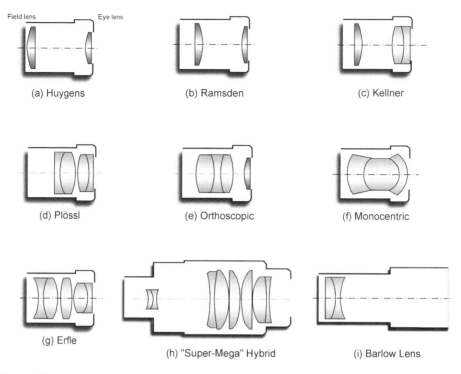

Field lens   Eye lens

(a) Huygens   (b) Ramsden   (c) Kellner

(d) Plössl   (e) Orthoscopic   (f) Monocentric

(g) Erfle   (h) "Super-Mega" Hybrid   (i) Barlow Lens

Figure 6.4 *The inner workings of several eyepiece designs from both yesterday and today. Eyepieces shown are (a) Huygens, (b) Ramsden, (c) Kellner, (d) Plössl, (e) Orthoscopic, (f) Monocentric, (g) Erfle, and (h) a hybrid. A typical Barlow lens is shown in diagram (i).*

first achromatic eyepiece (see Figure 6.4c). Based on the Ramsden, Kellner eyepieces replace the single-element eye lens with a cemented achromat. This greatly reduces most of the aberrations common to Ramsden and Huygens eyepieces. Kellners feature fairly good color correction and edge sharpness, little curvature, and apparent fields of view ranging between 40° and 50°. In low-power applications, Kellners offer good eye relief, but this tends to diminish as the eyepiece's focal length shrinks.

Although they are usable in small-to-medium aperture telescopes, Kellner eyepieces are becoming increasingly hard to find, as shown in Appendix B. Of the few still sold today, they cost about the same, or possibly even more, than comparable Plössl eyepieces, which tend to offer superior performance.

RKE

From Edmund Optics (not to be confused with Edmund Scientific Co.) comes this twist on the Kellner eyepiece. Instead of using an achromatic eye lens and a single-element field lens, the RKE (short for Rank-modified Kellner Eyepiece after its inventor, Dr. David Rank) does just the opposite. The computer-optimized achromatic field lens and single-element eye lens combine to outshine the Kellner in just about every respect. However, images are not generally as sharp as through orthoscopic or Plössl eyepieces.

Plössl

One of the most highly regarded eyepieces around today, the Plössl (see Figure 6.4d) features two close-set pairs of doublets for the eye lens and the field lens. The final product is an excellent ocular that is comparable to the orthoscopic eyepiece (discussed later) in terms of color correction and definition, but with a somewhat larger apparent field of view. Ghost images and most aberrations are sufficiently suppressed to create remarkable image quality.

Although it was developed in 1860 by Georg Simon Plössl, an optician living in Vienna, Austria, the Plössl eyepiece took more than a century to catch on among amateur astronomers. In 1980, Tele Vue Optics introduced a patented line of Plössls that was to start a new eyepiece revolution among amateur astronomers. They were an instant success. Today, many companies offer Plössls, although few actually manufacture them. Like many telescopes, most are imported from Taiwanese and Chinese sources, such as Guan Sheng, Nanjing Astronomy Equipment Factory, Kunming United Optics Corporation, and Synta. All are quite good, but some are perceptibly better than others. I prefer not to generalize, but in this case, you pretty much get what you pay for. Less-expensive Plössls cut corners that the premium brands do not. The more expensive Plössls typically maintain closer optical and mechanical tolerances, as well as include multicoated optics and blackened lens edges and threads. Little things like these can make the difference between seeing a marginally visible object and not.

Tele Vue Plössls, generally conceded as the finest, are available in eight focal lengths from 8 mm to 55 mm. Image quality, contrast, and brightness

---

CONSUMER CAVEAT: Plössls

Plössls come in a wide range of focal lengths, but you will quickly discover that the shorter focal lengths also have uncomfortably short eye relief values. Avoid them. Instead, get a moderate-focal-length eyepiece, something in the 10- to 12-mm range, and a high-quality Barlow lens, which is discussed later in this chapter.

Also avoid any 1.25-inch 40-mm Plössl. Although magnification is very low, their apparent fields of view are narrower than 32-mm Plössls with 1.25-inch barrels. In the end, the real fields of view (or how much sky actually squeezes into view) are actually the same, but the slightly higher magnification in a 32-mm Plössl will produce better image contrast. It is also difficult to look through a 40-mm Plössl for long because a very steady head position is needed to keep the image from blacking-out. Long-focal-length Plössls with 2-inch barrels are a different matter because of their larger lens elements.

are tops in the field, making them well worth the expense. All but the 55-mm have 1.25-inch barrels; the 55-mm requires a 2-inch focuser. Each has an integrated rubber eyeguard that helps prevent unwanted light infiltration and 50° apparent field of view, except for the 40-mm model, which has a 43° field. The 55-mm makes an especially nice low-power eyepiece if you own a Schmidt-Cassegrain, a Maksutov, or a long-focus refracting telescope.

For those working on tighter budgets, less-expensive Plössls are still worth consideration. All typically offer focal lengths between 4 mm and 40 mm, with apparent fields of view in the 40° to 50° range. Price-wise, Plössls from Discovery Telescopes, Celestron (called "E-Lux"), Hands-On Optics ("GTO," short for "Generic Telescope Optics"), and Orion ("Sirius") are among the least expensive of the lot, less than half the price of Tele Vues. Celestron's more upscale Omni and Orion's HighLight lines of Plössls are more expensive than the cheapest, but offer better quality control, superior barrels, and slightly wider fields of view.

Some eyepieces that are marketed as Plössls are actually variations on the traditional four-element design. For instance, Meade's Series 5000 Plössls are built around a five-element layout (six elements in the 5.5-mm model) that produces a wide, 60° apparent field of view and excellent contrast. Although they are very good in f/8 and slower telescopes, the Series 5000 Plössls show more field curvature toward the edge of the field than standard Plössls when used in faster instruments. Eye relief is comparable to traditional Plössls, which means viewing through the shorter focal length versions may prove uncomfortable. Their twist-up eyecups, however, are very comfortable. Personally, I prefer the 20- and 26-mm models.

Another eyepiece line that comes highly recommended is the Antares Elite Plössls from Sky Instruments of Canada. All use either five or seven optical elements to create 52° fields of view. The longer focal lengths are especially favored for their excellent light transmission, leading to bright, contrasty images. Edge sharpness is also acceptable when used in telescopes with focal ratios greater than about f/5. Below f/5, coma tends to take over. Parks Optical also markets an excellent line of five-element Gold Series hybrid Plössls. Images are very sharp and light transmission is excellent, but their comparatively high prices make them difficult to recommend.

Orthoscopic (Abbe)

Introduced in 1880 by Ernst Abbe, the orthoscopic eyepiece has become a perennial favorite of amateur astronomers. As shown in Figure 6.4e, it consists of a cemented triplet field lens matched to a single plano-convex eye lens. What results is a close-to-perfect eyepiece, hassled by neither chromatic nor spherical aberration. There is also little evidence of ghosting or curvature of field. "Orthos" offer flat views with apparent fields between 40° and 50° and moderate eye relief that is typically a millimeter or two shorter than the focal length. Color transmission and contrast are superb, especially when combined with today's optical coatings. With either a long focal length for low power, a short focal length for high power, or anywhere in between, orthoscopic eyepieces remain one of the best eyepieces for nearly all amateur telescopes. They yield equal or higher image contrast than many of the new-generation super-deluxe eyepieces, making them ideal for planet watching.

Orthoscopics are slowly making a comeback after several years of hiding in the shadow of Plössl eyepieces. There are generally two versions sold today, both of which are made in Japan. The more traditional *cone-* or *volcano-top* orthoscopics are imported from Kokusai Kohki (denoted by the company's Circle T logo on the barrels) and remain favorites among devout planetary and lunar observers. More recently, a new line of high-definition (HD) orthoscopics has appeared, recognizable by the flat-top barrels. Both have the same internal lens layout.

When comparing the two lines of eyepieces, three main differences are immediately apparent. The first and most obvious difference is appearance. The volcano-top orthoscopics have a built-in, solid eye cup, while HD orthos usually come with a removable, wing-style rubber eyecup. Most users prefer the latter, especially at short focal lengths when eye relief is so slim. The second difference, which some find a plus and others a minus, is that each HD barrel has an undercut to catch the eyepiece before it falls out of the focuser should the small retention screw loosen accidentally. (Many also find such undercuts annoying because snagging the undercut when switching eyepieces can cause an object to be lost.) Finally, the third difference, not readily apparent until the eyepieces are used, is that all HD orthos are approximately parfocal with one another.

After checking all the flat-top orthos through four different telescopes, I can report that image quality is indeed excellent, especially the 6- and 9-mm eyepieces. Image sharpness, edge correction, and freedom of false color are all noticeably better than through most Plössl eyepieces. If you are in the market for a good set of planetary eyepieces from the start, the flat-top orthoscopics, such as those sold through Helix Observing Accessories or Baader Planetarium, would be my choice. If, however, you already own volcano-top orthoscopics, save your money, as the improvement is not that dramatic.

Finally, there is a third line of orthoscopic eyepieces on the market. The 2.5- and 5-mm XO orthoscopics from Pentax are extremely sharp. Both would undoubtedly win best in breed, but I find it difficult to recommend them because of their exorbitant prices. Instead, I would go with one or more of the flat-top orthoscopics or, for the ultimate planetary contrast, a like-size mono-centric eyepiece from TMB Optical, discussed next.

Monocentric

Designed by the nineteenth-century German telescope maker Carl August von Steinheil, the monocentric eyepiece (see Figure 6.4f) consists of three thick lens elements, all ground and mated together so that they have a common center of curvature. Monocentric eyepieces produce excellent images but suffer severely from narrow fields of view and little in the way of eye relief. As a result, with the evolution of more comfortable eyepieces, the monocentric eyepiece design became fodder for any history of the eyepiece discussions and little else.

Recently, however, interest has been rekindled. Thomas M. Back of TMB Optics took Steinheil's original concept and applied modern-day multicoatings and high index glass to create what has been touted as the ultimate planetary eyepiece. The result is a set of eyepieces, dubbed Super Monocentrics, that have somewhat improved eye relief, better on- and off-axis image correction, and a slightly wider apparent field of view than the original Steinheil design. Even with the improvements, however, eye relief and apparent field of view are much more restrictive than other, far less expensive designs. They also suffer from noticeable field curvature in instruments faster than f/5, which takes away even more from their already narrow fields of view.

The question arises whether these eyepieces are worth the expense. One TMB Super Monocentric costs as much as three flat-top HD orthoscopic eyepieces. Are they that much better? If you are a fervent planet connoisseur who is striving to squeak the last little bit of contrast out of, say, Mars or Jupiter through your optimized apochromatic refractor, and as long as you don't have to wear eyeglasses when viewing, then I would say yes, absolutely. There is no question in my mind that Back's Super Monocentric eyepieces yield higher contrast images than any other eyepiece sold today. But if you are looking for a general purpose eyepiece that you will use on the planets, double stars, and deep-sky objects, then I would say no. These are specialized eyepieces designed for a clientele that is as narrow as the eyepieces' fields.

Erfle

The Erfle, the granddaddy of all wide-field eyepieces, was originally developed in 1917 for military applications. With apparent fields of view ranging from 60° to 75°, it was quickly embraced by the astronomical community as well. Internally, Erfles consist of either five or six elements; one variety uses two achromats with a double convex lens in between, while a second has three achromats, as shown in Figure 6.4g.

Erfles give observers an outstanding panoramic view of the deep sky. But the spacious view takes its toll on image sharpness, which suffers from astigmatism toward the field's edge, especially in telescopes with fast focal ratios. For this reason, Erfles are poor choices for lunar and planetary observations or any occasion that calls for higher magnification. In low-power, wide-angle applications, though, they are very impressive through f/8 and slower—that is, higher f-ratio—systems. Forget them in f/7 and faster telescopes.

Super-deluxe-extra-omni-ultra-maxi-mega-colossal

Whatever happened to the words *standard* and *regular*? Are they still in the dictionary? Apparently not, judging by today's advertising. Every product, from refrigerators to pet food, is extra special in some way. Top-of-the-line eyepieces are no different. Each is proclaimed by its manufacturer as something extraordinary.

The first super-duper hybrid eyepieces (see Figure 6.4h), introduced by Tele Vue in the 1980s, were intended to meet the demanding needs of amateurs

CONSUMER CAVEAT: Super Eyepieces

Super-deluxe-extra-omni-ultra-maxi-mega-colossal eyepieces offer exceptionally wide fields of view when compared with more traditional designs, but all are not created equally. You pretty much get what you pay for when it comes to these eyepieces. If you own an f/6 or faster telescope, most of the inexpensive super eyepieces discussed here will produce wider fields than orthoscopics or Plössls but with blurrier images, especially near the edges.

At the opposite end of the price scale, the finest super eyepieces offer excellent image quality, color correction, and freedom from edge distortions. Their designers are to be congratulated for creating superb eyepieces for the backyard astronomer. But if you want a premium quality super eyepiece, it's going to cost you. Some may actually cost more than your telescope! Are they worth the money? In all honesty, I have to say yes, especially if you own a fast—that is, low focal ratio—telescope. But at the same time, they are not absolutely necessary. Many hours of enjoyment can be yours with far less costly Plössl and orthoscopic eyepieces.

using Schmidt-Cassegrain telescopes and huge Newtonian reflectors, but they also work equally well in short-focus refractors and longer focal length reflectors. Most of these company-proprietary designs use multicoated lenses made from expensive glasses to minimize aberrations. Other pluses include huge apparent fields of view that let users not only see sky objects, but experience their grandeur firsthand. Most also offer longer eye relief than other classic designs, a big help for those forced to wear eyeglasses. The universe has never looked so good!

Celestron International. Ranging in focal length from 2.3 mm to 25 mm, each Celestron X-Cel six-element hybrid eyepiece has a 55° apparent field of view and 20 mm of eye relief. Couple those features with a foldable eyecup and rubber grip ring, and you have some very comfortable eyepieces to use. Each eyepiece in the series is physically large, although not so large that it could adversely affect balance. Images are quite bright through each X-Cel, although focus is a little soft when used in sub-f/6 telescopes. Planetary images will fall short of those provided by a good Plössl or orthoscopic costing a similar amount of money, but if eye relief is important, these are a good option.

Made in Japan, Celestron Axiom eyepieces are built around seven separate lens elements, all of which are multicoated to improve image contrast. Each ocular is nicely appointed with a rubber grip ring and a foldable rubber eyecup. Owners report good results, although they are slightly inferior in terms of contrast and edge sharpness to more expensive, premium eyepieces. Eyeglass wearers should also take heed that eye relief is surprisingly short in all but the 40- and 50-mm Axioms.

Guan Sheng Optics. Although its primary business focuses on making low-cost Newtonian reflectors, Guan Sheng also produces a line of eyepieces called GSO Superviews. Superviews are available in a wide range of focal lengths from 5 mm to 50 mm, each with five optical elements and boasting an impressive apparent field of view in the mid-60s (except the 50-mm, which has a stated field of 54°). Overall performance is quite good in f/7 and slower telescopes. Images appear sharp and contrasty nearly to the edge when used with an f/9 refractor, for instance.

Kokusai Kohki. The Wide Scan Type III 84° apparent-field eyepieces have received very good press from users. Most report the eyepieces hold a sharp focus for about two-thirds of the field, after which, the image softens quickly. As one respondent says, they are "not for large open clusters." Wide Scan IIIs come in four focal lengths. I favor the 13-mm model, especially in f/7 and slower telescopes. The 16-, 20- and 30-mm eyepieces are good as well, although edge sharpness is not their strong point. I wouldn't hesitate to recommend the 1.25-inch WSIIIs for binoviewer owners with limited budgets.

Meade Instruments. Meade's inexpensive Series 4000 QX line of five-element eyepieces is a good choice if your budget is low and your telescope's focal ratio is high. In an f/10 telescope, QX eyepieces perform surprisingly well considering their prices. Images stay sharp across all but the outer 15% or so of the field. Although there is some false color and image ghosting around bright objects, astigmatism and field curvature are well within reason. Cutting the telescope's focal ratio in half, however, causes an increase in both. Eye relief is also short for their respective focal lengths.

Around the same time it rolled out the QX eyepieces, Meade also introduced a new line of Series 5000 Super Wide Angle eyepieces. Available in six focal lengths ranging from 16 mm to 40 mm, Series 5000 SWAs include rubber-armored and metal housings rimmed by green accent stripes. All come with unusually designed eyecups that are similar in concept to those used on Pentax XW eyepieces. Although most eyecups extend around the eyepiece beyond the eye lens to shield the observer from stray light, SWA eyecups look like an upside-down funnel. To use them, twist the rubber housing counterclockwise to raise the eyecup. The observer then sticks his or her eye (which is slightly convex) into the eyecup. It takes a little getting used to, but the design works quite well.

Like the Series 5000 Super Wide Angles, Meade's Series 5000 Ultra Wide Angle (UWA) eyepieces remind me of mushrooms, with bulbous, rubber-armored housings protruding from chromed barrels. In practice, the Meade 5000 Ultra Wides, each with an 82° apparent field of view, prove to be very good, but not great, super eyepieces. I'll relegate that accolade to Tele Vue's Naglers and Pentax's XW lines, as well as to many of the old Meade Series 4000 Ultra Wide Angle eyepieces, which these replace. Viewing through the 5000 UWAs is comfortable and images are good for the most part. There is little off-axis color visible and good edge correction through all. Some users report issues with internal reflections, however, describing how moonlight "telegraphs" into the field when the Moon is placed just outside of the field. Of all models in the Ultra Wide line, I prefer the 6.7-mm and 8.8-mm eyepieces, while I find the 14-mm to be the poorest performer. I would also add a caveat about the big boy of the bunch, the 30-mm. That's a huge eyepiece, tilting the scales at nearly three pounds. I know of few telescopes that wouldn't be adversely affected by adding that much weight at the focuser. So, overall, I would grade the Meade 5000 Ultra Wide Angle eyepieces an A−, as compared with most Pentax XWs and Naglers an A+. Of course, the Meades are also about 30% cheaper than the other two lines, so if you just can't sink another penny beyond their cost, then I wouldn't hesitate recommending the Series 5000 UWAs.

A final caution about Meade eyepieces in general. For some reason, Meade uses a slightly different filter thread than any other manufacturer. As a result, many report difficulty screwing non-Meade filters onto their Meade eyepieces, possibly leading to cross-threading.

Orion Telescopes and Binoculars. Of all the lines of eyepieces that Orion imports, its Ultrascopics are probably the most highly regarded. Available in nine different focal lengths from 3.8 mm to 35 mm, Ultrascopics are similar to Celestron's top-rated Ultima eyepieces. All include five lens elements, except the 3.8-mm and 5-mm versions, which add a two-element negative achromat (in effect, a built-in Barlow lens), like many other hybrid eyepiece designs. Views are ghost-free with pinpoint stars seen across nearly the entire field. A common comment among owners is summed up by one who uses the eyepieces with a large-aperture reflector: "Ultrascopics are excellent; very good images to the edge of the field, even in an f/5.2 Newtonian." All Ultrascopics come with rubber eyeguards and produce a reasonable 52° apparent field of view (the 35-mm has a 49° apparent field of view).

For those on a budget, yet still lusting for a low-power, 2-inch eyepiece, Orion's 35-mm and 42-mm DeepView eyepieces are sure to attract attention. Built around a doublet field lens and single eye lens reminiscent of Edmund Optics' RKE line, DeepViews feature multicoated optics and foldable eyecups. The latter isn't quite long enough to prevent blackout, however, so you will need to hold your head back a bit from the eyecup to take in the view. DeepViews exemplify many characteristics that you'll read and read again in many of the inexpensive super eyepieces to come: good images when centered in the view, becoming blurrier as they move toward the edge. The problem increases as telescope focal ratio decreases.

Have a little more money to spend? Then, consider the three long-focal-length Optiluxe eyepieces from Orion. All have fully multicoated optics for improved light transmission. The longest focal length of the three, a monster 50-mm, is composed of five elements in what Orion describes as a modified Plössl design. The others, 40-mm and 32-mm, include four elements each. Light transmission and suppression of flare seem very good for such inexpensive eyepieces, although coma is quite pronounced in f/5 and faster telescopes. Also on the downside, all three Optiluxe eyepieces have excessive eye relief, which requires that you hold your head just so to see the field. Unfortunately, they do not have an eyecup, which would make positioning easier. Still, they produce respectable views, especially considering their comparatively low prices.

Orion's line of Stratus eyepieces looks like a twin of Vixen's Lanthanum Superwide line that was separated at birth. Not only do the specs read identically almost line-for-line, their outward appearance is also uncannily similar. Details even down to eyecup design and rubber grip rings make them difficult to distinguish from each other. All Stratus eyepieces have 20-mm eye relief, just like the Vixens, and 68° apparent fields of view (the Vixens have 65° fields, an insignificant difference). In a side-by-side test between the two, the Vixens beat the Stratus slightly in terms of image contrast and edge correction. The sky appears darker, and therefore images have a more distinct presence in the Vixens than through the Stratus, though only by a small amount. The Stratus line also shows more edge distortion and false color than the Vixens in my 18-inch f/4.5 reflector. But given that they cost half as much, Orion's Stratus eyepieces are worth strong consideration.

Orion's Epic ED-2 eyepieces incorporate some nice features at an equally nice price. Two of the six lens elements in each Epic ED-2 eyepiece are made of extra-low dispersion (ED) glass, which helps reduce unwanted false color in images. Light transmission is excellent, so images are also quite bright. As with many low-priced super eyepieces, the Epic ED-2s work well in slower instruments, such as my 4-inch f/10 refractor. Images are sharp right across the field. But when used in my 18-inch telescope, Epic ED-2 eyepieces soften toward the outer 15% of the field. Even though they may not be as crisp as some premium-priced eyepieces, Epic ED-2s nonetheless represent excellent values. And with 20 mm of eye relief, they are certainly superior to the standard Plössl eyepieces that come with many imported telescopes.

Finally, Orion's Expanse eyepieces look, feel, and perform just like their brethren, ScopeStuff's UltraWide eyepieces (and probably others, as well), all of which market the same series of imported wide-field eyepieces from the Far East under different names. Construction is adequate, although precise machining is not its strongest attribute. Optical quality is as you would expect for a cheap super eyepiece, with fair contrast and so-so image sharpness. Ghosting is a problem in the 6-mm and 9-mm models but not as obvious in the 15-mm and 20-mm. The latter two, however, suffered more from edge-of-field astigmatism than the two shorter focal lengths. All in all, they make great so-called sacrificial eyepieces. If you participate in public star parties, perhaps with a local astronomy club, then these are excellent choices. They are inexpensive, eye relief is comfortable, and image quality is adequate for the task. Best of all, your heart won't stop when a kid spills a soda and then grabs the eyepieces with sticky hands. Prices do vary from one importer to the next, so be guided accordingly.

Pentax. Renowned for its camera optics, Pentax makes several lines of oculars. In addition to its exceptional XO orthoscopic eyepieces discussed earlier, Pentax also offers premium-grade XW eyepieces. Replacing the highly regarded XL eyepieces, Pentax's XWs are available in eight focal lengths from

3.5-mm to 40-mm. Each has a 70° apparent field of view and 20 mm of eye relief. The 30-mm and 40-mm models have 2-inch barrels; the others fit only 1.25-inch focusers. (Don't be fooled by what appears to be a step-down barrel in photos of these eyepieces; the larger diameter is less than 2 inches.)

Although they bear a strong family resemblance to the predecessor XL eyepieces, XWs take a good thing and make it even better. Images are sharp right out to the edge, even in fast focal-ratio instruments, and they are easily the match of orthoscopics and Radians on the planets. Edge distortion, detectable in many Panoptic and Series 5000 SWA eyepieces, is absent in like-size XWs. Although none of the XW eyepieces has the field of view of either of the Tele Vue Naglers, their exceptionally crisp, well-corrected images both on- and off-axis, and long eye relief are second to none. In fact, I can really only see two drawbacks to Pentax XW eyepieces. First and foremost are their prices, which are among the highest on the market. The second drawback is their size. Like the Meade Super Wide and Ultra Wide lines, each XW's housing is too large to fit into binoviewers. The resulting weight may also prove troublesome for some finely balanced telescopes.

Sky Instruments. Hailing from Vancouver, British Columbia, Sky Instruments markets several different types of eyepieces, including Plössls and Erfles, under the brand name Antares. None attract more attention than its line of Speers-Waler eyepieces. Right off the top, it's clear that the made-in-Canada Speers-Waler eyepieces look nothing like anything else on the market. All have short, chromed 1.25-inch-diameter barrels, yet stand very tall. The 10-mm, for instance, is almost 6 inches long. Most of this is the black lens housing, attractively accented with orange lettering.

Unfortunately, most of the Speers-Waler eyepieces require so much focuser in-travel (turning the focuser in to the telescope) that none focus properly through many telescopes. In fact, the only telescope that all eyepieces focused in was a Schmidt-Cassegrain telescope, which has a much wider range of focus than instruments with conventional focusers. Ultra-low eyepiece adapters are offered by Astrosystems and are also available from Helix Manufacturing. Tim Hagan at Helix also kindly loaned me a set of Speers-Waler eyepieces for testing—thanks, Tim.

Performance-wise, the 82° 10-mm Speers-Waler is my favorite. It compares quite well to the Tele Vue 10-mm Radian, although it does not have the same long eye relief or produce images that are as sharp. There is also some minor lateral color in the Speers-Waler. On the plus side, however, the Speers-Waler has a wider apparent field of view and costs about half the Radian's price. I also think the 7.5-, 14-, and 18-mm eyepieces are very good, but I am not as enthusiastic about the 14-mm L (for "low-profile") version.

If Speers-Waler eyepieces are still too rich for your blood, then Sky Instruments' Antares W70 series might be more your cup of tea. None of the five 1.25-inch-diameter eyepieces, which include 5.8-mm, 8.6-mm, 14-mm, 19-mm, and 25-mm focal lengths, currently cost more than $80. Yet, each has a folding eyecup and rubber grip ring, as well as a stated 70° apparent field of view and 15 mm of eye relief. In actual use, all perform reasonably well through f/6 and longer focal ratios. Overall image quality is sharp throughout the inner 75% of the field; beyond that, some astigmatism causes stars to stretch into tiny lines that run parallel to the edge of the field. This is especially noticeable in the 19-mm and 25-mm W70s and is least conspicuous through the 8.6-mm. Some minor lateral color plagues each in the bunch, although the 8.6 seems least impacted here as well, making it my favorite eyepiece of the bunch.

Takahashi. Takahashi LE (for "long eye relief") eyepieces are another company-proprietary design. The LE line was recently expanded to include four new, short focal-length ED models for better false-color suppression, as well as five non-ED oculars. LE-ED eyepieces range from 2.8 mm to 7.5 mm, while non-ED focal lengths span 12.5 mm to 30 mm, all with 1.25-inch barrels. The monster 50-mm LE eyepiece, with its 2-inch barrel, caps the line. All are highly regarded by amateur astronomers. One respondent to the *Star Ware* survey described her 5-mm LE eyepiece as "the only lens that I use for planetary and lunar observing; very sharp, with no ghosting on bright objects like some of my other eyepieces; excellent eye relief."

Tele Vue. Not satisfied with the success of their Plössl eyepieces, Al Nagler and Tele Vue reinvented the eyepiece market by introducing oculars with extremely sharp, extremely wide apparent fields of view. As is customary, a new eyepiece design usually caries the name of its inventor, and so we know these as Nagler eyepieces. Not one to rest on his laurels, however, Nagler continues to evolve his design, replacing the original Nagler eyepieces with new and improved versions. First came the eight-element Type 2 design, then later, eight-element Type 4, six-element Type 5, and seven-element Type 6 Naglers. Three eyepieces make up the Nagler Type 4 line, all with the same dramatic 82° apparent field of view as the originals, but with better eye relief. The 12-mm Nagler Type 4 is designed to fit both 1.25- and 2-inch focusers thanks to its skirted barrel, while the 17-mm and 22-mm eyepieces fit 2-inch focusers only. These longer focal lengths are ideal for deep-sky observing, especially for viewing low-surface-brightness objects like diffuse nebulae and face-on spiral galaxies. The images delivered by the Type 4s are even better than the Type 2s, which suffered from minor image ghosting when viewing bright stars and planets. Those ghosts have been completely exorcized in the Type 4s. My only negative comment about the Nagler 4 line—and it's a very minor one—concerns the Instadjust eyecup, which is pulled in and out to just the right distance for the observer's eye. Some people love this feature, while others hate it. The problem is that unless the Instajust eyecup is adjusted just right, image blackout can be a problem. A tool is included to help newcomers get it right, although it still takes some getting used to. Me? I remain neutral, except to say that, either way, it doesn't seriously affect what has proven to be some outstanding eyepieces. Indeed, the 12-mm Nagler Type 4 is my most-used eyepiece through my 18-inch reflector.

The summer of 2001 saw the introduction of the first Nagler Type 5 and Nagler Type 6 eyepieces. Using an updated design and coating process, Type 5 and Type 6 Naglers take advantage of new optical glasses, while not sacrificing the spacewalk sensation that have made Naglers famous. All Nagler Type 5 and all Type 6 eyepieces feature foldout rubber eyecups. Type 5 Naglers are a mix of 1.25-inch and 2-inch barrels depending on focal length, while all Type 6 Naglers fit into 1.25-inch focusers. Leave it to Al Nagler to cram an 82° apparent field of view into a 1.25-inch format. Just amazing! If you own a telescope that only takes 1.25-inch eyepieces, or if you use a binoviewer attachment, then the Type 6 is for you. Just bear in mind that some purists have commented that Type 6 Naglers show minor ghosting around the Moon and bright planets, although images are still very sharp and contrasty.

Then there is the Termi-Nagler, as it has been called. The 31-mm Nagler Type 5 is the biggest, heaviest eyepiece yet to come from Tele Vue. The Termi-Nagler is tops in many areas, including the huge 82° apparent field of view, its surprisingly flat field even in fast f/ ratio telescopes, and, last, its cost, at more than $600. Ahh, but for those who can afford it, the views are just amazing. Looking at the Orion Nebula, for instance, is an experience that words cannot possibly capture. Contrast is excellent, despite the high number of lens elements, and there is no appreciable lens flare even when viewing bright objects.

A couple of words of caution about the Termi-Nagler, however. First, to reach focus, the 31-mm Nagler 5 requires more in-travel than many other eyepieces. If your telescope's focuser cannot be racked in sufficiently, focus will not be reached. Second, this is a heavy eyepiece, weighing more than 2 pounds. That could be enough to throw off a telescope's balance and cause a Dobsonian's front to drop to the ground. Just be careful when you first use it, especially if you are aiming the telescope fairly close to the horizon.

Tele Vue Panoptics are also heavyweights in the eyepiece arena, their six elements combining to yield unparalleled low-power views. Panoptics come in seven focal lengths, each sporting a 68° apparent field of view. The 15-, 19-, and 24-mm Panoptics feature 1.25-inch barrels, the 22-mm may be used in either 1.25-inch or 2-inch focusers, while the 27-mm, 35-mm, and 41-mm fit 2-inch focusers only. The 35-mm Panoptic used to be everyone's favorite wide-field eyepiece, but that title now goes to its big brother, the 41-mm Panoptic. Not only does the 41 squeeze in more stars than the 35, the images are also sharper across the field. The 35-mm Panoptic shows some minor pincushion distortion near the field's periphery, but the 41 is sharp to the edge. Coma may be a problem when Panoptics are used in f/5 and faster telescopes, although this can be solved with a coma corrector, which we will discuss later.

Recognizing that planetary observers have specific eyepiece requirements, and looking down the path of time to when most of us will need to wear eyeglasses, Tele Vue created its line of 1.25-inch Radian eyepieces. Each has a comfortable 20 mm of eye relief, allowing even those who must wear glasses to see all of its 60° apparent field of view at once. And what a view it is! Other designs may have wider fields of view, but they are not nearly as sharp. Stars snap into focus through Radians, making them especially nice for studying double stars and open star clusters. False color and ghost images are also well suppressed, while image brightness and contrast are both excellent. As good as Radians are on stars, they are just as adept on the planets, equaling the performance of nearly any other eyepiece on the market, including some exotic models sold in limited quantities. Like the Nagler Type 4s, all Radians include a click-stop Instadjust eyecup.

VERNONscope. Brandon eyepieces, offered exclusively by VERNONscope, yield sharp, crisp views with good image contrast across the entire field. Ghosting, prevalent in many lesser eyepieces, is effectively eliminated in the four-element Brandons thanks to precise optical design and magnesium-fluoride coatings. Overall performance is comparable to Plössl and ortho-scopic eyepieces on deep-sky objects, but superior on the planets. Brandons come in six focal lengths from 8 mm to 32 mm. All have 1.25-inch barrels and feature foldable rubber eyeguards. But take heed: for those who plan on using screw-in filters, Brandon eyepieces do not have standard threads to accept other manufacturers' filters. Instead, they can only use VERNONscope filters.

Vixen Optical. Each of Vixen's Lanthanum LV eyepieces, ranging in focal length from 2.5 mm to 30 mm, features a long 20 mm of eye relief. As a result, the observer can stand back from the eyepiece and still take the whole scene in, making life at the eyepiece much more comfortable. Based on the Plössl design, the Lanthanum LV adds a fifth lens element between the eye-lens and field-lens groups as well as an extra element(s) before the field lens. Apparent fields of view are in the 45° to 50° range. Image quality ranks among the high-est of any eyepiece, although some feel that the images through Lanthanums are dimmer than through some other designs.

Looking for a premium wide-field eyepiece, but prefer to save a little money for other accessories, or maybe just for food? Consider the Vixen Lan-thanum Superwides. Ranging in focal length from 3.5 mm to 42 mm, these eight-element oculars all boast 20 mm of eye relief and 65° apparent fields of view. The most pleasant surprise is that they show sharp, contrasty views across nearly the entire field, even through fast focal-ratio instruments. The 17-mm Lanthanum Superwide, for instance, is still one of my most-used eye-pieces with my 18-inch f/4.5 reflector, with edge correction that is comparable to Tele Vue Panoptics. All are designed to fit either 1.25-inch or 2-inch focusers, except the 42-mm, which fits 2-inch focusers only.

William Optics. Although best known for its exciting refractors, William Optics also markets a line of SWAN (an acronym for "Super Wide ANgle") wide-field eyepieces that feature good eye relief, lightweight construction, and a surprisingly low price. But like most lower-end super eyepieces, the SWANs are best in slower focal ratio systems. In an f/10 Schmidt-Cassegrain telescope, for instance, image quality is sharp across the inner two-thirds of the field. Light transmission is also very good, although images can't compare with the far costlier WO UWAN eyepieces, especially in fast telescopes. In an f/5 reflec-tor, for instance, the lens's sweet spot is restricted to only the inner third of the field; stars in the outer two-thirds appear soft and unappealing. One thing I do like about the SWAN eyepieces is that each barrel is tapered to keep the eyepiece from inadvertently falling out of a focuser if the thumbscrew is accidentally loosened. Personally, I prefer this technique to the more common machined-in barrel undercut, which I find frustrating.

WO's UWAN eyepieces are a different story. William and David Yang have really come up with something here! Currently limited to four focal lengths—4-mm, 7-mm, 16-mm, and 28-mm—UWAN (for "Ultra Wide ANgle") eyepieces work well in just about any telescope, from an f/15 refractor to an f/4.5 Newtonian. The three shorter focal lengths squeeze seven optical elements into their 1.25-inch barrels, while the UWAN 28 cuts the number to six elements in a 2-inch barrel. Each UWAN also includes a tapered barrel and a twist-out eyecup. Users report excellent results with each of the UWAN eyepieces. On-axis image contrast and resolution get high marks. In fact, the only negative word about these eyepieces, and it is minor criticism, is that some users complain about image blackout, especially when a UWAN is used in conjunction with a Barlow lens. Some also comment about minimal internal reflections when viewing the Moon, while the 16 shows a minor amount of astigmatism toward the edge in sub-f/5 scopes. But overall, the UWAN quartet adds four more members to the elite list of today's finest eyepiece choices.

Zoom Eyepieces

Some people love 'em, others hate 'em, but almost every amateur astronomer has an opinion about zoom eyepieces. Good or bad? Read the sidebar about zoom eyepieces for some thoughts.

Several companies—including Celestron, Meade, Orion, Tele Vue, and Vixen—feature zoom eyepieces that try to break the adage of "zoom is doom." All have a zoom range from 8 mm to 24 mm. Are they worth the price? The

---

CONSUMER CAVEAT: Zoom Eyepieces

Combining a wide variety of focal lengths into one package, zoom eyepieces claim to marry multiple eyepieces into one, saving the user time, money, and effort in the process. Twist the barrel and instantly go from low power to high. Sounds too good to be true? Unfortunately, it usually is! Most zoom eyepieces are compromises at best. For one thing, aberrations are frequently intensified in zoom eyepieces, perhaps due to poor optical design or because the lenses are constantly sliding up and down in the barrel.

Another problem is that their apparent fields of view are not constant over the entire range. The widest apparent fields occur at high power, but shrink rapidly as magnification drops. Many less-expensive zooms are also haunted by ghost images caused by internal reflections. Finally, many are not parfocal across their entire range, requiring you to refocus whenever the zoom eyepiece is zoomed. Inexpensive zoom eyepieces are almost always junk.

short answer is no in my opinion but with a qualification. Depending on who you talk to, these eyepieces are the greatest things since Lippershey invented the telescope or they are poor performers that are easily outclassed by conventional eyepieces. Those I have viewed through seem to vary in quality, leading me to question the inspections that they are put through. In the worst cases, images are dim and soft and contrast is blah. In the best cases, images are sharp but still dimmer than through a fixed focal-length eyepiece. All suffer from cramped apparent fields of view and are not precisely parfocal across their entire range. Of these, Tele Vue's 8- to 24-mm zoom is the best. It features detents at several intermediate focal lengths, which is a big convenience if you'll be using two of the zoom eyepieces in a binocular viewer.

Tele Vue sells two narrow-range, premium zoom eyepieces based on its Nagler design. Both the 2- to 4-mm and 3- to 6-mm Nagler Zooms can be adjusted to any point within their respective ranges, with click stops telling the user when each lens is at an exact focal length. Unlike all others, which vary in apparent field of view as the eyepiece is zoomed in and out, each Nagler Zoom maintains a 50° field regardless of focal-length setting. They also maintain an eye relief of 10 mm. But the most unusual thing about the Nagler Zooms is their appearance. Most zooms keep the sliding mechanism inside their housings, but the Nagler Zooms have the moving eye-lens assembly on the outside, creating a unique effect that can best be likened to a flattened mushroom. These are undoubtedly the best zooms on the market, and while they have many loyal users, I still believe that they are compromises. By their very nature, zooms will dim the view as they are twisted to their shortest focal lengths, and the Naglers are no exception. The Nagler Zoom, however, does offer the image quality of a fine Plössl with the convenience of being able to tune your magnification.

Eyepiece Accessories

In addition to eyepieces, amateur astronomers have a host of items to accessorize their telescopes. Probably the most popular item to buy for a telescope after eyepieces is a Barlow lens, which multiplies the system's effective magnification by either two or three times. But there are many more items to spend money on as well, including binocular viewing attachments, focal reducers, and coma correctors.

Barlow Lens

The Barlow lens was invented in 1834 by Peter Barlow, a mathematics professor at Britain's Royal Military Academy. He reasoned that by placing a negative lens between a telescope's objective, or mirror, and the eyepiece, just before the prime focus, the instrument's focal length could be increased (see Figure 6.4i). The Barlow lens, therefore, is not an eyepiece at all but rather a focal-length amplifier.

Depending on the Barlow's location relative to the telescope's prime optic and eyepiece, the amplification factor will change. For example, remember the 8-inch f/10 Schmidt-Cassegrain telescope and 25-mm ocular used to illustrate the concept of real field of view? If we insert the eyepiece into a 2× Barlow lens, and then place both into the telescope's star diagonal, the combination's magnification will increase by two times (80× to 160×). But if we insert the same Barlow before the star diagonal, the resulting amplification factor will be approximately three times, boosting magnification to 240×. The same change in amplification factor applies with any telescope that uses a star diagonal, such as a refractor or a Maksutov-Cassegrain.

Why use a Barlow lens? The first reason should be obvious. By purchasing just one more item, the observer effectively doubles the number of eyepieces at his or her disposal. But the benefits of the Barlow go even deeper than this. Because a Barlow stretches a telescope's focal length, an eyepiece/ Barlow team will yield consistently sharper images than an equivalent single eyepiece (provided that the Barlow is of high quality, of course). This becomes especially noticeable near the edge of the field. Another important advantage is increased eye relief, something that short-focal-length eyepieces always have in limited supply. In fact, several of the Super-Mega eyepieces get their long eye relief and short focal lengths by incorporating a Barlow-like lens (called a *Smythe lens*) right in the eyepiece's barrel.

When shopping for a Barlow lens, make sure that it has fully coated optics—fully multicoated are even better. Avoid so-called variable-power Barlows, because they tend to perform less satisfactorily than fixed-power versions.

There are several high-quality Barlows available in the 1.25-inch range, including those by Tele Vue, the Meade apochromatic Barlow, Orion's Shorty-Plus and Ultrascopic, and Celestron's Ultima. Many feel that the Celestron Ultima is the best in class, although it is made to tight mechanical specifications. Some users have reported problems when trying to insert some eyepieces that have barrels slightly greater than 1.25 inches in diameter. I have yet to experience any trouble personally, but I did notice an audible *pop* when some eyepieces are removed. Others prefer the Tele Vue 2× Barlow. Either will prove worthy of your finest eyepieces.

Should you buy a regular Barlow or a so-called shorty? That all depends on your telescope. Shorty Barlow lens are primarily designed for slipping into star diagonals, making them popular among owners of refractors and Cassegrain-style catadioptric telescopes. Although they are equally at home in Newtonian reflectors, don't feel obliged to buy a short Barlow. All other things being equal, a standard model will work just as well for you.

If you own 2-inch-diameter eyepieces, then clearly a 2-inch Barlow is your choice. There are several giant Barlows on the market today, including those sold by Apogee, Guan Sheng Optical, Hands-On Optics (Proxima), Orion, Parks, Sky Instruments (Antares), and Tele Vue. Of the group, Tele Vue's Big Barlow is easily the best of the bunch.

Astro-Physics sells a modular Barlow lens, designed primarily for refractors, that can be adapted to a number of different uses and situations. Called the Barcon, it can be disassembled into an extension tube and a threaded lens assembly, which can then be threaded into star diagonals and binoviewers. Inserting a conventional Barlow lens before a star diagonal also extends the physical length of the optical path as well as the focal length, which could lead to focusing difficulties. But since the Barcon threads into the star diagonal, the system's optical path remains the same.

Baader Planetarium's VIP Modular Barlow can also be taken apart and reconfigured depending on how you'll be using it. Extension tubes can be added or removed as required for attaching a camera or for visual use. Standard 1.25- and 2-inch filters can be threaded into the Barlow's housing behind the lens assembly, while the entire unit can be attached to Baader's dedicated Amici star diagonal to create an integrated assembly.

Another option is offered by Tele Vue's Powermate series. Not conventional Barlows in the strictest sense, Powermates are more appropriately classified as *image amplifiers*. Although traditional Barlows use two-element negative lens sets, the Powermate has both a negative achromatic doublet plus a positive pupil-correcting doublet to correct *vignetting* (darkening of the field edges) and other optical problems across a wider field than with a conventional Barlow. The net difference is that a Barlow stretches a telescope's existing focal length, while a Powermate effectively creates a new telescope. This may seem to be a fine point, but for the perfectionists in the crowd, especially those who own short-focus apochromatic refractors or fast Newtonians, and crave highly magnified views of the planets, the difference is noteworthy. Powermates all create nicely magnified views of their selected targets, without introducing false color, edge distortions, or other aberrations. Assuming you go into it with a high-quality telescope and eyepiece, the inclusion of a Powermate will only add to the system's magnification, and nothing else.

Powermates come in four magnification factors: 2×, 2.5×, 4×, and 5×. The 2× and 4× fits 2-inch focusers, while the others are for 1.25-inch focusers. All are very well made in typical Tele Vue fashion, with chromed barrels, black-anodized bodies, and captive setscrews for securing eyepieces.

Table 6.3 lists the most popular Barlow lenses available today.

Binocular Viewers

Research has shown that when it comes to viewing the night sky, binocular (two-eyed) vision is definitely better than monocular (one-eyed) vision. Our powers of resolution and ability to detect faint objects are dramatically improved by using both eyes, while any "floaters" that clog our vision are effectively eliminated. Although few experts will argue against the benefits of using binoculars over a same-size monocular telescope, the advantages of binocular viewing attachments for conventional telescopes are not as clear. Binocular viewers customarily either screw onto a telescope (as with Schmidt-Cassegrain telescopes) or slide into a telescope's focuser. Inside, we typically find a beam-splitter cutting the telescope's light into two equal-intensity paths, sending the light toward both eyepiece holders. Table 6.4 lists several popular models that are currently available.

Table 6.3 **Barlow Lenses**

Company/Model	Magnifying Factor	Barrel Diameter	Features
Apogee 1.75× Barlow	1.75×	2″	6″ long; 46.5-mm clear aperture
Apogee 3× Barlow	3×	1.25″	4.38″ long
Astro-Physics Barcon	2×	2″	Modular construction; fully multicoated optics
Baader VIP Modular Barlow	2×	1.25″ or 2″	Modular construction
Celestron Omni #93326	2×	1.25″	Short; fully multicoated optics
Celestron Ultima #93506	2×	1.25″	3-element optical design; 2.75″ long; fully multicoated optics
Guan Sheng Optical	2×	1.25″	Fully multicoated optics
Guan Sheng Optical	3×	1.25″	Fully multicoated optics
Guan Sheng Optical	2×	2″	Fully multicoated optics
Hands on Optics Short	2×	1.25″	2.88″ long; fully coated optics
Hands on Optics Proxima	1.5×	2″	42-mm clear aperture; multicoated optics
Meade #140 Series 4000	2×	1.25″	3-element optical design; multicoated optics
Orion #8711 Shorty	2×	1.25″	Multicoated optics; 3″ long
Orion #5121 Shorty-Plus	2×	1.25″	3-element optical design; fully multicoated optics; 2.75″ long
Orion #8725 Ultrascopic	2×	1.25″	3-element optical design; fully multicoated optics; 5.63″ long
Orion #8743 3-Element Barlow	2×	2″	3-element optical design; fully multicoated optics; 44-mm clear aperture
Orion #8704 Tri-Mag	3×	1.25″	Multicoated optics; 5″ long
Parks Optical 701-30010	2×	1.25″	26-mm clear aperture; multicoated optics
Parks Optical 701-30050	2×	1.25″	Short; multicoated optics
Parks Optical 701-30015	2×	2″	Multicoated optics
Sky Instruments B2S	2×	1.25″	Short; fully coated optics; 22-mm clear aperture
Sky Instruments UB2S	2×	1.25″	Short; multicoated optics; 27-mm clear aperture
Sky Instruments UB2SD	2×	1.25″	Short; 3-element apochromatic design; fully multicoated optics; 27-mm clear aperture
Sky Instruments UB2-3	2×–3×	1.25″	Fully multicoated optics
Sky Instruments UB3S	3×	1.25″	Short; multicoated optics; 27-mm clear aperture
Sky Instruments 2UBS	1.6×	2″	Fully multicoated optics; 41-mm clear aperture
Takahashi TBL0002	2×	1.25″	Fully multicoated optics; 3.5″ long

(continued)

Table 6.3 (continued)

Company/Model	Magnifying Factor	Barrel Diameter	Features
Tele Vue 2x	2x	1.25"	Fully multicoated optics; 3.75" long
Tele Vue Big Barlow	2x	2"	Fully multicoated optics; 4.5" long
Tele Vue Powermate	2x	2"	4-element optical design; fully multicoated optics
Tele Vue Powermate	2.5x	1.25"	4-element optical design; fully multicoated optics
Tele Vue 3x	3x	1.25"	Fully multicoated optics; 5" long
Tele Vue Powermate	4x	2"	4-element optical design; fully multicoated optics
Tele Vue Powermate	5x	1.25"	4-element optical design; fully multicoated optics
VERNONscope Dakin	2.4x	1.25"	Multicoated optics
Vixen 2x	2x	1.25"	Fully coated optics
Vixen 2x-DX	2x	1.25"	Fully multicoated optics
Vixen 3x	3x	1.25"	Fully coated optics

Binoviewers, rather costly accessories to begin with, must be used with two absolutely identical eyepieces, raising the total investment higher still. Even then, they do not work as well as true binoculars. This should come as no surprise when you stop and think about it. In essence, binoculars are two independent telescopes strapped together, while binoviewers must rely on the light-gathering power of a single telescope. As such, the images perceived will be dimmer than through the same telescope outfitted with a single, equal-magnification eyepiece.

It is important to consider a binocular viewer's clear aperture to maximize image brightness. If a clear aperture is too small, vignetting will eat into the edge of the field of view. As a result, I wouldn't recommend a clear aperture under 26 mm, especially if you will be using wide-angle eyepieces, like the Tele Vue Nagler Type 6. Note, however, that if your primary goal is medium- or high-power viewing, vignetting will not be a critical issue.

Another feature to look for in a binocular viewer is a diopter adjustment, as you would find with binoculars. A diopter adjustment allows you to fine-tune the focus for each eye. This is a must-have feature, because it is a rare person whose two eyes focus exactly the same.

Not all telescopes (especially Newtonian reflectors) have enough range in their eyepiece focusing mounts to accommodate the long path that light must follow inside a binocular viewer. To avoid this situation, some owners insert a 2× Barlow lens in between their telescopes and binocular viewers. A Barlow

Table 6.4 **Binocular Viewers**

Company/Model	Clear Aperture (mm)	Internal Light Path (mm)	Diopter Adjustment	Comments
Baader Planetarium Mark V	27.9	110	Yes	Includes 1.25× compensator lens
Burgess Premium Binoviewer Model C	22.0	100	Yes	Two eyepieces included; compensator lens sold separately
BW Optik Bino Fokus	22.0	120	Yes	
Celestron Stereo Binocular Viewer #93691	n/s	n/s	Yes	Adapters included for 1.25" and 2" focusers as well as threads for Celestron/Meade catadioptric telescopes
Denkmeier Standard	26.0	115	Yes	Options include 0.5× StarSweeper for SCTs; compensator lens for Newtonians and refractors; 2-position Power Switch
Denkmeier Denk II	26.0	115	Yes	Options include 0.5× StarSweeper for SCTs; compensator lens for Newtonians and refractors; 3-position Power Switch
Lumicon BinoViewer	22.0	120	Yes	Compensator lens sold separately.
Moonfish Binoviewer	18.0	100	Yes	No compensator lens offered; company recommends shorty Barlow instead
Orion Binocular Viewer	n/s	100	Yes	Includes thread-on 2× Barlow lens
Stellarvue Binoviewer	22.0	100	Yes	Two eyepieces included; compensator lens sold separately
Tele Vue BinoVue	27.0	130	No	Includes 2× compensator lens
William Optics Binoviewer	20.2	100	Yes	Compensator lens and two eyepieces included

extends the instrument's focal point, which cancels out the focusing problem, but then creates another. A 2× Barlow in front of a binoviewer will multiply the system's overall focal length not by 2× but by approximately 4×. So, a telescope-eyepiece combination that normally yields a low 40× is suddenly increased to a comparatively high 160×.

Here's a better idea. Several manufacturers offer, either as standard equipment or as an extra-cost option, a special parfocal compensator lens designed to overcome the focusing problem. These are a necessity for those using just about any telescope, except for catadioptrics that focus by moving their mirrors (for instance, Schmidt-Cassegrain telescopes); these have such wide focus ranges that auxiliary lenses are usually not necessary. But even with compensator lenses in place, focusing might still be a problem, as will be discussed later.

Of those listed in table 6.4, the Tele Vue and Baader viewers get high marks for quality by their owners. Image quality is excellent through both. The Baader is much more expensive than the Tele Vue but offers slightly larger prisms and a lower power compensator lens.

A common complaint often heard about binocular viewers is that the compensator lenses raise the system's magnification too much for wide-field viewing. To help compensate for the compensator, Denkmeier offers a unique add-on called a "Power × Switch" that inserts one (two-position Power × Switch) or two (three-position Power × Switch) auxiliary lenses into the optical path, actually demagnifying the view by as much as 34%, depending on the model and the telescope being used. When combined with the 2× compensator lens needed to bring images to focus, the Power × Switch brings the overall system magnification closer to what it would be without a binocular viewer. The Denkmeiers, especially the higher priced Denk II, also get high marks for optical and mechanical performance.

At the other end of the price scale are the economy-priced Burgess, Moonfish, Orion, Stellarvue, and William Optics binoviewers. Although we shouldn't expect them to compete with the Baader, Tele Vue, or Denkmeier models, all of which have larger, optically superior prisms; better coatings; and finer mechanics, each of these offers a way to try binoviewing and see if it's for you. But remember that unless you own a Schmidt-Cassegrain or other telescope with a large amount of focus in-travel, you will need that compensator lens.

But even with the compensator lens, focusing can still be a problem, as the binoviewer's barrel may not be long enough to hold the compensator far enough away. In that case, add a 1-inch barrel extension, available from ScopeStuff, as shown in Figure 6.5. That should do the trick, but if it is still not

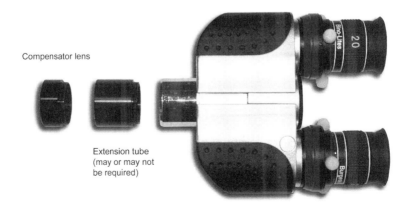

Compensator lens

Extension tube
(may or may not
be required)

Figure 6.5 *A binoviewer can deliver spectacular views of the Moon and the planets, but a special compensator lens as well as an extension tube may be required to bring things into focus.*

long enough, also add in an empty color filter ring to get a bit more length; just take the glass out of a cheap filter. This combination finally brought the Burgess binoviewer into focus through my 18-inch Starsplitter. Alternatively, if you are using a refractor, thread the compensator lens into the telescope side of the star diagonal to increase the distance. Although that will increase the effective magnification, it should help overcome focusing issues.

Binocular viewers come into their own when viewing the brighter planets and, especially, the Moon. Our satellite seems to take on an amazing although artificial three-dimensional effect that cannot be duplicated with a traditional monocular telescope. In addition, an increase in planetary detail can also be expected with a binocular viewer thanks to the aforementioned improvement in subtle contrast and color perception. As one owner summed up his experiences: "When I have compared Saturn [through an Astro-Physics 180EDT apochromatic refractor] with and without the Tele Vue binocular viewer at similar magnifications, I find the difference in illumination almost unnoticeable. Markings on the disk of Saturn, however, are decidedly more apparent with the binocular viewer than without. Admittedly, however, some dimming of faint objects is detectable." To my eye, the dimming amounts to about half a magnitude.

Focal Reducers

Although the f/10 focal ratios of most Schmidt-Cassegrain telescopes have great appeal for observers who enjoy medium-to-high-power sky views, their long focal lengths make it difficult to fit wide star fields into a single scene. To view or photograph large objects, such as the Andromeda Galaxy or the Orion Nebula, many amateurs use focal reducers to shrink a telescope's overall focal length by either 37% (from f/10 to f/6.3), 50% (f/10 to f/5), or 67% (f/10 to f/3.3), thereby increasing its field of view. Table 6.5 summarizes several of the more popular focal reducers sold today.

Lumicon's newly redesigned Rich-Field Viewer now comes in several varieties rather than its former one-size-fits-all model. The standard model is designed for 8-inch and 9.25-inch Schmidt-Cassegrains from either Meade or Celestron, reducing their f-ratios by 37%. Giant versions are sold separately for 10-inch and larger Meade and Celestron SCTs, and can vary their reduction depending on how they are set up. None of the Lumicons correct for edge distortions, making them less useful for photography than some others.

The StarSweeper from Denkmeier threads into the end of 2-inch star diagonals, unlike the Meade or Celestron reducer/correctors, listed next, which thread onto the small rear port of their telescopes. Ideally, it is intended to be mated with Denkmeier's binoviewer discussed earlier, but it can also be used alone if desired. Although the StarSweeper cuts the focal length of a Schmidt-Cassegrain in half, it only partially corrects for edge distortions. Stars toward the outer edge of the field are still distorted by field curvature and other optical irregularities.

The Meade and Celestron focal reducers do more than just compress the focal length; they also give pinpoint star images across the entire field of view. Although beneficial to all observers, this is especially attractive to

*Table 6.5 **Focal Reducers/Correctors***

Company/Model	% Reduction	Comments
Celestron International Reducer/Corrector	37%	Four-element design, fully multicoated optics
Denkmeier StarSweeper	50%	Designed to work with Denkmeier binoviewer but can also be used alone
Lumicon Rich-Field Viewer	37%	Designed for 8- and 9.25-inch SCTs
Lumicon Giant Rich-Field Viewer	Celestron version: 37% or 50% Meade version: 35% or 50%	Separate models for 10-inch and larger Meade and Celestron telescopes
Meade Instruments Series 4000 Focal Reducer/Field Flattener	37%	Four-element design; multicoated optics
Meade Instruments Series 4000 CCD Focal Reducer/Field Flattener	67%	Four-element design; fully multicoated optics; CCD cameras only; not designed for visual use or 35-mm film cameras
Optec NextGEN MAXfield 0.33X	67%	Three-element design; CCD cameras only; not designed for visual use or 35-mm film cameras

astrophotographers, as the reduction in focal length also cuts exposures. Not surprisingly, both Celestron and Meade offer focal reducers for their telescopes, each designed with its own brand in mind. Consumers should note, however, that both will fit each other's SCTs as well. They both work equally well, so the choice is yours.

Meade also manufacturers a 67% reducer/corrector, which cuts an f/10 SCT to a fast f/3.3. As discussed in the chapter 7, most CCD cameras have narrow fields as compared with conventional 35-mm film cameras. The Meade Series 4000 CCD Focal Reducer/Field Flattener helps to level the playing field a little. Like the others described earlier, the f/3.3 reducer/corrector threads onto the back of any Meade or Celestron Schmidt-Cassegrain. Overall performance of the Meade CCD reducer is very good, yielding reasonably sharp star images from edge to edge. Depending on the size of the CCD chip being used, some vignetting in the corners may be evident, however.

The Optec NextGEN MAXfield 0.33X reducer/corrector also shrinks an SCT's focal length by 67%, and with excellent results. The Optec's mechanical quality is superior to the Meade, while the price is competitive. Optical clarity and sharpness are also superior to Optec's original MAXfield, which came out more than a decade ago. In side-by-side tests, most agree that the Optec unit also has the edge over the Meade in terms of corner vignetting. Overall image sharpness is about comparable. Optec also makes reducer/correctors that shrink focal lengths by 38% and 50%.

Coma Correctors

One look through chapter 5 and it should be clear that Newtonians with low focal ratios have become immensely popular—and with them, unfortunately, so has coma. You'll recall that coma causes stars near the edge of an eyepiece's field to appear like tiny comet-shaped blobs instead of sharp points. Coma is an inborn trait of all telescopes, although it becomes objectionable only when the focal ratio is f/5 or less.

To help counter coma's deleterious effect, many amateurs use coma correctors. At first glance, coma correctors look just like Barlow lenses; in fact, that is just how they are used. Simply slip an eyepiece into the corrector's barrel, and then place the pair into the telescope's focuser. By refining the light exiting the telescope before it reaches the eyepiece, coma is effectively eliminated. As a side bonus, they also correct for off-axis astigmatism, another common problem with fast Newtonians. But bear in mind, coma correctors are not magic. They do nothing for bad optics. If your telescope has on-axis astigmatism, it is most likely caused by a poor-quality primary or secondary mirror.

Two companies, Tele Vue and Lumicon, manufacture general-purpose coma correctors, while Vixen Optical makes one specifically for its R200SS f/4 reflector (see Table 6.6). Although basically the same idea, Lumicon uses two single lenses in its design, while Tele Vue's Paracorr (short for "parabolic corrector") is based around a pair of achromatic elements for better correction.

The newest Paracorr has what Tele Vue calls a *tunable top*. Tele Vue found that an important aspect to optimizing coma correction is the distance between an eyepiece and the Paracorr lens. The Tele Vue Tunable Top Paracorr lets users adjust this spacing by loosening a thumbscrew and turning the top of the Paracorr, which raises and lowers the eyepiece. Some eyepieces work best with the Paracorr at its maximum height setting, some at minimum height, and some in between. Tele Vue provides recommended settings for its own eyepieces, but you'll need to experiment with your non-Tele Vue eyepieces for the best results.

Okay, are they worth it? Yes, but with a qualification. As mentioned earlier, they do nothing to help fix bad optics. If the telescope's optical system, including optical quality and collimation, is not in tiptop form, then the improvement will prove miniscule at best. Even then, it's been my experience that a coma corrector has a far greater impact when coupled with Plössl,

Table 6.6. **Coma Correctors**

Company/Model	% Focal Length Increase	Comments
Lumicon Coma Corrector	0%	Uses two single-element lenses; fits 2-inch focusers only
Tele Vue Paracorr	15%	Uses two achromatic lenses; fits 2-inch focusers only
Vixen AV-CC-3838	n/s	Designed specifically for R200SS reflector

orthoscopic, and other mere mortal eyepieces than it does with a top-quality eyepiece, such as a Pentax XW or a Tele Vue Nagler. These eyepieces are designed to eliminate internal astigmatism, which has the added benefit of improving off-axis performance. For prime-focus astrophotography through fast Newtonians, however, they are absolutely essential.

Reticle Eyepieces

For some applications, such as through-the-telescope guided photography, it can be useful (even necessary) to have an internal grid, or reticle, superimposed over an eyepiece's field of view. Reticle patterns, typically etched on thin, optically flat windows, come in a wide variety of designs depending on the intended purpose. The simplest have two perpendicular lines that cross in the center of view, while the most sophisticated display complex grids.

Most reticle eyepieces (see Table 6.7) come with strange-looking appendages sticking out one side of their barrels. These are illuminating devices used to light the reticle patterns for the observer. Most use red light–emitting diodes (LEDs). Ideally, an illuminator's light level should be adjustable so the reticle is just bright enough to be seen, but not so bright as to overpower the object in

Table 6.7 **Reticle Eyepieces**

Company/Model	Focal Length	Optical Design	Type of Reticle	Illuminated?
Celestron #94169	10 mm	Plössl	Double crosshairs	Sold separately
Celestron #94171 Micro-Guide	12 mm	Orthoscopic	Multiple scales	Yes, built-in, adjustable LED
Lumicon Illuminated Reticle Eyepiece	12.5 mm	Orthoscopic	Double crosshairs	Yes, built-in, adjustable LED; pulsating or constant brightness settings
Meade 12-mm Reticle	12 mm	Modified Achromat	Double crosshairs	Yes, two models: either cordless or plug into telescope power panel
Meade Astrometric	12 mm	Modified Achromat	Multiple scales	Yes, built-in, adjustable LED
Meade Series 4000 Advanced Reticle	9 mm	Plössl	Double crosshairs and two concentric circles	Yes, two models: either cordless or plug into telescope power panel
Orion #8484	12.5 mm	Plössl	Double crosshair	Yes, built-in, adjustable LED
Sky Instruments Antares Illuminated Reticle	12 mm	Kellner	Double crosshair	Yes, built-in, adjustable LED
Sky Instruments Antares Illuminated Reticle	27 mm	Kellner	Small red disk centered in view	Yes, built-in, adjustable LED
Sky Instruments Antares Illuminated Reticle	10 mm	Plössl	Double crosshair	Yes, built-in, adjustable LED

view. Some, rather than glowing steadily, alternately flash the reticle on and off. The advantage here is that by not keeping the reticle on continuously, fainter stars may be spotted and followed with greater ease. Rate and brightness are both adjustable by the user.

One of the most versatile reticle eyepieces is Celestron's Micro Guide. Built around a 12-mm orthoscopic eyepiece, the Micro Guide features a laser-etched reticle that includes a bull's-eye style target for guiding during astrophotography, and micrometer scales for measuring object size and distances, such as the separations and position angles of double stars. Meade's Astrometric eyepiece is similar, but based around a 12-mm Modified Achromat (Kellner) eyepiece. Note that the scales shown in these eyepieces are calibrated only for telescopes with 80-inch focal lengths. The Meade's performance, however, is inferior to Celestron's Micro-Guide. In fact, my advice here echoes what was written earlier in this chapter: if you are looking for sharp views, stick with either Plössl- or orthoscopic-based reticle eyepieces. Kellners are not adequate in this application.

Pieces of the Puzzle

Now comes the moment of truth: which eyepieces are best for you? It is always preferable to have a set of oculars that offers a variety of magnifications, since no one value is good for everything in the universe. Low power is best for large deep-sky objects, like the Pleiades star cluster or the Orion Nebula. Medium power is perfect for lunar sightseeing as well as for viewing smaller deep-sky targets, such as most galaxies. Finally, high power is needed to spot subtle planetary detail or to split close-set double stars.

Appendix B helps sort out the eyepiece marketplace by giving a blow-by-blow account of what eyepieces are sold by which companies. This table also looks at all of the important criteria that should be weighed when deciding what to purchase. In an ideal world, eyepiece quality, eye relief, field of view, and optical coatings would be our chief considerations, but in the real world, most of us must also factor in cost. That is why eyepieces have been grouped by their cost.

As an aid to guide your selection, Table 6.8 also offers three different possibilities for four of today's most popular telescope sizes. Each lists eyepieces according to the magnification they would produce (LP = low power, MP = medium power, HP = high power). The first is a dream outfit, where money is no object, the second offers a middle-of-the-road compromise between quality and cost, while the third represents the minimum expenditure required for a good range of eyepieces.

Let me offer some explanation for the choices. First, a Plössl listed in the "Good but Cheap" category is referring to the multitude of inexpensive eyepieces being imported from the Far East and sold by literally dozens of companies. Don't confuse that for a premium Plössl in the "Middle of the Road" category. The *premium* designation is restricted to costlier top-end versions, such as those sold by Tele Vue. Even though his Plössls are also imported from

Table 6.8. *Four Telescope/Eyepiece Alternatives*

	Dream Outfit	Middle of the Road	Good but Cheap
4" f/9.8 refractor	LP: 22-mm Tele Vue Panoptic	LP: 32-mm premium Plössl	LP: 25-mm Plössl
	MP: 11-mm Tele Vue Nagler Type 6	MP: 12-mm Baader Genuine Ortho	MP:—
	HP: 5-mm Tele Vue Radian	HP: 9-mm Baader Genuine Ortho	HP: 8.6-mm Antares W70
	2.5× Tele Vue Powermate	2× Celestron Ultima Tele Vue, or Orion Shorty-Plus	2× Celestron Ultima, Tele Vue, or Orion Shorty-Plus
8" f/6 Newtonian	LP: 24-mm Tele Vue Panoptic	LP: 25-mm premium Plössl	LP: 20-mm Guan Sheng Optical SuperView
	MP: 13-mm Tele Vue Nagler Type 6	MP: 9-mm Baader Genuine Ortho	MP: 9-mm Plössl
	HP: 8-mm Tele Vue Radian	HP: —	HP: —
	2.5× Tele Vue Powermate	2× Celestron Ultima, Tele Vue, or Orion Shorty-Plus	2× Celestron Ultima, Tele Vue, or Orion Shorty-Plus
18" f/4.5 Newtonian	LP: 31-mm Tele Vue Nagler 5	LP: 25-mm Orion Ultrascopic	LP: 20-mm Guan Sheng Optical SuperView
	MP: 12-mm Tele Vue Nagler 4	MP: 10-mm Speers-Waler	MP: 12-mm Plössl
	HP: 7-mm Pentax XW	HP: 7-mm Baader Genuine Ortho	HP: —
	2× Tele Vue Powermate	2.5× Tele Vue Powermate	2× Celestron Ultima, Tele Vue, or Orion Shorty-Plus
	Tele Vue Paracorr coma corrector		
8" f/10 Schmidt-Cassegrain	LP: 55-mm Tele Vue Plössl	LP: 40-mm Sky Instruments Classic Erfle	LP: 32-mm Plössl
	MP: 16-mm Nagler Type 5	MP: 15-mm Vixen Lanthanum LV	MP: —
	HP: 10-mm Pentax XW	HP: 9-mm Baader Genuine Ortho	HP: 9-mm Plössl
	2× Tele Vue Big Barlow	2× Celestron Ultima Tele Vue, or Orion Shorty-Plus	2× Celestron Ultima, Tele Vue, or Orion Shorty-Plus
	Reducer/corrector		

the Far East, Al Nagler's more exacting (and more expensive) design means improved performance. Also note that exact prices are not listed because they can change quickly and dramatically from dealer to dealer, so be sure to shop around.

Next, if your budget is tight, I recommend purchasing a high-quality Barlow lens rather than two inexpensive eyepieces. Any gaps in magnification, which are indicated by blank spaces in the table, can always be filled by combining the Barlow with another eyepiece in the collection.

Keep in mind that these represent only three possibilities for each telescope. You are encouraged to flip back and forth between eyepiece descriptions, Appendix B, and Table 6.8 to substitute your own preferences. And remember, the only way to learn which oculars are really best for your particular situation is to try them out first. Once again, I strongly recommend that you seek out and join a local astronomy club and go to an observing session. Bring your telescope along and borrow as many different types of eyepieces as possible. Take each of them for a test drive. Then, and only then, will you know exactly what is right for you.

7

The Right Stuff

Congratulations on making it through what might be thought of as telescope and eyepiece obstacle courses. Although the range of choices was extensive, you should now have a fairly good idea about which telescopes, binoculars, and eyepieces best meet your personal needs.

A telescope alone, however, cannot simply be set up and used. First, it must be outfitted with other things, such as a finderscope, maybe some filters, a few reference books, a star atlas, a red-filtered flashlight, some bug spray . . . well, you get the idea. So get out your wish list and credit card once again. It is time to go shopping in the wide world of "astro-phernalia"!

Let me just take a moment to calm your fears at the thought of spending more money on this hobby. Sure, it is easy to draw up a list of must-have items as you look through astronomical catalogs and magazines. Everything could easily tally up to more than the cost of your telescope in the first place, but is this truly necessary? Happily, the answer is no. Before you buy another item, we must first explore the accessories that are absolutely mandatory, those that can wait for another day, and the ones that can be done without entirely.

A quick disclaimer before going on. There are so many accessories available to entice the consumer that it would be impossible for a book such as the one you hold before you to list and evaluate every item made by every company. As a result, this chapter must limit its coverage to more readily available items. If, as you are reading this chapter, you feel that I have unjustifiably omitted something that you believe is the greatest invention since the telescope, then by all means, share your enthusiasm with me. Write your own review and send it to me in care of the address listed in the preface. I will try to include mention of that item in a future edition.

Finders

After eyepieces (and arguably even before), the most important accessory in an amateur astronomer's bag of tricks is a finder. There is nothing more frustrating to an amateur astronomer than not being able to locate an object through his or her telescope. That is why most instruments require an auxiliary device, called a *finder,* for pointing. Finders come in two different flavors nowadays: *finderscopes,* which magnify, and *unity finders* (sometimes called *reflex finders*), which do not.

Finderscopes (see Figure 7.1) are small, low-power, wide-field spotting scopes mounted piggyback on telescopes. Their sole purpose in life is to help the observer aim the main telescope toward its target. Although most telescopes come with finderscopes, all are not created equally! What sets a good finderscope apart from a poor finderscope depends on what it is going to be used for.

First, some words of advice for readers who will be using finderscopes only to supplement the use of setting circles or a GoTo computer for zeroing in on sky targets. If this applies to you, then the finderscope will probably be used only for locating alignment stars, possibly Polaris to align the equatorial mount to the celestial pole, and perhaps a few terrestrial objects. In this case, just about any will do.

If, however, star hopping is your preferred method for locating sky objects, then finderscope selection is critical to your success as an observer. The three most important criteria by which to judge a finderscope are magnification, aperture, and field of view. Finderscopes are specified in the same manner as binoculars. An 8×50 finderscope, for example, has a 50-mm aperture and yields 8 power. Most experienced observers agree that this is the smallest useful size for a finderscope. Under suburban skies, an 8×50 finderscope will penetrate to about 8th magnitude, which is roughly comparable with many popular star atlases.

Some telescopes come with right-angle finderscopes. In these, a mirror- or prism-based star diagonal is built into the finderscope to turn the eyepiece at

Figure 7.1 *This amateur telescope has the best of both worlds: an 8×50 straight-through finder as well as a Rigel QuikFinder unity finder. Note the 8×50 ring mount's silver spring-loaded piston discussed in the text.*

a 90° angle. Sure, a right-angle prism can make looking through the finder-scope more comfortable and convenient, but is it really a good idea? To my way of thinking: NO! There are two big drawbacks to right-angle finderscopes. First, although the view through a right-angle finderscope is upright, the star diagonal flips everything left to right. This mirror-image effect matches the view through a telescope using a star diagonal (Schmidt-Cassegrain, refractor, etc.) but makes it very difficult to compare the field of view with a star atlas. By comparison, straight-through finderscopes flip the view upside down but do not swap left and right. Personally, I find it easier to turn a star chart upside down to match an inverted view than to turn it inside out to match a mirrored image! Special prisms, called *Amici prisms,* can be used in place of an ordinary star diagonal to cancel out the mirroring effect, but these are supplied with rel-atively few finderscopes, probably because they also cause images to dim.

A straight-through finderscope also permits the use of both eyes when initially aiming the telescope. By overlapping the naked-eye view with that through the finderscope, an observer may point the telescope quickly and accurately toward the intended part of the sky. If you keep both eyes open when using a right-angle finderscope, one eye will see the sky, the other will see the ground!

Table 7.1 lists some of today's better finderscopes.

Although most of today's finderscopes are imported from sources in China or Taiwan, they all seem to follow the same basic design originated by Vixen in Japan. Focusing requires that a retaining ring around the objective lens be loosened and the objective's cell rotated. The eyepiece remains fixed throughout the process, which I find a little counterintuitive. The Antares find-ers from Sky Instruments in Canada have helical eyepiece focusers, which I much prefer.

Deep down inside, most observers agree that right-angle finderscopes are poor substitutes for straight-through designs. At least I assume they must, given the incredible popularity of one-power aiming devices, or unity finders, in recent years. These let observers keep both eyes open when aiming their telescopes by projecting a pattern onto a window. The observer then sights through the window, which acts as a beamsplitter, to see the reflected pattern as well as stars shining through from beyond. Unity finders are not necessarily meant to be used in lieu of a finderscope, but rather to supplement its use. Their biggest advantage is that they make aiming a telescope easy without needing to flip or twist star charts.

Unity finders have two disadvantages, however. One is that the window tends to dew over quickly in damp environs. But since it is not an optical device, the window may be wiped clear with a finger, sleeve, or paper towel. A more permanent solution to the dewing problem is to attach a shield over the window. The simplest is a large file card that has been painted black bent over the window and secured in place with masking tape; a project in chapter 8 offers a fancier alternative. The second disadvantage is that since a unity finder is not an optical device, it will not it show star-hopping guide stars that are dimmer than what can be seen by the naked eye—so, suburban and urban observers take heed.

Table 7.1 **Finderscopes**

Company/Model	Magnification	Aperture	Field of View (degrees)	RA or ST(1)	Features
Apogee, Inc. 50RA	8	50	6.0	RA	Interchangeable 1.25" eyepiece; mounting bracket sold separately
Apogee, Inc. 50ST	8	50	6.0	ST	Interchangeable 1.25" eyepiece; mounting bracket sold separately
Celestron #93785-8P	7	50	5.0	ST	Illuminated Polaris finder reticle; mounting bracket included
Celestron #93783-8	9	50	5.8	ST	Mounting bracket included
Discovery Telescopes #0608	8	50	n/s	ST	Mounting bracket included
Lumicon 50-mm Right Angle Finder System	8	50	5.7	RA	Mounting bracket and eyepiece included
Lumicon 80-mm Super Finder System	12	80	3.8	ST or RA	Mounting bracket and eyepiece included
Meade #825	8	25	7.5	RA	90° finder for ETX-90AT
Meade #544 (blue), #545 (white)	8	50	5.0	ST	Mounting bracket sold separately
Orion #7211	6	30	7.0	RA	Mounting bracket included
Orion #7212	9	50	5.0	RA	Mounting bracket included
Orion #7200	9	50	5.5	ST	Mounting bracket included
Sky Instruments Antares F750	7	50	7.0	ST	Mounting bracket sold separately; diopter focus; 1.25-inch eyepiece; available in white, blue, black
Antares FR750	7	50	7.0	RA	Mounting bracket sold separately; diopter focus; 1.25-inch eyepiece; available in white, blue, black
Antares FRE750	7	50	6.3	RA	Mounting bracket sold separately; 90° Amici prism; diopter focus; 1.25-inch eyepiece; available in white, blue, black

(continued)

Table 7.1 **(continued)**

Company/Model	Magnification	Aperture	Field of View (degrees)	RA or ST(1)	Features
Sky Instruments (cont)					
Antares F850	8	50	7.0	ST	Mounting bracket sold separately; diopter focus; 1.25-inch eyepiece; available in white, blue, black
Antares FR850	8	50	7.0	RA	Mounting bracket sold separately; diopter focus; 1.25-inch eyepiece; available in white, blue, black
Antares FRE850	8	50	6.3	RA	Mounting bracket sold separately; 90° Amici prism; diopter focus; 1.25-inch eyepiece; available in white, blue, black
Antares FR1280	12	80	4.3	RA	Mounting bracket sold separately; diopter focus; 1.25-inch eyepiece; available in white, blue, black
Antares FRE1280	12	80	4.3	RA	Mounting bracket sold separately; 90° Amici prism; diopter focus; 1.25-inch eyepiece; available in white, blue, black
Stellarvue F50	9	50	7.0	RA	Sold with or without bracket; 90° Amici prism; helical focusing 1.25-inch eyepiece; black or white
Takahashi TVF7500	7	50	6.2	ST	Mounting bracket and illuminated reticle sold separately
Takahashi TVF1170	11	70	4.2	ST	Mounting bracket and illuminated reticle sold separately
Vixen AV-FS-3546	7	50	6.8	ST	Mounting bracket sold separately
Vixen AV-FS-3887	10	50	4.0	ST	Mounting bracket sold separately

Notes: RA, right-angle; ST, straight-through; n/s, not supplied by manufacturer; n/a, not applicable.

CONSUMER CAVEAT: Finderscopes

The majority of finderscopes come with dew shields that are too short to do their jobs. As a result, the finder's objective will often dew over long before the telescope itself. If you find that to be the case, read the discussion on dew-suppression equipment later in this chapter. An alternative might be to fashion a dew shield extension for your finder. For instance, I found that the inside diameter of a can of Campbell's soup was just a little larger than the outside diameter of my 8 × 50 finder's dew shield. To take up the difference, I simply lined the inside of the can with a strip of thin, adhesive-backed foam insulation—the type used to seal around windows. Putting this on the finder at the beginning of an observing session keeps the finder clear the entire night, unless conditions are especially damp.

The most common of these new-generation sighting contraptions is the Telrad, invented by the late amateur telescope maker Steve Kufeld. Based conceptually on the World War II vintage Norden bombsight, the Telrad is described as a *reflex sight*. Using a pair of AA batteries and a red light–emitting diode (LED), a bull's-eye target of three rings (calibrated at 0.5°, 2°, and 4°) is projected onto a clear piece of glass set at a 45° angle. The brightness of the rings is controlled by a side-mounted rheostat that also acts as an on/off switch.

Rigel Systems offers its QuikFinder, which looks and works like a vertical Telrad. The QuikFinder, also shown in Figure 7.1, projects adjustable-brightness 0.5° and 2° red circles onto an angled window. Unlike any of the other unity finders, the QuikFinder's red circle can be set to pulse off and on. This allows the observer to keep the circle's brightness relatively high while aiming at dim stars, making it a helpful feature. The QuikFinder weighs just 3.4 ounces (as compared to the Telrad's 11-ounce weight) and stands about 4.5 inches tall. Its small footprint, just 1.38 inches, is the most compact of any unity finder currently made, although it also seems to imply a certainly frailty in that it might be easily snapped off. To prevent this from happening, the QuikFinder can and should be removed from its base and stored separately whenever the telescope is moved.

A new, third reticle finder was recently introduced by Apogee, Inc., under the name Galileo Circular Illuminated Reticle (CIR) finder. Shorter, cheaper, and lighter than the Telrad and QuikFinder, the Galileo CIR appears attractive at first, but it fails in use on several levels. The biggest problem is that the reticle pattern, consisting of three random circles rather than scaled references like the others, shifts from parallax depending on how you hold your head. As such, it is not recommended.

Several of the unity finders in Table 7.2, such as the Celestron Star-Pointer, the Orion EZ-Finder II, and the Tele Vue Qwik-Point, are direct adaptations of an aiming device that Daisy Manufacturing Company markets for its BB guns and pistols. All are made from black plastic and project single red

Table 7.2 **Unity Finders**

Company/Model	Type (target or dot)	Comments
Apogee Galileo Circular Illuminated Reticle	Reticle	Includes mounting base
Apogee Giant Mars-Eye Finder	Dot	Includes dimmer and mounting plate
Apogee Mars-Eye	Dot	Includes dimmer and mounting plate
Baader Planetarium Sky Surfer V	Dot	All-metal construction; includes dimmer and mounting plate
Celestron International Star-Pointer	Dot	Includes dimmer and mounting plate
Orion Telescopes EZ-Finder II	Dot	Includes dimmer and quick-release dovetail mount; glass window very dark
Rigel Systems QuikFinder	Target	Pulsing or steady 0.5° and 2° red circles; includes two dovetail mounting plates
Stellarvue Red Dot Finder	Dot	Includes dimmer and mounting plate
Tele Vue Qwik-Point	Dot	Includes dimmer and mounting plate
Tele Vue Starbeam	Dot	All-metal construction; includes dimmer and mounting plate
Telrad	Target	0.5°, 2°, and 4° red circles; includes dimmer and dovetail mounting plate
Vixen Red Dot Finder	Dot	Includes dimmer and Vixen/Synta compatible mounting base

dots onto partially reflective curved windows. The observer-side surface of the windows is coated with a clear reflective material, while the outer surface is uncoated. Both surfaces reflect images of the red dot, which merge into one when the observer's eye is on-axis. Power to the units is provided by thin, button batteries mounted on the undersides of the units. Most are adjustable in brightness. Of these, the twin Apogee Giant Mars-Eye Finder and the Stellarvue Red Dot Finder rate as best because of their larger, clear windows and lower prices.

Tele Vue's Star Beam is the leanest and sexiest red-dot finder of all. Like the others, the Star Beam projects a red LED dot onto a partially reflective window that the observer looks through when aiming the telescope. While the others are made of plastic, the lean body of the Star Beam is crafted from machined aluminum. Not surprisingly, the Star Beam costs about four times as much as the others. Beauty may be only skin deep, but its cost cuts right to the bone, so the Star Beam has never achieved the popularity of some of its ugly duckling cousins, despite its superiority. The Baader Sky Surfer V is also worth a look because of its more durable, metal construction. Like the Star Beam, however, it is expensive.

Finally, let's discuss finderscope mounts. Good finderscopes are secured to telescopes by a pair of rings with three or six adjustment screws. These thumbscrews allow the finderscope to be aligned precisely with the main instrument. On some, one of the thumbscrews has been replaced by a spring-loaded piston, which makes alignment quite easy. Let the buyer beware, however, that many smaller finderscopes (primarily inexpensive 5 × 24 models) use single mounting rings with only three non-spring-loaded adjustment screws. These are notoriously difficult to adjust, and even more difficult to keep in alignment.

Filters

For years, photographers have known the importance of using filters to change and enhance the quality and tone of photographs in ways that would be impossible under natural lighting. Today, more and more amateur astronomers are also discovering that viewing the universe can also be greatly enhanced by using filters with their telescopes and binoculars. Some heighten subtle, normally invisible planetary detail, others suppress the ever-growing effect of light pollution, while still others permit the safe study of our star, the Sun. Which filter or filters, if any, are right for you depends largely on what you are looking at and from where you are doing the looking.

Light-Pollution Reduction (LPR) Filters

Amateurs use many different filters, but the most intriguing—and most misunderstood—are light-pollution reduction filters (often abbreviated LPR) and nebula filters. Selections of each are shown in Figure 7.2.

Can we actually filter out light pollution? It turns out that many sources of outdoor lighting causing the problem do not shine evenly across the entire visible spectrum. Instead, they emit their radiation at only a few discrete wavelengths. For instance, the common high-pressure sodium streetlight used throughout North America radiates principally in yellow wavelengths. To do their magic, LPR filters suppress the broad portion of the visible spectrum that includes those wavelengths, while passing other wavelengths through.

The so-called *nebula filters* have a somewhat different mission. These are designed for the singular purpose of observing emission and planetary nebulae. The ionized atoms in these clouds are excited into fluorescence, causing them to glow. Because these objects are composed largely of hydrogen, their visible-light emissions correspond to that of ionized hydrogen, in the far red and blue ends of the visible spectrum. They also emit greenish light from doubly ionized oxygen. Most nebula filters muffle all wavelengths except those narrow portions associated with ionized hydrogen and doubly ionized oxygen. As a result, these filters are more correctly referred to as *narrowband filters*.

Other nebula filters suppress all visible light except for a specific wavelength, or line, that passes through. *Line filters* come in two varieties: oxygen-III (abbreviated O-III) and hydrogen-beta. As we will see, some narrowband and line filters are more appropriate for certain objects than others.

Figure 7.3 shows the band-pass profile of an LPR, a *narrowband,* and a *line filter.* All three work essentially the same way. Extremely precise, ultra-thin

Figure 7.2 *A collection of LPR and nebula filters, including (from left to right): Tele Vue Bandmate O-III, Astronomik O-III, Lumicon O-III, DGM Optics NPB, and Orion UltraBlock.*

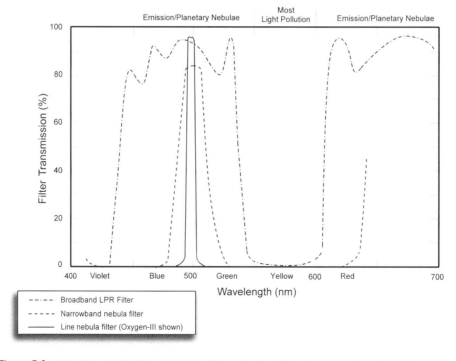

Figure 7.3 *Typical transmission characteristics of broadband, narrowband, and line (oxygen-III) filters. Chart based on data supplied by Lumicon, Inc.*

layers of nonorganic coating materials with known optical refractive characteristics, one with a high index of refraction and another with a low index, are deposited on glass using a vacuum chamber. As certain wavelengths of light strike the material, they are reflected back rather than passed through, as other wavelengths are.

There are many misconceptions about these filters. For instance, many amateurs mistakenly believe that LPR filters reduce all forms of light pollution. Seems like a natural assumption given the name, but that's just not the case. True, they cut certain wavelengths, but they do little to reduce the impact of car headlights, lights on buildings and houses, and other sources using incandescent bulbs, which shine at all visible wavelengths.

Therefore, in an effort to clarify their intent and usage, I'd like to propose dropping the term *light-pollution reduction filter* from our vocabulary. Instead, let's call them by their proper name of *broadband filters*, because they filter out a wide spread of wavelengths. Similarly, *nebula filters* should be referred to as either narrowband or line filters.

Another common fallacy is that these filters make deep-sky objects look brighter. Not so. Filters do not discriminate based on the source. They decrease all light at their specified wavelengths, dimming everything. In the process, however, the background sky is usually darkened more than the target, boosting contrast. This should make an object easier to spot but will not make it appear brighter.

Some amateurs believe these filters dim the view so much that they can't be used with smaller telescopes. That's just not true in practice either. Almost any astronomical instrument can draw some benefit from these filters. I've often used identical pairs of filters taped to the eyepieces of my binoculars to detect objects that were invisible otherwise.

Finally, there's the widespread myth that filters are only for urban and suburban observing. If you're observing far out in the country, you don't need them. Although experience shows that broadband filters make their biggest impact under severe light pollution, narrowband and line filters can enhance the view of emission and planetary nebulae no matter where you are.

To cut through the urban legends associated with these filters and to find out just how well many of them perform, I gathered together more than two dozen broadband, narrowband, and line filters for a review that appeared in the August 2005 issue of *Astronomy* magazine. Although most are available to fit either 1.25- or 2-inch eyepieces, I restricted my tests to 1.25-inch models only.

Examining the filters, it was clear that all were made to high optical standards. Coatings were uniform and metal housings were nicely machined, and they were either anodized or painted. Each was labeled for easy identification under dim lighting and screwed into my eyepieces effortlessly. Most were black, which could make them difficult to find at night if misplaced, even with a flashlight. Only the four Thousand Oaks filters were anodized with different colors.

Not surprisingly, the broadband filters did best against broadband objects, such as star clusters and galaxies. Although it was rare for a filter to show more detail than an unfiltered view or to reveal a target that was otherwise invisible, broadband filters did improve the visual aesthetics by darkening the background sky. Of the seven tested, the best overall results were delivered by the Orion SkyGlow, which added just enough contrast to make an object stand out nicely without sacrificing too much image brightness. I also judged the DGM Optics Very High Throughput (VHT), Meade 908B, and Thousand Oaks LP-1 filters to be effective. The VHT did a little better on nebulae, while the others were slightly better on star clusters and galaxies.

Of the nine narrowband filters, DGM Optics' Narrow Pass Band (NPB) filter was judged to be the best performer on emission and planetary nebulae. What sold me were the views I had of the Crab Nebula, planetary nebulae M76 and NGC 1514, and especially, the Orion Nebula. I was also impressed with the images delivered by Orion's UltraBlock and Tele Vue's Nebustar Bandmate filters.

Results were exceptionally close in the O-III round. I wasn't disappointed with any but found that the views I had through the Meade 908X were a little better than the rest. I especially enjoyed the remarkable details revealed in M42, as well as some otherwise invisible features in several planetary nebulae. Part of that difference might be due to Meade's filter passing more energy in the hydrogen-alpha portion of the visible spectrum than traditional O-III filters. In that sense, the Meade 908X is more of a narrowband filter than a true O-III filter.

Finally, *hydrogen-beta filters* are highly specialized accessories that produce positive results with only a handful of emission nebulae. Here, the Astronomik H-Beta filter rose above the others, but again only by the narrowest measure. All revealed the dim silhouette of the Horsehead Nebula against the subtle glow of the backdrop of IC 434 through my 18-inch reflector when there wasn't even a hint with any of the others. All three also showed finer structure in M43, the tiny appendage hanging off the northern edge of the Orion Nebula, than any of the O-III or narrowband filters, and even added new dimensions to M42 itself (although I still preferred the view through the O-III filters).

You can find my full review in the August 2005 issue of *Astronomy* magazine, but the bottom line is that light-pollution reduction filters are wonderful assets for the amateur astronomical community. With them, observers can spot nonstellar objects that were considered impossible to find just a few years ago. But remember, no filter can reveal the beauty of the universe as well as can a dark, light-free sky.

Lunar and Planetary Filters

Many seasoned observers agree that fine details on both the Moon as well as the five planets visible to the naked eye can be significantly enhanced by viewing through color filters. This improvement occurs for two reasons. First, they reduce irradiation, the distortion of the boundary between a lighter area and a darker region on the Moon's or a planet's surface. This effect usually is caused by either turbulence in the Earth's atmosphere or by the human eye being overwhelmed by the dazzling image.

Filters also help to increase subtle contrasts between two adjacent regions of a planet or the Moon by transmitting (brightening) one color while absorbing (darkening) some or all other colors. For instance, an observer's eye alone may not be able to distinguish between a white region and a bordering beige region on Jupiter. By using filters of different colors, the contrast between the zones may be increased until their individuality becomes apparent.

Table 7.3 compares different heavenly sights with the results that may be expected when they are viewed through a variety of color filters. The filters are listed according to their *Wratten number* as well as their color. The Wratten series of color filters was created by Eastman Kodak and contains more than 100 different shades and hues. Today, it is the industry's standard way to refer to a filter's precise color.

For those just starting out, choose basic colors, such as deep yellow (#15), orange (#21), red (#23), green (#58), and blue (#80A). You are free to use photographic filters sold in camera stores, however, most amateurs prefer color filters designed to screw into the field end of eyepieces. Companies that sell eyepiece color filters include Celestron, Lumicon, Meade, Orion Telescopes, and Sky Instruments. If you are going to buy a variable polarizing filter, be aware that some designs, such as those by Meade and Celestron, require that the eyepiece slide into a holder that sticks out of the focuser, like a Barlow lens. As a result, some eyepieces will not focus properly if the focuser has limited travel.

Sirius Optics of Kirkland, Washington, makes several filters for amateur telescopes, including the unique *variable filter system* (VFS). Rather than using one color filter at a time, the VFS is an enclosed filter assembly that inserts into a telescope's focuser. Eyepieces, in turn, slip into the VFS's eyepiece tube and are locked in place with a single thumbscrew. Inside the black-anodized housing, a piece of optical glass measuring 71 mm across has been coated with different thin films to produce varying color shades (technically speaking, *bandpass frequencies*) and mounted in a rotating metal ring. The VFS's colors are far more subtle than typical color filters.

To change from one color to the next, simply rotate the section of the filter's metal ring that protrudes from the side of the housing, then watch through the eyepiece as the target changes shades from red to green, yellow, orange, magenta, and all variations in between.

I tested the VFS filter using my 8-inch f/7 and 18-inch f/4.5 Newtonians, as well as my 4-inch f/9.8 refractor. Since the filter fits between the focuser and the eyepiece, the eyepiece protrudes about 1.5 inches farther out of the focuser than it would otherwise. This prevented images from focusing through my 18-inch without first inserting a shorty Barlow lens. My refractor and 8-inch reflector, with their greater focuser ranges, had no trouble. Note that Sirius Optics now offers the VFS with a built in 2× Barlow to addresses focusing issues with some (but not all) telescopes.

Although it is quite costly, the VFS is a very good choice for planetary viewing. Jupiter, for instance, presented an interesting change in appearance

Table 7.3 **Color Filters: A Comparison**

Object	Filter	Benefit
Moon	Moon filter (neutral density)	Reduces brightness of Moon evenly across the spectrum
	15 (deep yellow) 58 (green)	Enhances contrast of lunar surface
	80A (light blue)	Reduces glare
	Polarizer	Like the neutral density filter, reduces brightness without introducing false colors
Mercury	21 (orange)	Helps to see planet's phases
	23A (red) 25 (deep red)	Increases contrast of planet against blue sky, aiding in daytime or bright twilight observation
	80A (light blue)	Improves view of Mercury against bright orange twilight sky
	Polarizer	Darkens sky background to increase contrast of planet; helpful for determining phase of Mercury
Venus	25 (deep red)	Darkens background to reduce glare; some say they also help reveal subtle cloud markings
	80A (light blue)	Improves view of Venus against bright orange twilight sky
	Polarizer	Reduces glare without adding artificial color (especially helpful for viewing planet through larger telescopes)
Mars	21 (orange) 23A (light red) 25 (deep red)	Penetrates atmosphere to reveal reddish areas and highlight surface features such as plains; #21 best for 6-inch and smaller scopes, #23 best for 8- to 12.5-inch; #25 best for larger scopes
	38A (deep blue)	Brings out dust storms on surface of Mars
	58 (green)	Accentuates "melt lines" around polar caps
	80A (light blue)	Accentuates polar caps and high clouds, especially near the planet's limb

Table 7.3 (continued)

Object	Filter	Benefit
Jupiter	11 (yellow-green)	Reveals fine details in cloud bands
	21 (orange)	Accentuates cloud bands
	56 (light green) 58 (green)	Accentuates reddish features such as the Red Spot
	80A (light blue) 82A (very light blue)	Highlights details in orange and purple belts as well as white ovals
Saturn	15 (deep yellow) 21 (orange)	Helps to reveal cloud bands
Comets	80A (light blue)	Increases contrast of some comets' tails
Other	15 (deep yellow)	Helps block ultraviolet light when doing black-and-white astrophotography
	25 (red)	Reduces impact of light pollution on long-exposure black-and-white photographs taken from light-polluted areas
	58 (green)	Same as #25; works well for emission nebulae
	82A (very light blue) Minus violet	Suppresses chromatic aberration in refractors

as I slowly turned the filter wheel. I preferred the view in the filter's yellow-orange range, where there was a noticeable improvement in the contrast of the atmospheric belts, although many other variations had something to offer. Subtle features that were otherwise invisible, including several cloud swirls and dark markings, could be seen simply by rotating the filter wheel until I found just the right setting. Saturn also showed improvement, although it was more modest than with Jupiter. In both cases, magnifications between about 120× and 200× worked best. The VFS is also purportedly suitable for observing

CONSUMER CAVEAT: Color and Moon Filters

Color filters can be useful, but they are not magic. The improvement that they offer is subtle at best. Experience still plays the major role when it comes to seeing what is considered impossible detail; a filter itself will not do it. And as far as Moon filters go, save your money. Although they will lower overall brightness, I have never found them to be useful accessories.

nebulae, but in this case, I saw no advantage over narrowband and line filters outlined earlier.

Solar Filters

Monitoring the ever-changing surface of the Sun is an aspect of the hobby that is enjoyed by many. Before an amateur dares to look at the Sun, however, he or she must be aware of the extreme danger of gazing at our star. Viewing the Sun without proper precautions, even for the briefest moment, may result in *permanent vision impairment or even blindness.* This damage is caused primarily by the Sun's infrared and ultraviolet rays, the same rays that cause sunburn. Although it may take many minutes before the effect of sunburn is felt on the skin, the Sun's intense radiation will burn the eye's retina in a fraction of a second.

There are two ways to view the Sun safely: either by projecting it through a telescope or binoculars onto a white screen or piece of paper or by using a special filter. Sun filters come in a couple different varieties. Some fit in front of the telescope, while others attach to the eyepiece. *Never* use the latter—that is, the eyepiece variety. They can easily crack under the intense heat of the Sun (focused by the telescope as is the Sun's image), leading tragically to blindness. Happily, I know of no new telescope that is supplied with an eyepiece solar filter, but many were in the past.

Safe solar filters are made from either glass or a polymer, such as Mylar, and fit securely in front of a telescope or binoculars. Figure 7.4 shows one example. The filter must be placed in front of the telescope, so that it can filter out both the dangerously intense solar rays as well as the accompanying heat prior to entering the optical instrument. *Never place the filter between you and the eyepiece.* Be sure to use only specially designed solar filters; *do not* use photographic neutral-density filters, smoked glass, overexposed photographic film, or other makeshift materials that may pass invisible ultraviolet or infrared light.

The most popular solar filters are sold by Baader Planetarium, Thousand Oaks Optical Company, Orion Telescopes, and JMB. Thousand Oaks Optical, a

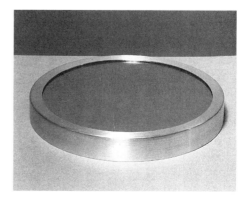

Figure 7.4 *A full-aperture white-light solar filter. Photo courtesy of Thousand Oaks Optical.*

well-established name in the solar-filter field, offers both glass and nonglass filters. Its Type 2 Plus filters are made from float glass, the same stuff used in window glass. Many people have expressed doubts about the quality of the Sun's image through glass solar filters due to the possibility of optical distortions. According to Pat Steele from Thousand Oaks, each piece of glass used for a Type 2 Plus filter is hand selected for its flatness, then coated with Solar II Plus, which is a mixture of chrome, stainless steel, and titanium. Thousand Oaks Type 3 Plus filters are made the same way, but they have a lighter density (Neutral Density 4) for photographic use. Type 3 Plus filters are *not* appropriate for visual observing.

JMB offers three different glass filters. Its advertising claims that all of the company's filters are made of machine-polished float glass. JMB Identi-View Class A filters, coated with an alloy of nickel, chromium, and stainless steel, are also sold by Orion Telescope Center. Premounted filters are available for telescopes ranging in aperture from 2 inches to 14 inches. The only difference between the Class A and less-expensive Class B is that the Class B's coating does not include stainless steel (and is, therefore, slightly less durable). Finally, like Thousands Oaks's Type 3 Plus filters, JMB's Class C filters have a lighter density that enable solar photographers to operate their cameras at faster shutter speeds when using slow, fine-grain films. In addition, the JMB cell is held in place by friction using a rubber strip, which is more secure than the felt used by Thousand Oaks.

Although Baader Planetarium in Germany is a relative newcomer to the field of solar filters, it has quickly established itself as a leader. Baader's AstroSolar solar filter is also made from a polymer film that is specially darkened to reduce internal reflection. The material is heat-annealed, like glass, to reduce internal strains, then it undergoes what is described as an ion implantation and metallization coating process. The Baader solar filter material is imported into the United States by Astro-Physics, which sells it only in raw sheets. In this case, the consumer must come up with his or her own cell to hold it in place, either by following the included instructions or by some other manner. Kendrick Astro-Instruments sells the Baader film in premade mountings, if you'd prefer, while Celestron also sells the material in cells made specifically for many of their instruments.

The most obvious difference between the filters is the color of the Sun's image, which depends on the type of coating used. The JMB and Thousand Oaks Polymer Plus filters show the Sun as a yellowish disk, while it appears yellow-orange through the Thousand Oaks Type 2 Plus. The Baader filter turns the disk white. None of these shades represents the Sun's "true" color, although the Baader is closest.

Of all the filters available today, the Baader AstroSolar filter impresses me both for its amazingly low price (as much as 60% lower than the others) as well as for showing the finest level of detail in sunspots. This was especially evident at higher powers when the other filters showed comparably dim images. Conversely, the image is bordering on too bright (but not dangerously so) at low magnifications. As a result, faculae are easy to see along the limb, but washed out elsewhere due to image brightness.

All of the solar filters just discussed might be thought of as broadband filters, because they filter the entire visible spectrum (and you can't get much broader than that). There are also special narrowband solar filters that allow observers to see our star in a completely different light! These filters block all of the light from the Sun with the exception of one distinctive wavelength: 656 nanometers, also known as hydrogen-alpha. Viewing the Sun at this wavelength with an H-alpha filter changes the view from a placid disk peppered with the occasional sunspot to a dynamic stellar inferno highlighted by ruby red solar prominences, sinuous dark filaments, and intricate surface granulation.

Hydrogen-alpha filters typically consist of two separate pieces: an energy-rejection filter (ERF) that fits over the front end of a telescope to prevent overheating and the hydrogen-alpha filter itself, which fits between the telescope and the eyepiece. The filters are available in several different, extremely narrow bandwidths (usually expressed in angstroms, where one angstrom equals 0.1 nanometer). The narrower the bandwidth, the higher the contrast but fainter the image. Bandwidths of 0.6 to 0.9 angstrom are the most popular because these offer the best compromise between brightness and contrast.

Unfortunately, most H-alpha solar filters are expensive. The lone exception, of course, is not just a filter but an entire H-alpha telescope. The Coronado Personal Solar Telescope (PST) has single-handedly brought hydrogen-alpha solar observing to the masses. Chapter 5 has a complete review of the PST as well as other Coronado solar telescopes.

Coronado, now a wholly owned subsidiary of Meade Instruments, makes add-on hydrogen-alpha filters as well. Each is sold as a set, consisting of an *energy rejection filter* that covers the telescope's objective, and a second part called a *blocking filter*. Coronado's least expensive filter, the SolarMax 40, is designed specifically for small refractors under 1,000-mm focal length. With a bandwidth of less than 0.7 angstrom, the SolarMax 40 shows solar granulation, filaments, sunspots, and erupting prominences all seen in vivid detail. Coronado's pricier SolarMax 60 is best for telescopes between 1,000- and 3,000-mm focal length, while the SolarMax 90 is designed for even larger instruments. Clearly, the larger the aperture, the greater the resolution and finer the detail, but regardless, the images are simply breathtaking.

DayStar Filter Corporation has been a leading source for these special accessories for more than a quarter century. Most expensive of the DayStar filters is the University model, with band passes from 0.4 to 0.8 angstrom available. As the name implies, it is geared toward the rigid requirements of professional institutions. The ATM filter, available in bandwidths from 0.5 to 0.95 angstrom, is designed with the serious amateur solar astronomer in mind. Both the University and ATM filters need to be heated to maintain thermal equilibrium and to function properly, and therefore require an external source of AC power. This can prove a major inconvenience with a portable telescope.

The least-expensive H-alpha filter of all DayStar models is the T-Scanner. Unlike the other two, the T-Scanner requires no external power supply and, therefore, is completely portable, making it more attractive. It is available in bandwidths from 0.5 to 0.8 angstrom.

Finally, Lumicon sells its Solar Prominence Filter, which works on the same no-heater principle as the T-Scanner. Its 1.5 angstrom band pass renders good views of prominences, but it is not suitable for viewing flares, filaments, and granulation. Complete with an energy rejection prefilter and all required adapters and hardware, the Lumicon filter is significantly cheaper than the T-Scanner and ASP-60, and is an excellent value for anyone living within a budget who wishes to get into this phase of solar observing with his or her own telescope.

Other Accessories
Collimation Tools

A telescope will deliver only poor-quality, lackluster images unless it is in proper optical alignment, or collimation. Typically, refractors should never need to be collimated, while reflectors and catadioptric telescopes should be checked regularly. Deep-dish reflectors should be checked each time they are set up.

Several companies manufacture collimating tools for amateur telescopes, with some more useful for certain optical designs than others. The least-expensive collimation tool sold today, at less than $10, is the Aline from Rigel Systems. The Aline is simply a black plastic cap with a centrally drilled peephole and a reflective surface on the cap's inside surface. Place the Aline into your telescope's 1.25-inch focuser and look through the hole. Adjust the telescope's diagonal and primary mirrors until both appear centered. The Aline is a handy tool for slower focal ratio reflectors, but faster instruments require some of the more sophisticated tools discussed next.

The three most popular collimation tools remain the sight tube, the Cheshire eyepiece, and the Autocollimator. The sight tube is helpful for approximating the collimation of just about any telescope. At one end lies a small peephole used by the observer to check collimation, while at the other, a pair of thin crosshairs serve as a reference for centering optical components. (Collimation procedures are outlined in detail in chapter 9.) Tectron Instruments, AstroSystems, Inc., and Orion Telescopes all sell 1.25-inch-diameter sight tubes that will fit most amateur telescopes. The second edition (1998) of this book contained plans for making your own, which you can find today in the *Star Ware* section of my Web site www.philharrington.net.

The Cheshire eyepiece (see Figure 7.5), invented eighty years ago by Professor F. J. Cheshire at a British university, is intended primarily for in-the-field adjustments of fast Newtonians. Not an optical assembly, a Cheshire eyepiece is a variation on the sight tube with a part of one side cut out and a 45° mirrored surface inside. A light is shone through the cutout, reflects from the diagonal to the primary mirror, back to the diagonal, and out through a hole in the center of the eyepiece. The telescope is properly collimated when the reflected image of the Cheshire's center hole is centered in the primary mirror.

Tectron, AstroSystems, Celestron, and Orion Telescopes all sell 1.25-inch Cheshire eyepieces. Those sold by Celestron, Orion, and AstroSystems are

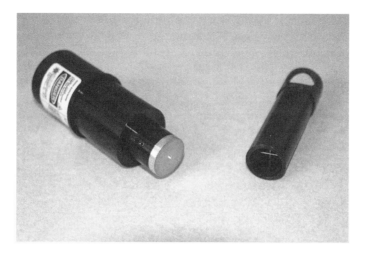

Figure 7.5 *Two tools for checking telescope collimation: a laser collimator (left) and a combination sight tube/Cheshire eyepiece (right).*

actually all-in-one sight tube/Cheshire eyepiece combinations. Although they are less expensive than Tectron's separate tools, I find them more difficult to use. Jim Fly, Inc., sells a similar tool called the CatsEye, which uses a red reflective triangle to mark the center of the primary, rather than the more typical white ring. Another difference is that the user must shine a light onto the mirror from the front of the tube, rather than through the cutout in the side of the tube, as with a traditional Cheshire eyepiece. Which is better? Fly's reflective triangle is a very good idea, but his system also makes it possible to drop the flashlight into the telescope tube. All things being equal, both a traditional Cheshire eyepiece and the CatsEye work about the same.

The Autocollimator is the most sensitive of the three tools for collimating a fast Newtonian reflector. At first glance, it looks like a short sight tube—that is, a hollow tube capped with a piece of metal that has a tiny hole drilled in the center. The difference only becomes evident when you look inside the Autocollimator. There, encircling the peephole and perpendicular to the eyepiece mount's optical axis, a flat mirror serves to reflect all light from the primary mirror back down the telescope.

Are all three necessary? The sight tube is a *must* for anyone who owns a reflecting or catadioptric telescope and is concerned with optimizing its performance. I also consider the Cheshire eyepiece a requirement for anyone who owns a deep-dish (f/6 or faster) Newtonian reflector. Don't leave home without it! The Autocollimator, although a handy tool for fine-tuning the alignment of fast Newtonians, is not absolutely necessary. But beware, some of you Newtonian owners out there: I have seen some imported Newtonian reflectors with 2-inch focusers where the 1.25-inch adapter wobbles sloppily in the focuser's drawtube. That will cause invalid collimation readings. No one brand is guilty of this more than another, but if your focuser has this problem, speak with the company about a repair.

Laser collimators have become all the rage in certain circles. Lasers emit precise, narrow beams of light that spread minimally as they travel away from their source. Within the past few years, small, low-power units have been created and incorporated into many diverse applications, including telescopes.

The principal behind a laser collimator is simple enough. Place the laser into the eyepiece holder and turn on the beam. If the telescope's optics are properly collimated, the beam should bounce from the diagonal mirror to the primary mirror, and then back exactly on itself. You then look through the front of the telescope tube at the bottom of the collimator protruding out of the focusing mount. If you see an off-center image of the reflected beam on the collimator, then the optics are not properly aligned. Table 7.4 lists current offerings.

Of those listed in Table 7.4, some project simple dots, while others generate intricate grid patterns. Personally, after trying both types through my 18-inch f/4.5 reflector, I find the latter to be more precise. They seem to make it easier to determine which way the mirrors must be tilted to achieve optimal alignment. But they are also more expensive, which may be a consideration.

Cost is one of the drawbacks to laser collimators, although those from Orion Telescopes and Helix Observing Accessories are well within reach of most observers. The two units from Orion, imported from the Far East, are also rebadged and sold as their own by several companies. I am hesitant to mention them, as even the Orion units vary in mechanical quality (notably barrel diameter, which can wobble and lead to spurious results). Orion has an excellent return policy, however, so if at first you don't succeed . . .

That raises the question of whether the laser collimator itself is collimated. Those listed in Table 7.4 should be, because all are made from metal

CONSUMER CAVEAT: Laser Collimators

Do laser collimators work as advertised? According to some of the advertising hype, they are absolutely necessary for checking a telescope's collimation. How did we ever get along without them?!

Okay, time for a little truth in advertising. The short answer is yes, they do work, but with some qualification. First, a laser collimator is not the only collimation tool you need. Every amateur should have a sight tube in his or her eyepiece case. A Cheshire eyepiece should be right next door if you own an f/6 or faster Newtonian reflector. Once you become skilled with using those, you may not even need a laser collimator, except to check results. That's how I use mine, following the collimation procedure detailed in chapter 9.

Also be aware that most laser collimators are not designed to be used with refractors or catadioptric instruments; they are really only intended (and needed) for deep-dish Newtonians. If your Newtonian has a slower focal ratio than this, then the collimation tools mentioned earlier in this chapter are probably all you'll need. Save your money and buy another eyepiece instead.

Table 7.4 **Laser Collimators**

Company/Model	Pattern	Diameter	Power Source
Helix Laser Collimator	Dot or several holographic projections	1.25″	Two LR44 batteries
Glatter Laser Collimator	Dot or 9 × 9 square grid	Either 1.25″ or 2″	One 123A lithium battery
Kendrick 2062	Dot	2″	Two AAA batteries
Kendrick 2063	Dot	1.25″	Two AAA batteries
LaserMax TLC-202N	Crosshairs with graduated concentric circles	1.25″ and 2″	Three AAA batteries
Orion LaserMate	Dot	1.25″	Three SR44 batteries
Tech 2000 Laser Collimator	Dot	1.25″	Three SR44 batteries

that has been machined to very tight tolerances. Some other laser collimators might be less expensive but may have plastic housings, which are far more prone to flexure and warping over time. Even with the collimators listed here, there is always the possibility that a defective unit may slip through quality-control measures. That's why you will find Craig Stark's Lazy Laser Collimator Collimator, detailed in chapter 8, so handy to make and use. If you find that something is amiss with a laser collimator, contact the manufacturer about having it fixed.

A third quandary is that the snout of the collimator may not protrude enough to make it easily visible in some focusers. This is usually not a problem with the so-called low-profile focusers included with many Newtonian reflectors, especially open truss-tube models. But it can be a showstopper for others that are so tall that the telescope's solid tube blocks the view from the side. If you can't see the bottom of the collimator through the focuser's tube, it can't be used.

This last problem isn't a concern to the LaserMate Deluxe from Orion or the laser collimators from Kendrick Astro Instruments. The tall body of the Orion LaserMate Deluxe includes a rear-view port that sticks above the focuser's drawtube, making it possible to see the reflected laser beam without having to look inside the front of the telescope. The Kendrick collimators are long enough to extend out from the bottom of all but the longest Newtonian focusers. Kendrick has adapter tubes with cutouts on one side through which the user can see the reflected laser beams (instead of having to look down the front of the telescope tube, as with others).

Star Diagonals

Rather than make observers crane their necks to look straight through the eyepiece, most refractors and Cassegrain-style instruments come supplied with either a 45° or a 90° star diagonal that diverts the optical path to a more comfortable angle. Although they certainly make life at the eyepiece much more enjoyable, some star diagonals work much better than others. Indeed, I have seen some supplied with beginners' telescopes that are so far out of alignment that they make focusing impossible. Others, because of relatively poor optics, dim images greatly. To owners of those telescopes, your best bet, if the telescope is acceptable otherwise, is to replace the original star diagonal.

There are two types of star diagonals, based on their angles: 45° and 90°. Traditionally, 45° prisms are associated with terrestrial rather than astronomical viewing. They are not nearly as easy to view through when a telescope is aimed at, or anywhere near, the zenith. Most 45° diagonals also generate upright images. To do this, the image must flip one additional time, dimming it further. My best advice is that if your telescope has such a star diagonal (perhaps it would be better called a terra-diagonal), you should consider replacing it with a real, 90° star diagonal. But which one?

Some 90° diagonals use prisms to redirect the light, while others use flat mirrors. In general, assuming all other things are equal, a high-quality mirror diagonal will outperform a prism diagonal. The problem is that some light is dispersed and lost as it enters the prism, even if the prism is fully coated. More is lost as it exits. A mirror diagonal, however, has only one optical surface, resulting in a brighter, sharper image. The best mirror diagonals, like other telescopic mirrors, are coated with enhanced aluminizing for even brighter images. Also check that the diagonal's nosepiece, which inserts into the telescope's focuser, is threaded for filters. That way, you can change eyepieces without having to swap the filter from one to the next.

Should you own a telescope that takes only subdiameter eyepieces, consider purchasing a hybrid star diagonal. These fit into 0.965-inch focusers but accept standard 1.25-inch eyepieces. Unless you are using some exceptional 0.965-inch eyepieces, the change will be quite an improvement.

Some of the most popular 90° 1.25-inch star diagonals on the market today are sold by Celestron, Lumicon, Meade, Orion Telescopes, Sky Instruments, Tele Vue, and William Optics. The same companies also sell 2-inch diagonals, as does Astro-Physics. If ordering a 2-inch model, check that it comes with a 1.25-inch adapter, as well.

Rubber Eyecups

To help shield an observer's eye from extraneous light, many eyepieces sold today come with collapsible rubber eyecups, and they certainly can make a big difference! In chapter 6, I tried to point out whenever an eyepiece came with a built-in eyecup. But just because an eyepiece is cupless, doesn't mean it is not worth consideration. Most can be retrofitted with aftermarket eyecups that work just as well. Orion Telescopes, Tele Vue, and several other companies sell rubber eyecups designed to fit both their own particular lines of oculars as well as others. Always match the proper size eyecup to your eyepiece before ordering, since a too-tight or too-loose eyecup isn't of much value.

Eyepiece Cases

If properly cared for, any of the equipment listed in this book should outlive you. Proper care is critical, especially when it comes to sensitive optical surfaces. Although most of us treat our telescopes with kid gloves, eyepieces are often handled and stored far more haphazardly. Any eyepiece, whether a $20 Ramsden or a $600 Nagler, should be protected from dust and dirt whenever it is not in use. To do that, several companies, such as Sky Instruments, ScopeStuff, and SmartAstronomy offer eyepiece carrying cases and individual storage packs. Orion Telescopes, for instance, sells several plastic and aluminum briefcase-style cases, each lined with cubed high-density foam that is easily customized to accommodate your accessories. Their smaller, plastic cases are fine to start with, but if you really get into the hobby, you will need to expand to a larger aluminum unit as your eyepiece collection inevitably grows.

You may have a great eyepiece and accessory case right in your neighborhood. Where, you ask? Is there a Wal-Mart, a Lowe's Home Center, or a Home Depot nearby? If so, drop by and browse their tool cribs. My local Lowe's sells an aluminum tool attaché case that is identical to Orion's large aluminum case but at a substantial savings. Unlike Orion's precut foam, the Lowe's case comes with cubed foam so you can customize the layout.

Another brand name to look for is Doskocil. The company makes a wide variety of hard-shell cases that are used for transporting anything from pets to golf clubs. It offers a series of foam-lined cases that are ideal for eyepieces, photographic equipment, and entire telescopes. Many have low-profile pressure release valves, which is an important feature for airline travel. More about that in chapter 10.

Tele Vue has adapted its small telescope case into a nicely padded eyepiece tote, complete with a handle, an adjustable over-the-shoulder strap, and compartments for up to thirteen 1.25- and 2-inch eyepieces. High-density foam protects the contents from shocks and keeps the gray nylon case from sagging. As we have come to expect from Tele Vue, its eyepiece carrying bag is very well made but expensive.

Books, Star Atlases, and Periodicals

Not long ago, I read that no pastime has more new books published about it each year than amateur astronomy. Although that is bad news for us authors (too much competition!), it is good for the hobby. In fact, with so many excellent books and periodicals available, it is difficult to draw up a short list. Here is a brief listing of some of the better offerings that are in print today.

Periodicals

Even though our electronic age affords an excellent opportunity to stay informed on all late-breaking discoveries and announcements in the astronomical world, there is still something nice about going out to the mailbox every month or so and finding a copy of an astronomy magazine waiting for you. The following magazines cater to the amateur astronomer and come highly recommended. And for those readers living outside the United States, I have included several international magazines. I would strongly recommend subscribing to at least one, to stay in touch with your local universe.

Amateur Astronomy, 5450 NW 52 Court, Chiefland, FL 32626; www.amateuras tronomy.com. Quarterly aimed toward observers and amateur telescope makers.

Astronomy, P.O. Box 1612, Waukesha, WI 53187; www.astronomy.com. Very good monthly general-purpose magazine, offering both technical and observational articles, including a monthly binocular column written by yours truly.

Astronomy and Space, Astronomy Ireland, P.O. Box 2888; Dublin 5, Ireland; www.astronomy.ie. Very good monthly general-purpose magazine, offering both technical and observational articles with an emphasis on events of interest to amateurs in Ireland.

Astronomy Now, P.O. Box 175, Tonbridge, Kent, TN10 4ZY, United Kingdom; www.astronomynow.com. Very good monthly general-purpose magazine, offering both technical and observational articles with an emphasis on events of interest to amateurs in England.

Griffith Observer, Griffith Observatory, 2800 East Observatory Road, Los Angeles, CA 90027; www.griffithobs.org. Bimonthly geared more for the armchair astronomer.

Mercury, Astronomical Society of the Pacific, 390 Ashton Avenue, San Francisco, CA 94112; www.astrosociety.org. Bimonthly that features articles of general interest as well as basic articles on observing.

Sky & Space, PO Box 1690, Bondi Junction, NSW, 1355, Australia; www.skyandspace.com.au. General-purpose bimonthly magazine, offering both technical and observational articles with an emphasis on events of interest to amateurs in Australia.

Sky & Telescope, P.O. Box 9111, Belmont, MA 02178; www.skyandtelescope.com. Very good monthly general-purpose magazine, offering both technical and observational articles. An Australian edition, appropriately named *Australian Sky & Telescope,* is available for amateurs in Oz.

Sky Calendar, Abrams Planetarium, Michigan State University, East Lansing, MI 48824; www.pa.msu.edu/abrams. Not a magazine per se, but rather a monthly flyer that details naked-eye events in a calendar format.

Sky News, Box 10, Yarker, Ontario K0K 3N0 Canada; www.skynews.ca. Very good bimonthly general-purpose magazine, offering both technical and observational articles with an emphasis on events of interest to amateurs in Canada.

Stardate, 1 University Station A2100; Austin, TX 78712; www.stardate.org. Bimonthly that features articles of general interest as well as basic articles on observing.

Annual Publications

Although most of the periodicals listed above highlight monthly sky events and goings-on, it is often nice to know of things to come further in advance. For this, there are several annual publications that focus primarily on the year's major happenings. These include:

Astronomical Calendar, Ottewell, G.; Sky Publishing Corporation
Astronomical Phenomena, U.S. Naval Observatory and Her Majesty's Nautical Almanac Office; U.S. Government Printing Office
Exploring the Universe, from the editors of *Astronomy* magazine; Kalmbach Publishing
Observer's Handbook, Gupta, R., et al; Royal Astronomical Society of Canada
Sky Watch, from the editors of *Sky & Telescope* magazine; Sky Publishing Corporation

Star Atlases

Just as we need road maps to plan our summer vacations, so, too, do astronomers need star maps to find their way around the sky. The following star atlases are suitable for both telescopes and binoculars, but with certain caveats. First, note that each atlas's limiting magnitude—that is, the faintest stars plotted—is listed, as is the number of deep-sky objects. If you are new to astronomy or still learning your way around the sky, you would be best advised to stick with the simpler atlases that plot fewer objects. Those who are more accustomed to the sky should look for the more sophisticated atlases.

Cambridge Star Atlas, Tirion, W. Cambridge University Press, third edition, 2001. Stars to magnitude 6.5, 866 deep-sky objects. Recommended for beginners.
Herald-Bobroff Astroatlas, Herald, D. and P. Bobroff. Lymax, 2004. The spiral-bound atlas consists of six separate atlas series, labeled *A* through *F.* The A-series charts show the distribution of various objects across the entire sky; the B-series charts plot stars to magnitude 6.5, over 2,900 deep-sky objects; the C-series shows stars to magnitude 9.0 and more than 13,000 deep-sky objects to magnitude 14; supplemental D-, E-, and F-series charts show overly crowded regions of the sky, plotting stars as faint as magnitude 14 and deep-sky objects to magnitude 15. Highly recommended for intermediate and advanced amateurs.
Norton's Star Atlas and Reference Handbook, Ridpath, I., et al. Longman Scientific and Technical, twentieth edition, 2003. Includes stars to magnitude 6.49 and 600 deep-sky objects, as well as extensive discussions on telescopes and observing techniques. Recommended for beginning and intermediate amateurs.
Sky Atlas 2000.0, Tirion, W. Sky Publishing Corporation, second edition, 1998. Stars to magnitude 8.5, 2,700 deep-sky objects. The *Sky Atlas 2000.0* is

available in three editions: an unbound field edition (showing white stars on a black background), desk edition (black stars on white), and a spiral-bound deluxe color edition (black stars, red galaxies, etc.). Highly recommended for intermediate and advanced amateurs.

Uranometria 2000.0, Tirion, W., B. Rappaport, and G. Lovi. Willmann-Bell, 2001. This second edition includes 22 lower-detail finder charts to help orient the user, while the main charts show more than 280,000 stars to magnitude 9.75 as well as over 30,000 deep-sky objects. Sold in two volumes, which may be purchased separately. Volume one covers the northern sky from declination +90° to –6°, while the second includes +6° to –90° (the overlap is intentional). Highly recommended for intermediate and advanced amateurs.

Introductory Books

Astronomy Hacks, Thompson, R., and B. Thompson. O'Reilly Media, 2005. This "insider's book" into amateur astronomy gives a wealth of tips and tricks for you, your equipment, and your time at the eyepiece that would take years to accumulate otherwise.

Backyard Stargazer, Price, P. Quarry Books, 2005. This friendly introduction to the hobby for anyone who is brand-new to stargazing includes many suggested activities and helpful resources.

Cambridge Encyclopedia of Amateur Astronomy, Bakich, M. Cambridge University Press, 2003. Clear prose coupled with easy-to-understand illustrations make this a well-organized and user-friendly addition to any astronomy library.

Nightwatch, Dickinson, T. Firefly Books, 2006. Highly recommended for beginners. Considered by many to be the Bible for new stargazers.

Peterson's Field Guide to the Stars and Planets, Pasachoff, J., and D. Menzel. Houghton Mifflin, 1999. Recommended for beginners, although the small format makes reading the star charts difficult.

365 Starry Nights, Raymo, C. Fireside Press, 1992. Excellent night-by-night format to introduce the night sky to all.

Observing Guides

Note that while many of these guides include finder charts, most are best used with a separate star atlas.

Atlas of the Moon, Rukl, A. Sky Publishing, 2004. This is the definitive observing guide to our nearest neighbor in space. The Moon's surface is visible in exquisite detail even through the smallest telescopes, but it takes this comprehensive visitor's guide to show you which crater is which.

Burnham's Celestial Handbook, volumes 1, 2, and 3, Burnham, R., Jr. Dover, 1978. A classic reference, listing just about every celestial object visible through small- and medium-aperture telescopes. Facts and figures are becoming a bit dated.

Celestial Sampler, French, S. Sky Publishing, 2005. A collection of sixty "Small-Scope Sampler" articles penned by columnist Sue French in *Night Sky* magazine. Excellent for someone with a 3- to 4-inch telescope.

Deep-Sky Wonders, Houston, W., and S. O'Meara. Sky Publishing, 1998. Superb compilation of Walter Scott Houston's "Deep-Sky Wonders" column, which appeared in *Sky & Telescope* magazine for nearly fifty years.

Exploring the Moon through Binoculars and Small Telescopes, Cherrington, E. Dover, 1984. An excellent guide to the lunar surface, although the photo maps are a bit cramped.

Night Sky Observer's Guide, volumes 1 and 2, Kepple, G., and G. Sanner. Willmann-Bell, 1998. A monumental work that combines observations of 5,500 deep-sky objects, made by the authors as well as many contributors. Highly recommended for intermediate and advanced deep-sky observers.

Sky Atlas 2000 Companion, Strong, R., and R. Sinnott. Sky Publishing, 2000. This second edition contains data on each of the 2,700 objects plotted on the *Sky Atlas 2000.0.* Objects are itemized twice, first alphabetically, then by chart number.

Star Watch, Harrington, P. John Wiley & Sons, 2003. A star party in a book, with instructional chapters and charts on observing the Sun, the Moon, the planets, and all of the Messier objects.

Touring the Universe through Binoculars, Harrington, P. John Wiley & Sons, 1990. Are you an intermediate or advanced amateur who is looking to get the most out of your binoculars? This guide lists more than 1,100 binocular targets for both small and large glasses.

Turn Left at Orion, Cosmolmagno, G., and D. Davis. Cambridge University Press, 2000. An excellent introductory guide to the night sky for beginners, with instructions for finding 100 sky objects.

Year-Round Messier Marathon Field Guide, Pennington, H. Willmann-Bell, 1998. An excellent guide to the Messier objects, complete with maps and directions. Organized by area of sky rather than numerically.

Astrophotography

The Art and Science of CCD Astronomy, Ratledge, D., et al. Springer Verlag, 1999. Excellent, though somewhat dated, introduction to this ever-advancing aspect of astrophotography.

Astrophotography: An Introduction to Film and Digital Imaging, Arnold, H. Firefly Books, 2003. Well organized, written at a level that can be appreciated by both amateur photographers who are unfamiliar with astronomy and amateur astronomers who are unfamiliar with photography.

Astrophotography for the Amateur, Covington, M. Cambridge University Press, second edition, 1999. The best general-purpose book on film-based astrophotography in print today. Includes some discussion on digital imaging, although this is not the book's strength.

Digital Astrophotography: The State of the Art, Ratledge, D., et al. Springer-Verlag, 2005. A rapidly changing field is discussed by a collection of experts who have authored and assembled this anthology that discusses digital equipment, techniques, and image processing.

Introduction to Digital Astrophotography, Reeves, R. Willmann-Bell, 2004. A comprehensive introduction to digital astro-imaging featuring detailed images, illustrations, and screenshots that will be valuable for both novice and expert astrophotographers.

The New CCD Astronomy, Wodaski, R. Self-published, www.newastro.com, 2001. This comprehensive book, arguably the best on the subject, includes a one-year subscription to the author's Web site, which includes a wealth of extras. Good, practical instruction throughout.

Photoshop Astronomy, Ireland, R. Willmann-Bell, 2006. A detailed book that helps astrophotographers master the powerful tools in Adobe's Photoshop software. Includes a tutorial DVD.

Wide-Field Astrophotography, Reeves, R. University of Alabama Press, 1999. Practical guide that will be a great help for beginners interested in trying some basic astrophotography.

Telescopes, Optics, and Telescope Making

Amateur Telescope Making, Tonkin, S., et al. Springer-Verlag, 1998. Great reference for those looking to make a variety of astronomical projects. Each chapter is written by a different author who has actually designed and built the project discussed.

Backyard Astronomer's Guide, Dickinson, T., and A. Dyer. Firefly Books, 2002. Excellent guide to the aesthetics of backyard astronomy along with a good dose of practical advice.

Build Your Own Telescope, Berry, R. Willmann-Bell, 2001. Contains complete plans and photos for building several excellent telescopes at home.

Choosing and Using a Schmidt-Cassegrain Telescope, Mollise, R. Springer-Verlag, 2001. The bible for owners of this popular breed of instrument, written by one of the subject's leading authorities.

The Dobsonian Telescope: A Practical Manual for Building Large Aperture Telescopes, Kriege, D., and R. Berry. Willmann-Bell, 1997. Required reading for anyone who owns a large-aperture Newtonian reflector.

Making and Enjoying Telescopes, Miller, R., and K. Wilson. Sterling, 1997. Like the *Build Your Own Telescope,* this book contains complete plans and photos for building several telescopes.

Making Your Own Telescope, Thompson, A. Dover, 2003. A classic book that has probably led tens of thousands to make their own optics and telescopes since it was first published half a century ago. Easy to follow and understand.

Perspectives on Collimation, Menard, V., and T. D'Auria. Self-published, 1993. The most thorough treatment of this critical topic. Order directly from the authors (Vic Menard, 2311 23rd Avenue, West Bradenton, FL 34205 or Tippy D'Auria, 1051 NW 145th Street, Miami, FL 33168).

Scientific American: The Amateur Astronomer, Carlson, S., ed. John Wiley & Sons, 2000. A compilation of the best amateur astronomy projects published over the years in *Scientific American* magazine. If you're a tinkerer, this is a must-have.

Star Testing Astronomical Telescopes, Suiter, H. Willmann-Bell, 1994. Although written at a fairly high level, this book offers a thorough treatment of this very revealing topic.

Cloudy-Night Reading

Alvan Clark and Sons: Artists in Optics, Warner, D., and R. Ariail. Willmann-Bell, 1995. The story of nineteenth-century America's first family of telescope makers.

The Planets, Sobel, D. Viking, 2005. The author of best sellers *Galileo's Daughter* and *Longitude,* Dava Sobel has created an eclectic tour of the solar system that views each world from a perspective that other authors have never even considered.

Seeing in the Dark, Ferris, T. Simon & Schuster, 2003. Entertaining book profiling some of today's most prolific amateur astronomers.

Starlight Nights, Peltier, L. Sky Publishing, 1999. The autobiography of America's most famous amateur astronomer. If anyone ever asks why you are interested in astronomy, tell them to read this book. No one has ever expressed the fascination we all feel as eloquently as Leslie Peltier.

The Electronic Age

As with just about every other aspect of our lives, amateur astronomy is becoming more sophisticated thanks to tremendous advances in electronics and computerization. Here is a short review of some of the more popular astronomical software and electronic gadgets on the market today.

Computer Software

Astronomical software can be divided into three basic categories based on their intent: observing programs (which include planetarium simulations and, often, telescope motion control), image-processing programs (typically used in conjunction with CCD cameras but may also be used to enhance scanned-in conventional photographs), and special-purpose programs (covering a wide range of subjects, from optical design to predicting eclipses). Since most amateur astronomers are interested primarily in observing the sky, I have restricted the listing here to some of the best observing programs.

Many of the programs detailed here can be purchased directly from their companies via the Internet. Most also offer downloadable demonstration versions that give users a chance to try before they buy.

AstroPlanner (CD-ROM; Windows or Macintosh), iLanga, Inc. A program that helps an observer set up a night's worth of astronomical targets, steer a compatible GoTo telescope to each, and then log what was seen. Outside, the field of view of the telescope is displayed in the form of a star chart, making it easy to compare with the actual view through the telescope.

Cartes du Ciel (free download from www.astrosurf.org/astropc). A free charting program created by French amateur astronomer Patrick Chevalley that should be on every amateur's Windows computer. CdC plots millions of stars and deep-sky objects, and combines complete planetarium function with exceptional charting capability.

Earth-Centered Universe (CD-ROM; Windows), Nova Astronomics. From David Lane in Canada, ECU graphically plots all objects in its database of over 15 million stars, the planets, Sun, Moon, comets, over 19,000 asteroids, and more than 10,000 deep-sky objects.

Deepsky Astronomy Software (CD-ROM; Windows), Steven S. Tuma. An integrated planetarium/telescope control program as well as searchable spreadsheet database listing more than 18 million stars and 426,000 deep-sky objects.

Guide (CD-ROM; Windows), Project Pluto. An excellent program for the deep-sky observer at a terrific price! Guide plots over 18 million stars, over 20,000 asteroids, hundreds of comets, and 75,000 deep-sky objects from the Messier, NGC, IC lists as well as many comparatively unknown catalogs.

MegaStar (CD-ROM; Windows), Willmann-Bell. I consider MegaStar to be the best program available for the diehard deep-sky observer. MegaStar plots over 15 million stars to magnitude 15, 110,000 deep-sky objects, and more than 13,000 asteroids and comets. Printed charts also cross-reference chart numbers in the Uranometria 2000 atlas.

SkyMap Pro (CD-ROM; Windows), SkyMap Software. Many consider this one of the best planetarium-type software packages available. Created by British amateur Chris Marriott, SkyMap Pro plots more than 15 million stars as faint as magnitude 15 and more than 200,000 deep-sky objects, as well as over 11,000 asteroids and comets. Updated asteroid and comet catalogs can be downloaded free of charge from the SkyMap Web site.

Starry Night (CD-ROM; Windows or Macintosh), Imaginova. One of the most widely acclaimed astronomy programs offered, Starry Night is available in five levels. The Pro version accurately plots more than 17 million stars and deep-sky objects and displays the sky from any planet or moon in the solar system at any time. The program's ability to aim and control computerized telescopes, however, is more limited than some others listed here.

TheSky (CD-ROM; Windows or Macintosh), Software Bisque. *TheSky* is one of the finest sky-simulation, star-charting, and telescope-control programs available. With this program, you can show the Sun, the Moon, and the planets against a myriad of background stars for any time, any date, any place on Earth. GoTo telescope control is excellent, but some simple functions prove unnecessarily difficult. For instance, users must scroll through multiple dialog boxes just to change the date and time.

Voyager (CD-ROM; Windows or Macintosh), Carina Software. An outstanding sky simulation program, now available for both Windows and Macintosh users. Voyager displays 259,000 stars; 50,000 deep-sky objects; and many comets, asteroids, and Earth-orbiting satellites. Animation compresses time to show eclipses or planets orbiting the Sun. Computer control is available using the optional Sky Pilot plug-in.

Digital Setting Circles

One of the biggest complaints that amateur astronomers have had for years is how useless the setting circles are that come with many equatorial mounts. Because of their small size and gross calibration, they are little more than decoration. All this changed with the invention of electronic digital setting circles. Digital setting circles (DSCs) make it possible to aim a telescope accurately to within a small fraction of a degree of hundreds or thousands of objects that are listed in their built-in data libraries. And forget polar aligning (although it is still usually required to use the clock drive); just aim the telescope at two of the many stars stored in memory, tell the unit which ones they are, and the built-in computer calculates the rest. This ease of operation means that the setting circles can be attached to just about any kind of telescope mount—either equatorial or alt-azimuth. Don't know what to look at? That's fine, because many of these units come with a whole catalog of thousands of sky objects from which to choose. Simply move the telescope until the LED prompt announces that you have hit the preselected target, look in the eyepiece, and there it is!

Each unit listed in Table 7.5 can be mounted unobtrusively to just about any telescope. Two encoders attach to the instrument's axes and are connected to the "brain" of the outfit by a pair of thin wires. Once properly secured and calibrated, the digital setting circles will automatically keep track of the passage of time as well as where the telescope is aimed. The LED typically reads to within an accuracy of 10 arc-minutes in declination and 1 minute of right ascension.

The biggest differences between the models lie in their onboard libraries of objects. The simplest units contain data on slightly more than 200 objects, while the most advanced include data on more than 12,000 sky sights. The latter can also be connected to a personal computer, typically via an RS-232 serial port, which allows the observer to link to various software packages like *Megastar, SkyMap Pro,* or *TheSky.*

Each of these computerized telescope-aiming systems require that a pair of encoders be mounted to the telescope's axes. Some come with them, while others sell them separately. Keep that in mind when checking prices. Many also offer customized packages for attaching the encoders to older fork-mounted Celestron or Meade Schmidt-Cassegrains simply by unscrewing and replacing one screw on both axes. Other telescopes, however, may not have it

Table 7.5 **Digital Setting Circles**

Company/Model	Objects		Catalogs	RS-232 port
	Deep-sky Objects*	User Defined		
Jim's Mobile, Inc.				
NGC-microMAX	200	28	M, more	No
NGC-miniMAX	3900	28	M, some NGC and IC, more	No
NGC-MAX	12000	28	M, NGC, IC, more	Yes
Lumicon				
Sky Vector I	200	0	M, some NGC, more	No
Sky Vector II	3500	25	M, most NGC, some IC, more	No
NGC Sky Vector	12000	25	M, NGC, some IC, more	Yes
Sky Engineering				
Sky Commander	9000	120	M, NGC, IC, more	Yes
Vixen				
SkySensor 2000	15000	60	M, NGC, IC, more	No
Wildcard Innovations				
Argo Navis†	29000	1100	M, NGC, IC, more	Optional

*Rounded to the nearest 100.
†The Argo Navis system is also available from Jim's Mobile, Inc., (JMI) as the NGC-superMAX.

CONSUMER CAVEAT: Digital Setting Circles

Here are two warnings about digital setting circles. First, watch those object numbers in Table 7.5. They do not necessarily mean unique objects. Some manufacturers count the same object more than once if it belongs to more than one of the catalogs in their database. For instance, all but a couple of the Messier objects are also listed in the New General Catalog (NGC). Not fair counting them twice, but some do. Be sure to find out which catalogs are listed. At the very least, I would only recommend units that include all of the objects in the Messier and NGC lists.

A second caveat has to do not with the DSCs themselves but rather the telescope to which they will be attached. I have corresponded with several unhappy amateurs who went through the toil of attaching multiple brackets, encoders, and assorted hardware to their telescopes only to find that the digital setting circles didn't work correctly. The fault ultimately was found to lie not with the DSCs but with the telescope mount. The onboard calculations assume that the mount's axes are perfectly perpendicular to one another. If they are not, possibly due to warped wood or sloppy assembly, aiming accuracy will be seriously hampered.

so easy, requiring some drilling and tapping to fit the encoders to their mountings. I strongly urge you to contact the manufacturer to find out what is required to attach the encoders to your particular telescope before you purchase any of them.

Of those listed, the Argo Navis from Australia's Wildcard Innovations rises above the rest in terms of sophistication. Part of that sophistication lies in the unit's design, which includes not one, but two microprocessors for improved pointing accuracy. The built-in database, which includes information on more than 29,000 objects and can accept up to 1,100 user-defined objects, is the largest of any digital setting circle sold today. The Argo also has a built-in heater to ensure that the two-line LCD readout can be seen even in cold temperatures, addressing a problem that other units cannot. Finally, the Argo can interface with many charting programs, including TheSky6, SkyMap Pro, Cartes du Ciel, and MegaStar, as well as StellarCAT's ServoCAT and Starmaster's StarTracker GoTo controllers. Encoder adapter kits are available for more than thirty brands of telescopes.

Astrophotography Needs

One of amateur astronomy's most popular, and most expensive, pastimes is trying to capture the night sky on film. As most soon discover, however, there is a lot more to taking a good picture of the night sky than one might suspect at first. Patience is the most important requirement of the astrophotographer, followed closely by equipment. It's tough to bottle the former for sale, but there are lots of companies looking to sell you the latter!

Here's a look at some of the equipment available today to tempt you into the world of astrophotography.

Cameras

Naturally, without a good camera, astrophotography is impossible. But there are so many cameras sold, which is best? That depends on where your interests lie. Today's astrophotographic camera market is divided into two major segments based on media—film-based and digital—with several subdivisions within each. Let's try to break it down further.

Film Cameras. Of all the styles of film cameras sold today, 35-mm single-lens reflex (SLR) cameras are the most popular for astrophotography. Single-lens reflexes allow the photographer to look directly through the lens of the camera itself, a critical feature for aligning the image, especially when photographing through a telescope (with most other cameras, the photographer is viewing through a separate viewfinder).

But times change. Once the very backbone of astrophotography, SLRs are no longer the only game in town. Part of the problem lies with the cameras themselves. For astrophotography, a camera must have a removable lens with a manually adjustable focus, provisions for attaching a cable release to the camera and the camera to a tripod, a manually set mechanical shutter with a "B" (bulb) setting, mirror lockup, and interchangeable focusing screens. Unfortunately, few of today's 35-mm SLRs fit this bill. In an attempt to attract more weekend photographers, most camera manufacturers offer cameras with automatic everything—from focus to exposure to flash control. All of these features are nice for taking pictures of the family picnic, but they are of no use to astrophotographers. The long exposures required for astrophotos (usually measured in minutes, even hours), however, will quickly drain the power from expensive camera batteries. When that happens, the camera shuts down and becomes useless until a fresh set of batteries is inserted.

Which cameras are best for astrophotography? Table 7.6 lists several alternatives, both past and present.

Expensive does not necessarily mean better for astrophotography. All of these cameras will work well for wide-field constellation shots as well as through-the-telescope photos of the Moon and the Sun (the latter requiring safety precautions outlined earlier in this chapter), but their differences will become more apparent when taking long-exposure telescopic shots. Here, the benefit of interchangeable focusing screens and mirror lockup will become apparent. Most subjects photographed through telescopes are very faint, making it difficult to line up and focus the shot when viewing through most standard focusing screens. A simple ground-glass screen will provide the brightest

Table 7.6 **Suggested Film Cameras for Astrophotography**

Model	Operates without Batteries	Manual Focus	Interchangeable Focus Screen	Mirror Lockup
Today's Best				
Nikon FM3A	Y	Y	Y	N
Nikon FM10	Y	Y	N	N
Promaster 2500PK	Y	Y	N	N
Vivitar V3800N	Y	Y	N	N
A Few of Yesterday's Best				
Canon F-1	Y	Y	Y	N
Canon FTb	Y	Y	N	Y
Minolta SRT-101*	Y	Y	N	Y*
Nikon F3HP	Y	Y	Y	Y
Nikon FM2N	Y	Y	Y	Y
Nikon F	Y	Y	Y	Y
Olympus OM-4TI	Y	Y	Y	N
Olympus OM-2	Y	Y	Y	Y
Olympus OM-1	Y	Y	Y	Y
Pentax LX	Y	Y	Y	Y
Pentax K1000	Y	Y	N	N

*I thought that all Minolta SRT-101s had mirror lockup until I bought a used one recently that did not. Be sure to check before purchasing.

possible images, a great aid in focusing and composing. Mirror lockup is recommended for reducing *mirror slap,* which occurs every time the shutter is tripped and the camera mirror pivots out of the way. Swinging the mirror out of the way before the shutter is opened eliminates most vibration, reducing the chances for blurred images.

Consumer Digital Cameras. The biggest single change in our hobby of astronomy since the first edition of *Star Ware* was published in 1994 is not in the area of telescopes or eyepieces or binoculars. The biggest change—literally an upheaval of astronomical proportions—is in the area of digital astrophotography. "Way back then," digital astrophotography was the sport of a few, well-to-do amateurs using expensive CCD imagers coupled to state-of-the-art computers set up next to their electronically erudite telescopes.

Those telescopes and imagers are still alive and well, and detailed later in this section. But the revolution that I'm referring to here has to do with the advances made in inexpensive digital cameras. Today, for less than $100, you can buy a digital camera that you can use to take some amazingly clear images of the Moon and the Sun (with proper filtration) through just about any telescope listed in chapter 5. All it takes is a little practice.

Trying to list specific camera models here, however, is impractical. There are so many suitable cameras out there that listing them all would fill dozens of pages. Instead, let's look at a few general guidelines outlining features needed for astrophotography, and wind up with a few specific suggestions.

You will often see a camera's resolution referred to as being so many *megapixels.* Rather than a piece of film capturing and retaining photons of light, a digital camera uses a silicon chip made up of thousands of light-sensitive areas called *pixels,* which convert light into electrons during an exposure. At the end of the exposure, each pixel converts its stored electrons into a digit. Each value is then converted into a color that is ultimately combined into an image. A camera advertised as, say, 3.2 megapixels typically has a silicon chip measuring 2,048 pixels by 1,536 pixels (2,048 times 1,536 equals 3,145,728 pixels, rounded up to 3.2 megapixels).

In general, the higher the number of megapixels, the better the results. In today's market, you would be hard-pressed to find a camera with fewer than 3.2 megapixels. That should be set as a minimally acceptable value, since fewer will cause images to break down into a series of tiny squares (the pixels themselves) when enlarged. Images captured with cameras rated at 5, 6, or more megapixels can be enlarged more before a picture *pixelates.* By comparison, a piece of 35-mm film has an equivalent resolution between 2 and 16 megapixels depending on the film's speed.

A digital camera's settings should also be manually adjustable. The user must be able to override a camera's automatic shutter speed, aperture, focus, ISO speed setting, and flash in order to take successful astrophotos. Many lower priced point-and-shoot digital cameras do not have this flexibility, and so should be avoided.

Although it is possible simply to hold a digital camera up to the eyepiece of a telescope and snap a shot of the Moon that way, it's far easier to use a threaded adapter. More about those later, but first, make sure that the camera you are considering can accept threaded lens accessories. Also make sure that the camera has a self-timer or, better still, a wireless remote shutter control. Using either means that you won't be touching the camera when the shutter opens, lessening the chance of blurring. Finally, check for a threaded tripod socket on the bottom of the camera.

Cables to connect the camera to your computer and two sets of rechargeable batteries are also musts, although you'll likely have to buy the second set separately. You will also probably need to buy a memory card (or two), because most cameras come with minimal or no cards. Size matters here, since the larger the card's capacity, the greater the storage capacity. Most cameras have several resolution settings, but in general, you'll want to shoot at the highest, such as RAW or Hi JPEG, in order to record the finest detail. For example, using a lower resolution or *compressed* setting could store 200 or more shots on a single 128-megabyte card, while in *raw* mode, it might only be able to hold 15 or 30 images.

Consumer digital cameras can be broken into two subcategories: digital compacts and digital single-lens reflex (DSLR). Things to look for in a digital compact camera include an LCD viewscreen on the back of the camera. This is an absolute necessity for composing the shot and checking focus. The lens itself should be physically small in diameter. Although this sounds counterintuitive, smaller lenses vignette less and allow the use of a wider variety of eyepieces when shooting through a telescope. This will let you zoom in less when trying to eliminate vignetting (some vignetting is almost inevitable with these cameras). Finally, speaking of zooming, it is much easier to frame and eliminate vignetting using a camera with an internal zoom lens rather than one with a lens housing that moves outward from the body when zoomed.

Some of the better digital compact models these days are found among Nikon's Coolpix 800 and 900 series, the Olympus C series, Canon Powershot G series, and Sony Cyber-shot DSC series. Of course, this hot market is not likely to stay still long, which means that these models will probably be superseded quickly by newer, better, and sexier cameras.

Although more expensive, DSLRs enjoy many advantages over point-and-shoot cameras, including direct, through-the-lens viewing and removable lenses. This means they can be hooked up directly to a telescope in the same way that a 35-mm SLR can, by using commonly available adapters. Shutters are still electronically controlled, however, so expect batteries to need recharging or replacement at least once during a long night.

The majority of DSLRs also have a *35-mm lens factor,* the result of their sensor chip being smaller than a 35-mm film frame. For example, the Canon Digital Rebel XT has a lens factor of 1.6. This means that if it were hooked up to a telescope with a 1,000-mm focal length, the effective focal length would be 1,600 mm. Focal ratio remains unaffected.

Among DSLRs, the 10.1-megapixel Canon Digital Rebel XTi leads the bunch in terms of value for the dollar. Nikon's 10.2-megapixel D200 is also an exceptional camera. Higher-end models with strong consideration include the Canon 20D (see Figure 7.6) and 30D (both use the same 8.25 megapixel sensor,

Figure 7.6 *One of the finest digital single-lens reflex cameras is the Canon 20D, shown here with a universal camera-to-telescope adapter. Photo by Brian Kennedy.*

the Pentax K10D [10.2 megapixels]), and the Fujifilm FinePix S3 Pro (stated at 12.3 megapixels, but image quality appears comparable to 7 or 8 megapixels).

Webcams. Looking to get into lunar and planetary imaging on the cheap? Then a webcam is the only way to go. Although they must be connected to a computer right at the telescope, inexpensive webcams are producing some amazing results. The user takes a short burst of video exposures with the webcam, and then uses specialized software to extract the sharpest frames in the stream while eliminating the rest. The final image is then produced by stacking the best individual frames.

There are many webcams out there today, but only a few are suitable for astrophotography. The Phillips ToUCam Pro II (also sold as the ToUCam 840k) is the best of the best as this is written, although it is comparatively hard to find here in the United States. It's worth the search, however, since it uses a CCD chip rather than a comparatively insensitive CMOS chip. Other webcams that are suitable for astrophotography include Logitech's Quickcam 4000 and the 3Com Home Connect. All connect to a computer via a USB cable.

It takes more than a camera and computer to be a successful webcam astrophotographer, however. Like the other types of cameras discussed earlier, none of these webcams are designed to be mounted directly to a telescope without an adapter. Most webcam adapters look like empty 1.25-inch eyepiece barrels that screw into the tiny threads of the webcam lens. A Barlow lens or the Tele Vue Powermate will also be required to increase image size. Finally, an infrared filter must be screwed into the adapter to compensate for the filter that was removed with the camera's lens to attach the adapter in the first place. Astronomik in Germany manufactures the requisite infrared filters, while an especially popular adapter comes from Australia's Webcaddy.com.au.

Grabbing the image is only half the battle, however. The real magic is in the image processing. This typically involves using a computer program to analyze, isolate, align, and "stack" the sharpest frames from a short series taken through the telescope. There are several programs available to do this; one that is especially popular is a freeware program called *RegiStax* created by amateur astronomer Cor Berrevoets. Find it on the Web at registax.astronomy.net (note: no www in the URL). Another favorite program is *K3CCD* by Peter

Katreniak. You can download a free trial version at www.pk3.org/Astro. Other popular imaging applications include *Iris* (www.astrosurf.org/buil/us/iris/iris .htm) by Christian Buil, *Registar* by Auriga Imaging (www.aurigaimaging .com), and Axel Canicio's *Astro-Snap* (www.astrosnap.com). You'll find updated links to each of these programs in the *Star Ware* section of my Web site, www.philharrington.net.

Recently, Meade, Celestron, and Orion Telescopes have gotten into the modified webcam business. The Meade Lunar Planetary Imager (LPI) is a complete package that comes with the camera itself, as well as image-processing software, a too-short USB cable, adapters, and a parfocalizing ring, all for less than $200. The beauty of the LPI, shown in Figure 7.7, is that it's all in one package; no need to hunt down and purchase the components needed from several different sources. The LPI is also designed with astro-imaging in mind, so it has some nice touches, like the parfocalizing ring. Put the ring onto any eyepiece's barrel, insert it into your telescope's focuser, aim at a bright star, and bring it into sharp focus. Without moving the telescope, remove the eyepiece/ring combination and reinsert the LPI. The image is still in focus.

Overall results with the LPI are quite good, although not as good as the league-leading ToUCam Pro II. The difference lies in the ToUCam's superior light sensitivity, which allows for shorter exposures. Even though the Meade runs at 0.25-second per exposure, that time is still long enough for atmospheric turbulence to blur the image. The ToUCam freezes the action at 0.04-second per exposure. However, the LPI can take single exposures as long as 15 seconds, while the ToUCam is limited to only 0.04 second, which means that the LPI can also dabble into the world of deep-sky photography.

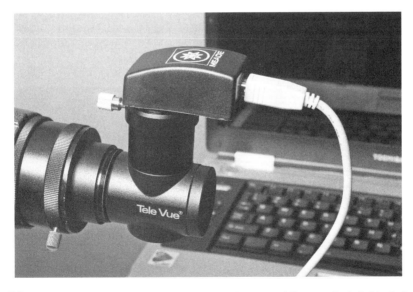

Figure 7.7 *The Meade Lunar Planetary Imager (LPI) is one of the smallest, lightest digital astro-imagers sold today. Photo by Brian Kennedy.*

Orion's StarShoot Solar System imager also uses a CMOS chip, which in this case, measures 0.33 inch across and is rated at 640 × 480 pixels. Given today's high-resolution computer monitors, that is going to leave some people wanting more. The images are fine for Web posting and e-mailing to friends, but it is unlikely that you'll be able to make enlargements without blurring the images due to pixelation. Exposures can range in length from .001 second to 0.5 second. Like the LPI, the StarShoot connects to a computer via the hard-wired USB cable that sticks out of the cylindrical plastic body. Also included in the deal is *MaxIm DL Essentials Edition* image-processing software. With the software installed on a PC that can accept a USB 2.0 connection, images can be displayed in real time or captured for later stacking or processing. A parfo-calizing ring is not part of the package, so focusing must be done by viewing the computer screen.

Finally, there's Celestron's NexImage. How does it stack up against the LPI, StarShoot SSI, and ToUCam Pro II? It certainly has the right stuff. For openers, the NexImage uses the Sony ICX098BQ color CCD chip, the same as in the ToUCam. The NexImage also includes a CD-ROM with RegiStax image-processing software, but like the StarShoot SSI, no parfocalizing ring. Cele-stron is, however, the only one to offer a screw-on reducer lens to double the imager's field of view. Overall, NexImage users report very good results on Jupiter, Mars, Saturn, and the Moon. Most agree that images are comparable to those captured by the venerated ToUCam, and have better color fidelity than the CMOS chips. Images are sharp and clear once the good frames are weeded out from the bad and blurry, and then combined with RegiStax. Some owners comment, however, that exposure control is tough to master due to the limits of the included video capture software, but still, it proves the best of the bunch for convenience and image quality.

CCD cameras. Finally, we come to the pièce de résistance in the world of astrophotography: thermally cooled, supersensitive charge-coupled devices, or

---

CONSUMER CAVEAT: Webcams and CCD Cameras

Although the images we see taken through webcams and dedicated CCD astrocameras are truly amazing, they are the result of very patient, very tal-ented amateur astronomers who have taken months, even years to develop their craft. One of the first things most beginning webcammers complain about is the difficulty involved in getting the target into view. The problem is that the chips inside webcams, as well as in most CCD cameras dis-cussed in the next section, are very small. Depending on your telescope's focal length, it's almost like trying to thread a needle while wearing a pair of gloves. Even if you are able to get the target into view, it will whoosh out of the field very quickly unless the telescope is tracking the sky. That's not to say that images of the Moon and the planets can't be taken with an un-guided telescope, but you need to have an infinite supply of patience.

CCDs for short. These make it possible to capture deep-sky objects using exposures many times shorter than is possible with conventional film.

But there is much more to astrophotography than just buying a CCD camera and hooking it up to your telescope. Besides having to buy a computer as well as a sturdy equatorial mounting, you also need to have some idea of what kind of imaging you are interested in doing. The ideal setup for planetary work may not necessarily be appropriate for deep-sky imaging, and vice versa. Planetary images are typically bright and small, requiring short exposures. The amount of sky coverage is usually not critical, but fine resolution is. Deep-sky objects, on the other hand, may cover wide expanses and are comparatively very dim, requiring larger fields and longer exposures.

Earlier, the statement was made that the higher the number of pixels, the better the results. Although that is still true, the size of those pixels must also be taken into account. Pixel size is always measured in microns (μ), where 1 micron = 0.001 millimeter, and varies from less than 5μ to more than 20μ. You need to strike a balance between the two competing factors, with the ideal telescope-CCD combination depending on what you are imaging. The larger the pixel, the more sensitive the camera will be in a given telescope. The smaller the pixel size, the less sensitive the camera, which necessitates longer exposures. Think of it like selecting the right film for the subject. The planets and the Moon are so bright that they need only very short exposures, while deep-sky objects need considerably longer exposures. For lunar and planetary photography, you would choose a slower speed film, since it has finer grain and better resolution. Although resolution is still an important factor in deep-sky photography, film speed is usually the overriding factor when selecting film. In general, the optimal deep-sky setup will have each pixel on the CCD chip covering between 1.5 and 2.5 arc-seconds of sky, while sky coverage between 0.4 and 0.7 arc-second produces the best planetary results. To find out just how much sky will be covered by each pixel, plug numbers into the following formula:

Image scale = (pixel size in microns ÷ focal length of telescope in mm) × 206.27

or

Image scale = (pixel size in microns ÷ focal length of telescope in inches) × 8.12

For instance, let's consider which CCD chip is best for planetary imaging through an 8-inch f/10 SCT, with a focal length of 2,032 mm. Because planets are small-scale objects, image detail is more important than a large field of view. Without getting into a long discussion of Nyquist sampling theory, suffice it to say that the general rule is that pixel resolution should be twice that of your telescope. Therefore, plugging numbers into the previous formula shows that a chip with pixels measuring between 5 and 7 microns corresponds to between 0.6 and 0.7 arc-second per pixel, which is a good range. This would also work well for small-scale deep-sky objects, such as planetary nebulae. Exposure time, however, would need to be comparatively long.

As exposure grows, so does the background electronic noise, or dark current, generated by the CCD camera. Although not a concern if exposures are restricted to the Moon and the planets, dark current poses a problem with the long exposures required for deep-sky imaging. Electronic noise can be reduced by cooling the CCD chip, but some chips require more cooling than others to bring this noise level down to workable levels.

To decrease exposures, astrophotographers use a technique called *binning.* Binning effectively increases pixel size while also increasing sensitivity. By digitally combining a two-by-two square of pixels into a single pixel, only one-quarter of the exposure time is needed to record an image (since binning is combining the energy recorded by four pixels into one). Of course, resolution is also decreased proportionally. Binning 3×3 increases sensitivity by a factor of 9.

The field of view with that same 8-inch f/10 telescope-camera combination would cover only a total field between 7 and 15 arc-minutes, which is too small for most deep-sky objects. For deep-sky imaging, a total field of 25 to 60 arc-minutes is preferred, with a pixel image scale in the range of 2 to 3 arc seconds. Here, a chip with pixels measuring 9 to 14 microns would serve well. Alternatively, a telecompressor or reducer/corrector could be used to shrink the telescope's focal ratio between 37% and 50%, decreasing exposure times and also expanding the field proportionally, albeit with some vignetting possible.

Some CCD chips are listed as 8 bit, some as 12 bit, and others as 16 bit. These numbers refer to the analog-to-digital (abbreviated A/D) converter, which transforms the signal stored in each pixel on the CCD chip into a digital value that corresponds to a specific shade of gray. This value is then combined with all the other pixel values to create the image. Simply put, the larger the number, the greater the dynamic range in image brightness, and, therefore, the better the image quality. Cameras with 8-bit A/D conversion are few and far between these days, and should be avoided. A 12-bit converter has 4,096 shades of gray, while a 16-bit has 655,535 levels. With the greater range comes the need for more computer memory, so make your choice to match your needs and equipment.

Antiblooming is another important feature to consider when selecting a CCD. Blooming occurs when a bright object oversaturates the chip, causing an excess charge to bleed down a column of pixels. The result is a bright shaft of light above and below the offending star. Most chips nowadays feature antiblooming, although it can also reduce the chip's sensitivity and amplify dark current.

Table 7.7 lists many of today's most popular CCD chips, including their dimensions, size of pixels, and which imagers they can be found in.

Not every telescope is CCD-friendly. The focuser, for instance, is critical. It must be solid enough to support the CCD imager without changing its position. Ideally, it should have a locking thumbscrew, just to make sure that everything stays put. Also, it must not shift when that thumbscrew is locked down. Even a slight movement will blur the image, leading only to wasted time and ever-growing frustration.

Table 7.7 **Popular CCD Chips**

Sensor Chip	Pixel Array Size (mm)	Pixel Size (microns, μ)	Imagers*
E2V CCD47-10	1024 × 1024	13.0	FLI MaxCam CM2-1
E2V CCD77	512 × 512	24.0	FLI MaxCam CM1-1
Kodak KAF-0261E	512 × 512	20.0	Apogee KX260E, FLI MaxCam CM9, SBIG ST-9XE
Kodak KAF-0401E	768 × 512	9.0	Apogee KX1E, FLI MaxCam CM7
Kodak KAF-0402ME	765 × 510	9.0	SBIG ST-402ME, SBIG ST-7XME, ST-7XMEI, FLI MaxCam ME2
Kodak KAF-1001E	1024 × 1024	24.0	Apogee AP6E, FLI IMG1001E
Kodak KAF-1401E	1340 × 1037	6.8	Apogee KX14E, FLI IMG1401E
Kodak KAF-1602E	1536 × 1024	9.0	Apogee KX2E, FLI MaxCam CM8
Kodak KAF-1603ME	1530 × 1020	9.0	SBIG ST-8XME, SBIG ST-8XMEI
Kodak KAF-16801E	4096 × 4096	9.0	Apogee AP16E, FLI IMG16801E
Kodak KAF-3200E	2184 × 1472	6.8	Apogee KX32ME, FLI MaxCam CM10, SBIG ST-10XE, SBIG ST-10XME
Kodak KAF-4200	2048 × 2048	9.0	Apogee KX4
Kodak KAF-4202	2048 × 2048	9.0	Apogee AP4, FLI IMG4202
Kodak KAF-4300E	2084 × 2084	24.0	FLI IMG4300E
Kodak KAF-6303E	3072 × 2048	9.0	Apogee AP9E, FLI IMG6303E, SBIG STL6303
Kodak KAI-2020M	1600 × 1200	7.4	SBIG ST-2000XM
Kodak KAI-11000M/CM	4008 × 2672	9 × 9	SBIG STL-11000M/CM
Micron MT9M001	1280 × 1024	5.2	SAC4-IID
Sony ICX055BK SuperHAD	500 × 290	9.8 × 12.6	Starlight Xpress MX5C
Sony ICX055BL SuperHAD	500 × 290	9.8 × 12.6	Starlight Xpress MX516
Sony ICX083AL SuperHAD	752 × 580 (high-res mode) or 376 × 290 (high-sensitivity mode)	11.6 × 11.2 (high-res mode) or 23.2 × 22.4 (high-sensitivity mode)	Starlight Xpress MX916
Sony ICX084AL HyperHAD	660 × 494	7.4	Starlight Xpress HX516
Sony ICX085	1300 × 1030	6.7	Apogee KX85
Sony ICX098AK SuperHAD	640 × 480	5.6	SAC7
Sony ICX249AK EXview	752 × 580	8.6 × 8.3	Starlight Xpress MX7C
Sony ICX249AL EXview	752 × 580	8.6 × 8.3	Starlight Xpress MX716

(continued)

*Table 7.7 **(continued)***

Sensor Chip	Pixel Array Size (mm)	Pixel Size (microns, μ)	Imagers*
Sony ICX254AL ExView HAD	640 × 480	9.6 × 7.5	SAC8.5
Sony ICX259AK ExView HAD	752 × 582	6.5 × 6.25	Orion StarShoot Deep-Space Imager
Sony ICX262AQ SuperHAD	2140 × 1560	3.45	SAC10
Sony ICX285AK Exview HAD	1392 × 1040	6.45	Starlight Xpress SXV-H9C
Sony ICX285AL Exview HAD	1392 × 1040	6.45	Starlight Xpress SXV-H9
Sony ICX405AK SuperHAD	500 × 580	9.8 × 6.3	Starlight Xpress SXV-M5C
Sony ICX405AL SuperHAD	500 × 580	9.8 × 6.3	Starlight Xpress SXV-M5
Sony ICX406AQ SuperHAD	7225 × 5375	3.125	Starlight Xpress SXV-M8C
Sony ICX413AQ SuperHAD	3024 × 2016	7.8	Starlight Xpress SXV-M25C
Sony ICX423AL SuperHAD	752 × 580	11.2 × 11.4	Starlight Xpress SXV-M9
Sony ICX424AL SuperHAD	660 × 494	7.4	Starlight Xpress SXV-H5
Sony ICX429AK EXview	752 × 580	8.4 × 8.2	Starlight Xpress SXV-M7C
Sony ICX429AKL EXview HAD	752 × 582	8.3 × 8.2	Meade Deep-Sky Imager II
Sony ICX429AL EXview HAD	752 × 580	8.4 × 8.2	SAC9, Starlight Xpress SXV-M7
Sony ICX429ALL EXview HAD	752 × 582	8.3 × 8.2	Meade Deep-Sky Imager Pro II
Sony SuperHAD (chip unspecified)	510 × 492	9.6	Meade Deep Sky Imager
Sony ExView HAD (chip unspecified)	510 × 492	9.6	Meade Deep Sky Imager Pro
Texas Instruments TI-237	640 × 480	7.4	SBIG STV
Thomson THX7899M	2048 × 2048	14.0	Apogee AP10

FLI: Finger Lakes Instruments; SBIG: Santa Barbara Instruments Group.

Most of the CCD cameras listed in Table 7.7 image in black and white only, yet color CCD photography is possible with a variation on the tricolor photographic technique introduced by James Maxwell in 1861. Three separate exposures are made with red, green, and blue filters over the camera, and then combined electronically to produce a true-color image. Some CCD cameras have optional built-in color-filter wheels, while others require a separate attachment. A notable exception is the C line from Starlight Xpress (that is, the SVX and MX models with the suffix *C*, such as the MX716C), as well as the ST-2000XMC from Santa Barbara Instrument Group.

In addition to the considerations cited earlier, a camera or a combination of imaging equipment that includes *autotracking* is practically a necessity if you want to keep your sanity. Back in the Dark Ages of film astrophotography, the photographer would have to sit with his or her eye glued to the eyepiece of a side-mounted guidescope throughout the entire exposure, making sure that a preselected star stayed centered in the guidescope's crosshairs. Often, the exposure would exceed an hour in length. But many of today's CCD imagers include autotracker technology that ensures a preselected star stays centered by automatically adjusting the tracking. Different companies call this feature by different names and accomplish it in various ways, but the net result is the same.

Several companies market CCD cameras for today's amateur astronomer. Some of the most highly regarded are manufactured by Santa Barbara Instrument Group, or SBIG for short. SBIG's least-expensive model, the 16-bit ST-402ME, uses Kodak's popular KAF-0402ME chip and includes a feature that SBIG calls "Track and Accumulate." This means that even though the ST-402ME cannot autoguide and take an image at the same time, a sequence of images can be captured and automatically aligned and stacked. Although not as good as a true autoguider, it beats having to track the sky manually. Still, the ST-402ME is an excellent entry-level CCD camera; just be sure to pay the extra $200 or so for the internal color wheel.

SBIG's ST-7XME series is built around the same KAF-0402ME imaging CCD, but it also includes a separate Texas Instruments TC-237H chip that is used as an autotracker. (A side note: Isn't it funny how the TC-237H used to be an excellent imaging chip but now is relegated to autotracker duty only?) That alone puts the ST-7XME in an entirely different league than the ST-402ME.

Just because SBIG's ST-8XME through ST-10XME cameras are more expensive does not necessarily mean they are better. Remember pixel size. The ST-9XE, for instance, uses the Enhanced KAF-0261E detector from Kodak, with 512×512 pixels, each 20 microns square. This suits it well for wide-field deep-sky imaging through telescopes with focal lengths in excess of 2,500 mm, but not less. Last, SBIG's ST-2000XMC is the company's first single-shot color CCD camera. Rather than relying on a color wheel like the other STs, the ST-2000XMC uses a red-green-blue (RGB) filter matrix over the pixel array to produce real-time color images. Results are very impressive.

Without a doubt, the two SBIG cameras that have created the most interest are their large-format cameras, the STL-11000 (11 megapixels) and the

STL-6303 (6.3 megapixels). They are a revolution in CCD imaging that yields images the size of a 35-mm film frame. Now, rather than having to combine multiple shots to create mosaics of larger sky objects, these two imagers let astrophotographers capture the big picture in one fell swoop. For those interested in absolutely state-of-the art equipment, here it is. Each includes an integral autoguider chip and internal shutter control. The monochromatic versions of each have internal filter wheels for tricolor imaging, while one-shot color imaging is also possible with each imager's CM model. In the months and years to come, the STL-6303 and the STL-11000 will likely be yielding the most impressive (and probably the most costly) images ever taken by amateur astronomers.

Note that each SBIG ST camera connects to a computer via a USB cable, speeding up image downloads as compared with older cameras, which used parallel ports. Each ST-7 through ST-10 model is available in two or three levels. The least expensive of each, designated with an I, includes only an imaging CCD chip, and not a second chip for autotracking. Unless you already own an autotracker or your telescope's mount cannot accept an autotracker, you would be well advised to spend the extra money.

Let me interject a word of caution here from Kevin Dixon, an accomplished astrophotographer from Somerdale, New Jersey. He writes, "The internal guide chip is not very effective with many of the filters in use today, often including the blue filter from a red-green-blue, or RGB, set. Some H-alpha filters will work, but there is no way you could use an oxygen-III filter. You simply cannot find a usable guide star. This is where a guidescope is invaluable."

Finger Lakes Instrumentation (FLI) sells a number of top-of-the-line CCD imagers for the amateur astronomy market, all using CCD chips from E2V, Kodak, or Thomson. Its least-expensive unit, the MaxCam ME2, includes the Kodak KAF-0402ME chip that is also found in SBIG's ST-402ME and ST-7XME. Like those, the MaxCam ME2 includes a USB computer connection for efficient image downloading, as well as a stainless-steel shutter capable of exposures as short as 0.02 second, a power supply, but no case or color wheel. An external color wheel is available separately.

Some of the other CCD cameras from Finger Lakes Instrumentation are among the most sophisticated—and expensive—in the field, and so are really designed for those who have already earned their CCD wings with less-expensive equipment. Although users report excellent results, they also caution that the documentation that comes with the IMG and MaxCam imagers is minimal. Read that as meaning you need to have a pretty good handle on what you are doing beforehand. Also note that while the FLI CCD cameras can be used for imaging as well as autoguiding, they cannot do both simultaneously.

From England, Starlight Xpress sells some of the smallest CCD imagers on the market. Starlight Xpress is also notable as being the company that pioneered single-shot color CCDs. Least expensive of the lot is the MX-5C, which takes exposures through different color filters simultaneously. The trick is that yellow, magenta, cyan, and green microscopic filters are distributed over the pixel array. The software supplied with the MX-5C decodes the intensity data from the pixels and reconstructs a full-color image in one step. Pretty clever, although image scale is limited.

To answer the demand for a larger imaging area, Starlight Xpress came out with the MX-7C imager. The same physical size as the MX-5C, the MX-7C also uses a four-filter internal system to generate some very impressive results. Inside, the Sony ICX249AK CCD measures 6.47 × 4.83 mm, as compared to 4.9 × 3.6 mm in the MX-5C, and has smaller pixels, making the MX-7C suitable for fast focal ratio instruments.

To help keep things on track, Starlight Xpress sells autoguiding software called *S.T.A.R.*, which is short for *Simultaneous Track and Record*. This Windows-based software relies on the imager's CCD chip for tracking information, while image-processing software gathers data for the final photograph. To use the software, an interface box is placed between the serial port of the computer and the autoguider port of the telescope mount using an RJ11 connector.

All MX imagers connect to their computers via a parallel cable, which slows download time significantly compared with a more modern USB connection. A USB adapter is available separately and strongly recommended.

Starlight Xpress also offers a series of sophisticated USB monochrome and single-shot color SVX cameras. Each SVX camera has a built-in autoguider interface and output that is compatible with its optional external guide camera. One end of the autoguider plugs into the back of the SVX camera body, while the autoguider's separate CCD chip fits into a standard 1.25-inch focuser. This arrangement necessitates a separate guidescope. Leading the pack is the SVX-M25, which uses Sony's ICX413AQ 6-megapixel CCD chip. Like the SBIG STL-11000 and the STL-6303, these large-format CCD imagers are the wave of the future, taking away the pains felt by other CCD imagers with small chips that capture tiny areas of the sky.

Perhaps your needs are more basic. If you are looking for an inexpensive CCD imager, just to test the waters, Meade's Deep-Sky Imager (DSI) is for you. For less than $500, you can buy a complete, ready-to-run digital imaging setup, complete with Meade's AutoStar image-processing software. Some Meade advertisements imply that anyone can take spectacular deep-sky images his or her first night out. That might be a little optimistic, however, the DSI does work quite well within its design parameters.

The compact Meade DSI is neatly housed in a 3.25 × 3.25 × 1.25-inch aluminum box that weighs only 10 ounces. A 1.25-inch adapter that threads into the housing is used to insert the imager into a telescope's focuser, while a side-mounted jack accepts a USB cable that runs to a computer. Depending on the computer being used, the DSI can generate either 8-bit or 16-bit images but with a caveat. The more common file formats, including JPEG and bitmaps, are available only in 8 bits, while FITS images record in 16 bit. With exposure times ranging between $1/10{,}000$ of a second to one hour, the DSI is just as adept at photographing the planets as it is deep-sky objects.

The AutoStar's image processing will help control many features of the DSI, including exposure, focus, dark-frame subtraction (needed to eliminate noise and so-called hot pixels), telescope aim, image stacking and alignment, and other functions. Some users report problems installing the software on their computers, with the most common error message saying "DSI imager not found" when trying to take a photo. The Meade Web site offers patches that usually help to resolve the problem.

One of the features controlled by AutoStar is guiding, but don't confuse that with the autoguiding found in SBIG's cameras, for instance. SBIG's two CCD chips allow for simultaneous guiding and imaging. With the DSI's single chip, you can do one or the other. Of course, given the comparatively low price, I suppose buying two DSIs—one for imaging and another for autoguiding through a separate guidescope—isn't totally out of the question.

Meade's Deep-Sky Imager Pro costs a little more and is more involved to use, but it pays off with better chip sensitivity and better ability to control final image color. That's because the DSI Pro is not a one-shot color camera like the standard DSI. Instead, it uses a set of color filters that requires three separate exposures, one through each of the three colors. Although this approach is common in high-end CCD imagers, most of those control filter insertion electronically. The Meade DSI Pro forces the user to slide the filters back and forth manually, which could knock the target out of the camera's small field if the telescope's mount isn't rock steady. Despite that apparent disadvantage, the DSI Pro is capable of capturing superior images to the one-shot DSI in side-by-side tests. On-the-fly image processing, done automatically by the included *Autostar* software, can yield very satisfying results provided the telescope is polar aligned, the mount is sturdy enough for the task, and you're reasonably familiar with long-exposure photography. Is the DSI Pro worth the extra cost over the DSI? For beginners, I would say no. The DSI is more than capable of letting beginners get their feet wet without spending more money than needed. It can provide years of fun and produce images suitable for e-mailing and posting on Web sites. The DSI Pro is really best for someone who is experienced with film-based long-exposure photography, and who wants to see if digital imaging is for him or her. The DSI Pro really shines by incorporating features found in more-expensive cameras, thereby giving a taste of things to come if and when they graduate to an advanced imager. One warning about the DSI Pro is that the color filters are always exposed, so users need to be careful not to damage them during use.

Early in 2006, Meade rolled out its next generation of deep-sky imagers, appropriately called the Deep-Sky Imager II and Deep-Sky Pro Imager II. Both emulate their first-generation counterparts in terms of overall size, weight, and design in that the DSI-II uses a color chip, while the DSI Pro-II uses a monochromatic chip with a tricolor filter bar. Improvements under the skin include CCD sensors that are almost twice as large, greater sensitivity, higher resolution, and lower thermal noise. Early results in the hands of DSI veterans look promising.

In late 2005, Orion Telescopes introduced its economy-priced StarShoot Deep-Space Color Imager. Like the Meade DSIs, the StarShoot DSCI is intended as an introductory tool for novices venturing out into the universe of digital photography. Priced comparably, the Orion imager has several notable design features. Perhaps the most noteworthy is its single-shot color capability, which eliminates the possibility of knocking the telescope off target while fumbling with filters. There's a downside to that too, however. Monochromatic cameras that use tricolor filtering usually produce more realistic colors than one-shot color images, in large part because of the photographer's ability to control how the colors mesh during processing. Much of that advantage is lost in one-shot imagers.

Another design feature worth mentioning is the StarShoot's thermal quality. Rather than rely on convective cooling like the Meade DSIs, the StarShoot incorporates active cooling that is powered by a separate 12-volt source. The net effect is diminished electronic noise, and therefore better images. But again, there is a downside, which is the need to carry along a 12-volt battery to power the camera's cooling circuit. The camera itself is powered by its USB connection to the computer. Early images with the StarShoot DSCI are very encouraging, and it remains to be seen what can be done with this great tool after it's been in the hands of users for a while.

Other econo-imagers include several models from SAC Imaging. All connect via a USB 1.1 port, speeding up image download times as compared to earlier SAC imagers that connected via a computer's parallel port. The company's least-expensive model, the SAC4-IID, uses the Micron MT9M001 CMOS chip. Its 5.2-micron pixels make it suitable for planetary and lunar photography but not for capturing deep-sky targets. In general, users report very good results. The SAC4-IID is an excellent beginner's camera, although the included documentation leaves a little to be desired for someone just starting out. Still, at a cost of under $300, it's a good value.

SAC Imaging's flagship camera, the color SAC-10, costs under $1,000. Designed with standard camera T-mount threads, the SAC-10 can be used in a variety of configurations, including prime focus and eyepiece projection through the telescope as well as with any T-mounted telephoto lens. It's also worth noting that the SAC-10, as well as the less expensive SAC-8.5, include software for both Windows and Macintosh computers, a rarity.

Although still more expensive than 35-mm SLR cameras, CCD cameras continue to drop in price just as CCD-based home video camcorder prices have fallen as the technology becomes cheaper to produce. Of course, all imagers require a personal computer and high-resolution monitor, adding further to the start-up cost, yet many amateurs think this is a small price to pay for the outstanding quality that can be achieved with CCDs.

Unfortunately, the area of CCD photography is far too complex to address adequately in the small space provided here. Although the discussion here may be enough to whet your appetite, if you are new to this burgeoning field, I recommend up front that you consult one of the books listed earlier in this chapter before purchasing anything.

Video Cameras. Adirondack Video Astronomy sells several terrific video cameras that are ideal for mating with a telescope. Each of its AVA Astrovid CCD video cameras is small enough to fit into a jacket pocket, yet powerful enough to get some terrific views of the Moon and the brighter planets through just about any telescope on any mounting. All come complete with power supply, cables, C-mount adapter (needed to couple the camera to a telescope), and instructions. All you need to add is a telescope and a video recorder (for field use, it might be best to attach the video camera to a camcorder rather than a conventional VCR).

Weighing only 6.6 ounces (190 grams), the color AVA Astrovid PlanetCam features a sensitivity of 1 lux and a horizontal resolution of 480 lines, which is fine for lunar and planetary shots. A deluxe version is also available that offers complete image control with the supplied software and EEPROM (electrically erasable programmable read-only memory) features. Both use Sony's ICX0208AK CCD chip, which measures 4.5 mm × 3.8 mm (pixel array measures 768 × 494, individual pixels cover 4.75 × 5.55 microns). The result is sharp images across a very small field of view.

Interested in videotaping deep-sky objects? Until recently, that was considered an impossible dream. After all, it takes many minutes of exposure time just to capture a still image of a favorite cluster or nebula, let alone a video. Enter the AVA StellaCam II with its 0.00002 lux CCD sensor (at f/1.4). The monochromatic (black-and-white) StellaCam II weighs an incredible 5.3 ounces (150 grams, less than many eyepieces), yet offers higher resolution than the Planetcam, with 600 lines of horizontal resolution. The StellaCam II is breeze to use, even if the manual is a little sparse in terms of details. Just plug the camera into a video monitor (or video recorder to record images for posterity), power it up by plugging its 12-volt DC power supply into a 110-volt AC outlet, and you're ready to image. The StellaCam is designed to take about 8.5 seconds worth of exposures and then automatically generate a final image on the monitor. The more individual images you stack, the more striking the results.

SBIG offers a completely different kind of CCD imager that combines the best of an autoguider with the convenience of a supersensitive black-and-white video camera. The result is the first, completely self-contained CCD imager, called the ST-V. Unlike the CCD imagers from SBIG or any of the others discussed earlier, the ST-V does not require a separate computer. Instead, the ST-V consists of a camera head and a control box, connected to each other by a 15-foot cable. The system is designed to run on 12-volts DC. A standard version includes both the imager and control box, but needs to be attached to a video monitor to be able to monitor exactly what the imager is seeing. The control box supplied with the deluxe ST-V includes a 5-inch LCD display, making a separate monitor unnecessary. Although this unit works well as a guider, few people seem to use it as an imager.

Photographic Accessories
Lenses

Just as all cameras are not suitable for photographing the sky, neither are all lenses. But before an educated choice can be made, the photographer must first determine what he or she wants to photograph. For wide-field photography, either with the camera attached to a fixed tripod or guided with the stars, the standard lens may be all that is needed. Most 35-mm SLR cameras are supplied with lenses of 50 mm to 55 mm focal length. These cover an area of sky $28° \times 40°$. If a wider field is desired, then either a 28-mm or 35-mm lens would be a good choice. They cover $50° \times 74°$, and $40° \times 59°$, respectively. On the other hand, if a magnified view is what you want, try an 85-mm, 135-mm, or 200-mm telephoto lens, with $16° \times 24°$, $10° \times 15°$, and $6.9° \times 10.3°$ fields of view, respectively.

In general, an astrophotographer wants to use as fast a lens as possible, since the faster the lens, the shorter the required exposure. Quite simply, the faster the lens, the lower the f-number. Holding the focal length constant, the only way to lower the focal ratio is to increase the lens' aperture. For instance, most 50-mm and 55-mm camera lenses have f-ratios between f/2 and f/1.2, resulting in apertures that range between about 1 and 2 inches. The larger the aperture, the greater the light-gathering ability of the lens, and therefore the shorter the required exposure.

Years ago, lens quality varied dramatically from one company to the next, but today, design and manufacturing procedures have been so perfected that most lenses will produce fine results (of course, every photographer thinks his or her brand is best). Flare and distortion, two of the biggest problems in older lenses, have been all but eliminated thanks to optical multicoatings. In general, most lenses made by reputable companies (Nikon, Canon, Pentax, to name a few, as well as those by Vivitar, Sigma, and Tamron) are fine for astrophotography. A *big* exception is the zoom lens. Although extremely popular for everyday photography, zoom lenses almost always produce inferior results to their fixed-focal-length counterparts.

Tripods

If you will be affixing your camera to a tripod (as opposed to shooting through a telescope), then pay close attention to the tripod you will be using. Many less-expensive tripods sold in department stores and other mass-market outlets are just not sturdy enough to support a camera steadily for any length of time. If the tripod is shaky, then the photographs will be hopelessly blurred. It makes no sense mounting a camera outfit costing hundreds, even thousands, of dollars on a cheap tripod!

Of all the tripods made, some of the sturdiest are manufactured in Italy by Manfrotto and marketed in the United States under the Bogen name. For instance, the Bogen model 3036 is sturdy enough to hold a 4-inch refractor even with its legs partially extended. Lesser tripods would collapse under such a load.

Gitzo tripods are certainly equal to, or arguably better than, those from Bogen/Manfrotto but can cost considerably more. Gitzo's Pro Studex Performance Tripod, which rises to nearly 80 inches above the ground, is a good choice for vertical viewing. Tripods from Slik, Velbon, and Vivitar are less expensive than Manfrotto and Gitzo models, but most are also far flimsier.

Camera-to-Telescope Adapters

Coupling an SLR camera, either film or digital, to a telescope requires a two-piece T-ring/adapter combination. The T-ring attaches to the camera in place of its lens, while an adapter attaches to the telescope. The ends of the adapter and the T-ring are then screwed together to form a single unit.

Different cameras require different T-rings. For instance, a T-ring for a Minolta will not fit a Canon. Likewise, different adapters are required for different telescopes. In the case of most catadioptric telescopes, an adapter called a *T-adapter* screws onto the back of the instrument in place of the visual back that holds the star diagonal and eyepiece. Most refractors and reflectors, on the other hand, use a different item called a *universal camera adapter,* which is inserted into the eyepiece holder.

Positive-projection astrophotography, commonly used when shooting the planets or lunar close-ups, requires that an eyepiece be inserted between the lensless camera and the telescope. Most camera adapters, such as the one shown in Figure 7.6, come with an extension tube for this purpose. The eyepiece is inserted into the tube, and the tube then screwed in between the adapter and T-ring.

Celestron, Meade, Questar, and some other telescope manufacturers offer camera adapters that custom-fit onto their telescopes. Aftermarket brands, often less expensive, are also available. For instance, Orion Telescopes sells several different adapters to fit most popular telescopes. None are supplied with camera T-rings, however, which can be purchased separately. Better photographic-supply stores carry T-rings for most common single-lens reflex cameras, as do many astronomical mail-order companies.

Flip Mirrors

One of the best things about a single-lens reflex camera is that the photographer is actually looking directly through the camera lens when peering through the viewfinder, enabling the shot to be composed exactly as it will appear in the final photograph. As the shutter is depressed, the mirror that has been directing light to the viewfinder flips out of the way, letting it pass to the film frame that lies behind the shutter. How do we compose a shot using a CCD imager? Some rely on the dim, ill-defined image appearing on their computer screen, which can make centering the target difficult, if not impossible.

Some astrophotographers prefer to use a device called a *flip mirror.* In effect, a flip mirror combines the mirror of a single-lens reflex camera into a star diagonal. With the mirror down, it redirects light to an eyepiece for focusing and centering. Flip it up, and the mirror swings out of the way so that light passes onto the CCD chip.

Several companies offer flip mirrors, including Meade Instruments, Murnaghan Instruments, True Technology, Taurus Technologies, and Van Slyke Engineering. Make sure you select one that is compatible with both your CCD imager and your telescope. In addition, some companies sell beam splitters, which let users view through the eyepiece and image using the CCD at the same time. In general, these result in dimmed images that are often inferior due to optical aberrations introduced by the beam splitter itself. Just be aware that flip mirrors may not be usable with some telescopes that have limited back-focus.

Star Wear

An area that few manuals of amateur astronomy address is the environment around the observer. Sure, most books complain about excessive light pollution and the need for good sky conditions, but there is so much more to enjoying the night sky than just the sky.

Baby, It's Cold Outside

The old saying that clothing makes the man (excuse me, person) is certainly true in astronomy. Nothing can take the enjoyment out of observing faster than weather-related discomfort. Although usually not a problem during the summer, it certainly can be at other times of the year. Even the sturdiest telescope mount will wobble if the observer using it is shivering!

It goes without saying that the clearest nights occur after a high-pressure weather front sweeps the atmosphere of clouds, haze, and smog. Unfortunately, the clear atmosphere also causes the Earth to lose a great deal of the heat that it has built up during the day. Many amateurs decide to sit these cold nights out, but by doing so, they are missing some of the clearest skies of the season. Others try to brave the cold, wearing their usual overcoat and a thin pair of gloves, but soon return inside, teeth chattering and fingers numb. Is this any way to enjoy the wonders of the universe?

Most hardy souls agree that layering clothes works best. For temperatures above about 20°F, wear (from the inside out) a T-shirt, flannel shirt, sweater, and parka on top, while underwear, long underwear, and heavy pants round out the bottom. In colder temperatures, or when the wind is howling, replace the sweater with a one-piece work suit, snowmobile suit, or insulated coveralls. These provide a good, windproof barrier between you and the cold, cruel world.

Although these items should keep you warm enough in moderately cold conditions, they also can make you stiff as a board. The multilayered look can make it difficult to pick up that pencil that was dropped on the ground, or even to bend for a peek through the eyepiece. There must be a better way! Happily, today there is.

Thanks to modern synthetic fabrics, it is now possible to stay outside even in subzero temperatures in relative comfort and with full freedom of movement. Increasing numbers of observers are joining other outdoor enthusiasts who wear clothing made of advanced materials such as Dupont's Thermax and Thinsulate. Both have amazing heat-retention properties, yet are thin and light enough to permit the wearer to bend with ease.

The best selections of cold-weather apparel are found at either local sporting goods retailers or in national mail-order catalogs such as Eastern Mountain

Sports, REI, L. L. Bean, and Campmor. Unlikely as it may sound, I have gotten much of my cold-weather clothing either from local bike shops or from two mail-order outlets: Bike Nashbar and Performance Bicycle Shop (I'm an avid cyclist by day).

Keeping the extremities warm is the most critical part of your cold-weather assault. In less extreme temperatures, a pair of thick socks and work boots for the feet, a hat and ear muffs for the head, and a pair of gloves for the hands should do the trick. Under colder conditions, the head is best protected with a silk or wool balaclava. Looking like a full-face ski mask, a balaclava is thin enough to hear through yet warm enough that it may make a heavy hood or hat unnecessary.

For the hands, try a pair of ski mittens stuffed with Thinsulate or a similar material. Unfortunately, although they are warmer than gloves, mittens can make it difficult even to focus the telescope. Some mittens come with a thin insulating glove that may be worn separately, handy when changing eyepieces but still a problem when taking notes or making drawings. One other possibility is to wear a separate thin glove liner inside each mitten. Once again, many outdoor outlets sell mittens and glove liners that work quite well.

Nothing is more painful than frozen feet. I have seen some people walk out in 10°F weather wearing a heavy parka, a down vest, long underwear, thick hood, heavy gloves, and a pair of sneakers! They didn't last long. Work boots, Sorel brand boots, and moon boots have excellent heat-retention qualities but only when used with a good pair of socks. The two-pair strategy usually works best. Wear a thin pair on the inside and a thicker, thermal pair on the outside. For truly frigid weather, even the best-insulated gloves and boots may not do the trick. Although some outdoorfolk use heated hand warmers that run on cigarette lighter fluid, I get nervous keeping an incendiary device near clothing. (Do you smell something burning?)

No doubt about it, battery-powered socks will keep your feet warm, yet they fail on several accounts. First, all of those that I have tried tend to warm my feet like a microwave oven warms a bowl of leftover spaghetti—unevenly! Invariably, part of my foot was too hot, while another part was too cold! Then, too, you must decide what to do with the battery pack. In most cases, it must be strapped onto your leg, adding to the discomfort. Some people love electric socks, which is fine, but I prefer another approach.

A safer means for fighting the cold are nontoxic chemical hand- and feet-warming pads, such as those sold by Grabber-Mycoal of Grand Rapids, Michigan, and Heat Factory in Vista, California. Once removed from their packaging and exposed to the air, these nontoxic heaters maintain a temperature of about 120°F for several hours. It is important to note, however, that the warmer must be wrapped in a cloth or other protective material before use. Burns can result if these heaters are left in direct contact with the skin for long. I have found the advertised claim that they last seven hours or more to be a little optimistic. From actual use, four to five hours is probably a better estimate. Afterward, simply discard the pad.

Die-hard winter hikers also tell me that treating the feet with antiperspirant keeps them warm by keeping the feet dry.

Don't Bug Me!

All that talk about cold weather makes me long for spring and warmer nights. But of course, with warmer weather come things that go *buzz* in the night. Mosquitoes, gnats, and black flies can prove more annoying than the cold. Can anything be done to ward off these nighttime pests? Different solutions, ranging from voodoolike rituals to toxic chemical brews, have been advanced over the years with varying degrees of success.

The best way to avoid insect bites is a combination of long-sleeved clothing, an observing site that is high and dry, and a good insect repellent. Studies show that the most effective brands use N, N-diethyl-3-methylbenzamide, better known as DEET. Most commercial repellents specify a DEET concentration of no more than 25 to 30 percent because higher amounts are potentially harmful. Common repellents that use DEET include Cutter, Off!, Deep Woods Off!, and 6-12 Plus.

Exercise caution whenever applying an insect repellent. DEET may be applied directly to the skin and clothing, but be sure to use it far from your telescope or other equipment. Although it works well at warding off insects, DEET also acts as a wonderful solvent when sprayed onto vinyl, plastic, painted surfaces, and optical coatings! DEET should not be used on infants or young children.

A repellent growing rapidly in popularity is actually not a repellent at all. For years, Avon has been selling Skin-So-Soft lotion and bath oil as a way to promote youthful skin; according to modern-day folklore, it works as a mosquito repellent, too. Tests in *Consumer Reports* magazine and by others have found that it works marginally, if at all. Maybe that's why so many astronomers look so young!

Still More Paraphernalia
Flashlights

Every astronomer has an opinion of what makes a flashlight astronomically worthy. Some prefer pocketable penlights, others like focusable halogen models, and a few favor dual-bulb models (which provide a built-in replacement in case one bulb burns out). Most agree that the best are small enough to fit into a pocket but large enough not to get lost at night. White or brightly colored housings are also preferred to black or dark models, because they are easier to find if dropped.

Regardless of the style or design, a flashlight must be covered with a red filter to lessen its blinding impact on an observer's night vision. There are many different ways of turning a white light red. Some of the more common methods include painting the bulb red with fingernail polish, or using red tissue paper or transparent red cellophane from stationery and party supply stores. These tend to chip, tear, or fade with time, forcing repeated filter renewal or replacement. A more permanent solution is to use red gelatin filter material sold in art-supply and camera stores. One or two layers of Wratten gelatin filter number 25 (yes, the same classification system as eyepiece filters) works especially well. But perhaps the most versatile red filter material is sold by auto parts stores as repair tape for car taillights. Sold in rolls of several feet, its adhesive backing makes it ideal for sticking onto a flashlight.

Flashlights that use red light–emitting diodes (LED) are much better than those using conventional bulbs thanks to the LEDs' low power drain. Many astronomical suppliers sell LED astronomy flashlights, with my favorites being the Skylite by Rigel Systems and Orion's DualBeam. Both can be switched between red and white LEDs, are adjustable in brightness, and run on single 9-volt batteries that will last for years.

Dews and Don'ts

One of the most frustrating things that we, as amateur astronomers, are forced to deal with is dew. The formation of dew on a telescope objective, a finder, or an eyepiece can end an observing session as abruptly as the onset of a cloud bank.

Dew forms on any surface whenever that surface becomes colder than the dew point temperature, which varies greatly with both air temperature and humidity. To illustrate this, think of a can of cold soda. In the refrigerator, the can's exterior is dry because its surface temperature is above the dew point of the surrounding refrigerated air. Now, take the can out of the refrigerator. Almost immediately, its surface becomes laden with moisture because the can is now colder than the warmer air's dew point. Under a clear sky, objects radiate heat away into space and soon become colder than the surrounding air.

Nothing—neither telescopes, binoculars, eyepieces, star atlases, nor cameras—is impervious to the assault of dew, but there are ways to slow the whole process down (no, one way is *not* to give up and go inside—although there have been times when I was tempted!). One option is to install a dew shield on the telescope. A dew shield is a tube extension that protrudes in front of a telescope tube to shield the optics from wide exposure to the cold air, thus slowing radiational cooling. Binoculars, refractors, and catadioptric telescopes stand to gain the most from dew shields because their objectives and corrector plates lie so near the front of the telescope tube.

Reflectors usually do not need a dew shield because their primary mirrors lie at the bottom of the tube (the tube itself acts as a dew shield.) The only exceptions to this are if the secondary mirror dews over (only in exceptionally damp conditions) or if the reflector has an open-truss tube. The former situation can be slowed by installing battery-powered heater elements around the secondary, detailed later. For the latter, many amateurs wrap the truss with cloth, effectively shielding the mirror from radiational cooling.

Dew shields merely slow the cooling of an objective lens or corrector plate, and, therefore, only delay dew from forming. Depending on the humidity, this may be enough, but to be effective, a dew shield must extend in front of the objective or corrector at least 1.5 times (preferably two to three times) the telescope's diameter. Although most refractors are supplied with dew shields, most binoculars and catadioptrics are not (the Questar being an exception). Owners must therefore either purchase or construct their own.

Most dew shields are made of molded plastic and are designed to slip on and off the telescope as needed. For instance, Orion Telescopes supplies Flexi-iShield dew shields. Made of thin, durable ABS plastic, FlexiShields are sold as flat rectangles. They are formed into tubes by wrapping the ends around and sealing the full-width, permanently attached Velcro closure. After the observing session, simply pull the Velcro seal apart and the FlexiShield lies flat for easy storage. FlexiShields are only effective in relatively dry environments. Kendrick Astro-Instruments and Astrozap sell similar dew shields made from ABS plastic, but also come lined with black felt for better absorption of moisture and extraneous light. Finally, the dew shields sold by Just-Cheney Enterprises and Dewzee Astronomy Products are made from a thicker, padded material with a black velvet interior and a reflective exterior. These last two are the most effective of all unheated shields against dewing, but they are also less convenient to store.

Astrozap also sells a collection of Flexi-Heat heated dew shields, designed to ward off heavy dew. Each felt-lined shield comes with an integral heating element. By warming the corrector plate and the air inside the cap ever so slightly by convection, dewing can be effectively prevented even in high humidity. The cap's heating element emits between 10 and 20 watts of heat, depending on telescope size, and requires a 12-volt power source, such as a car battery, DC power supply (useful only if a 110-volt AC outlet is nearby), or a rechargeable battery with at least 5 (preferably 10 or more) amp-hour capacity.

Heater straps are also popular among owners of refractors and catadioptric telescopes. These electrically powered heaters wrap around the front end of the telescope tube to heat the corrector plate by conduction to just above the air's dew point. Although these may be used independently of a dew shield, they are more effective with one in place. Heater straps are sold by several companies, including the Kendrick Dew Remover System, the Dew Zapper from Orion Telescopes, Dew-Not's Dew Remover, and Astrostraps from Astrozap. The Orion Dew Zapper plugs directly into a 12-volt DC cigarette-lighter style jack, while the Kendrick, Dew-Not, and Astrozap units come with cords designed to plug into a controller box (such as those sold by Dewbuster and Kendrick Astro Instruments, mentioned next). Of these, Dew-Not's straps seem to be designed the most efficiently in terms of power draw, which means that a given battery will last longer before needing a charge.

Kendrick Astro Instruments sells four heater control units for setting and adjusting the heaters' operating temperatures. The original Kendrick control box plugs into a 12-volt cigarette lighter–style jack and has four RCA-style inputs for heaters. Once the heaters are plugged into the controller, turning a knob adjusts the time that a heating element is on; it does not vary the element's temperature, however. The other Kendrick control boxes work basically the same way, although grow rapidly in sophistication. The most complex controller, the Premiere, has six input jacks, and can be used to digitally set and control a wide variety of 12-volt accessories, including cooling fans, digital setting circles, illuminated eyepieces, motor drives, as well as heater straps. In all cases, Kendrick recommends using, at a minimum, a 12 amp-hour battery. The system may also be run off house current by plugging the controller into an AC/DC converter of adequate amperage.

The Dewbuster heater control unit is different from the Kendrick in that turning its adjustment knob can vary an element's temperature, like the thermostat in your house. A scale on the controller's front panels lets the user select how many degrees the element will be maintained above ambient air temperature. The Dewbuster can control up to eight heaters simultaneously. Like the Kendrick controllers, the Dewbuster isn't cheap. But it is very effective—many believe more effective than any of the Kendricks—in even the dampest environments, provided a suitably strong battery is powering it.

A third way to wage war against dew is to use a low-power hair dryer or a heat gun. By blowing a steady stream of warm air across an optical surface, dew may be done away with, albeit temporarily. If an AC outlet is within reach, a small portable hair dryer at its lowest setting makes a good dew remover, but if you are out in the bush away from such amenities, use a heat gun designed to run off a car battery (DC heat guns draw too much power to use with rechargeable batteries). A wide variety of sources sell basically the same 12-volt heat gun at a wide variety of prices for a wide variety of uses. Auto parts stores, for instance, have windshield defrosters and camping equipment outlets offer mobile hair dryers. Call them what you will, all are pretty much the same. The gun puts out about 150 watts of heat; not exactly enough to dry hair or defrost a windshield but plenty to remove dew from an eyepiece.

Although I never go observing without one, heat guns have some drawbacks. First, although they are fine for undewing finderscopes and eyepieces, their small size limits their effectiveness for objectives and corrector plates. (If the lens or the mirror is much larger than about 3 inches, it is likely that the entire surface will not be cleared before a portion becomes fogged again.) A second shortcoming is that, sadly, the dew will return as soon as the surface cools below the dew point, making it necessary to halt whatever you are doing and undew the optic all over again.

Right about now, if you listen carefully, you can hear the traditionalists in the crowd jumping up and down, screaming that all this is blasphemy. Many amateurs believe that imparting any heat to a telescope will cause optical distortion. True enough, heat will upset the delicate figures of optics. This is why telescopes must be left outside for an hour or more before observing, to let their optics adjust to the outdoor temperature. In practice, however, the disturbance from contact heaters will be minimal as long as the heating is done in moderation. By definition, the dew point can only be less than or equal to the air temperature, never greater (under most clear-sky conditions, the dew point is going to be much less than the air temperature). The purpose of a contact heater is not to raise the telescope's temperature above that of the air, only above the dew point. Therefore, the telescope should never feel warm to the touch and overheating should never become a problem.

If dew is a big problem from where you observe, heater straps are hard to beat. Kendrick's system is the best of the bunch in terms of control and effectiveness, although the others also work reasonably well. I would recommend against those that just plug directly into a battery. Even though they are less expensive, their heat output is difficult to control; more often than not, it is either too much or too little.

Drier environs can probably do with a dew shield and heat gun, but regardless of where you are observing from, never wait for dew to form. Before observing begins, always turn the heater on, put the dew shield in place (remember the finderscope, too), and have the heat gun holstered at the ready to clear any fog that may form on eyepieces. By following this three-step program, optical fogging should be minimized, if not eliminated.

Observing Chairs

It is a proven fact that faint objects and subtle detail will be missed if an observer is fatigued or uncomfortable, yet many amateurs spend hours outside at their telescopes without ever sitting down. Observing is supposed to be fun, not a marathon of agony!

Observing chairs help relieve the stress and strain associated with hours of concentrated effort at the telescope. The best chairs have padded seats and may be raised or lowered with the eyepiece (not always possible with long-focus refractors or large-aperture Newtonians).

Musician's stools or drafting-table chairs make good observing chairs. Check the offerings at local music shops and drafting/art supplies stores. Orion Telescope's standard chair adjusts between 19 and 25 inches and differs little from a piano stool. The Tele Vue Air Chair is also similar to a heavily padded piano stool and includes a handy pneumatic height adjustment. This makes setting the height in the dark much easier than with the Orion standard chair; with that model you have to get off the seat, flip a small lever, and then pull or push the seat up or down. The limited seat height range for both models proves very limiting with most telescopes, however.

A much more popular design features a seat pad that slides along a pair of rails. To adjust the pad's height off the ground, simply lift it up and slide it along the rails to where you want it, and then release it. Friction holds it in place. The most popular example of this design is the Starbound observing chair, which is also sold by Orion as its Deluxe Observing Chair. The design has a couple of basic flaws, however. The most common complaint is that the seat doesn't stay still but slides down the chair rails unless you are sitting, literally, on the edge of your seat. Although much more flexible than stools, the Starbound/Orion chair is also considerably more expensive. Shop around, however, since prices for the Starbound chair can vary by as much as $50.

Kendrick Astro Instruments sells a nicely designed aluminum observing chair that includes provisions for an optional accessory tray that hooks over the chair frame. The Ultimate Observing Chair III is similar in concept to the Starbound chair, although it doesn't seem to share its friction problem. If you opt for the optional tray, which is also made of aluminum, be sure to add some foam padding (perhaps a pair of trimmed mouse pads) to prevent damage to eyepieces that may tip and roll. Like the Starbound chair, the Kendrick chair folds flat for easy storage.

Many lament that the Catsperch chair from Jim Fly is no longer available. Made of wood, it had a comfortable seat pad, an adjustable footrest, and a good range in height. For those who have some skill at woodworking, plans for the chair are still available for sale through the Jim Fly Web site. Consult Appendix C for contact information.

If you prefer a wooden chair but don't possess the tools or the skills to make your own, the BEER Chair offers an alternative. No, not a bar stool adapted for astronomy, the BEER (short for better, easier, effortless, relaxed) Chair from Insight Technology is made from maple that is stained and sealed against the elements. It has a carrying capacity of 325 pounds according to the manufacturer. Like the metal chairs mentioned earlier, the seat pad slides vertically along four round rails. Release the seat and it stays put due to friction. The BEER Chair also offers an optional footrest and a seat heater. Although more expensive than the metal chairs discussed earlier, the better friction coefficient of wood means that the seat pad doesn't slip down the rails as easily.

The Stardust chair from Hands-On Optics doesn't rely on friction to keep us in our seats. Instead, this chair's pad is held to the frame with small hooks that grab onto cross members on the chair frame. Although this means the chair's height must be set in increments, it's still a great idea. The pad is also spring-loaded, allowing the user to lean into the eyepiece while still staying squarely seated. Overall, the Stardust chair is very comfortable, although it requires two hands to move the seat pad up and down, a minor inconvenience.

If the thought of spending $200 or more on an observing chair is too much for your budget, consider the Astro Chair from R.F.S., Inc. (better known as buyastrostuff.com). It doesn't seem to suffer from the slipping-seat syndrome, weighs less than either the Starbound or Kendrick chairs, supports up to 300

Figure 7.8 *The Starmaster observing chair. Photo courtesy of StarMaster Telescopes.*

Table 7.8 **Observing Chairs**

Company/Model	Material	Seat Height Off Ground
BEER Chair	Maple hardwood, padded seat	9″ to 35″
Hands-On Optics StarDust Observing Chair	Metal frame, padded seat	13″ to 35″
Kendrick Astro Instruments Ultimate III Observing Chair	Metal frame, padded seat	9″ to 32″
Orion Telescopes Standard Observer's Chair	Metal frame, padded seat	21″ to 27″
Deluxe Adjustable Chair	Metal frame, padded seat	9″ to 32″
R.F.S. (buyastrostuff.com) Astro Chair	Metal frame, padded seat	18″ to 32″
Starbound Viewing Chair	Metal frame, padded seat	9″ to 32″
Starmaster Telescopes StarStep	Wood	6″ to 29″
Tele Vue AirChair	Metal frame, padded seat	21″ to 28″

pounds, and as of this writing, costs just under $100. The height range isn't as wide as the others, however, nor is the frame as rigid, which could be a concern on uneven ground.

Finally, Starmaster Telescopes offers the unique StarStep chair/stepladder combination seen in Figure 7.8. Two aluminum side rails, each standing 38 inches tall and placed at right angles to each other, support a triangular platform of oak, which in turn, holds the side rails together. The platform can be locked in four height positions; the standard chair sits the observer as high as 29 inches. Unlike the others described here, the StarStep chair may also be used as a short stepladder, although the manufacturer warns that this should only be done with the platform set in either of its two lowest positions. The entire chair disassembles for flat storage.

Table 7.8 lists some facts and figures for each of these observing chairs discussed in this chapter.

Binocular Mounts

After purchasing a pair of giant binoculars, most people suddenly come to the realization that the binoculars are too heavy to hold by hand and must be attached to some sort of external support. The favorite choice is the trusty camera tripod, but not all tripods are sturdy enough to do the job. Although all the brands mentioned earlier are strong enough to support binoculars, most are too short to permit the user to view anywhere near the zenith comfortably (the bigger Bogen/Manfrotto models being an exception to this). What are the alternatives? Basically, the only alternative is to use a special mount designed specifically for the purpose.

Four companies currently offer parallelogram binocular mounts: Blaho Company, Orion Telescopes, T&T Binocular Mounts, and Universal Astronomics. All range in price from $175 to $300, with very quick delivery times.

The Blaho Company offers two mounts, the Stedi-Vu and Stedi-Vu Junior, for different sizes of binoculars. The Stedi-Vu is designed to support up to 80-mm giant binoculars, while Junior is good up to 50 mm. Both offer sleek, black-anodized finishes, the nicest of any binocular mount sold today. Although the other models are made from aluminum square tubing, Blaho uses solid aluminum bar stock. A neat feature of the Stedi-Vu is the integrated counterweight used to balance the binoculars. The others mount their counterweights on long shafts. The only drawback to the Stedi-Vue is that binoculars can only be aimed directly in line with the parallelogram, meaning that the mount and the tripod are always in front of the observer. This makes it less convenient to view while sitting down, because the observer's legs can easily hit those of the tripod.

Orion's Paragon-Plus binocular mount is available with or without its Paragon tripod. The pair works well together, although the binocular mount can also be supported by any heavy-duty photographic tripod. The Paragon-Plus mount is made from square aluminum tubing that is painted semigloss black. Large knobs on the altitude axis can be adjusted to apply just enough pressure so that the binoculars can be aimed and released without constantly having to be loosened and tightened, which is a nice convenience. Motions are very smooth, although in critical tests, I find that the Paragon-Plus is only marginally suitable for binoculars larger than 70-mm models. Like the Blaho mounts, the Paragon-Plus only moves in two directions—azimuth (left to right) and altitude (up or down)—and as a result, observers face the same problem of having to dance around the tripod legs. Several of the mounts described next let the binoculars pivot to either side of the parallelogram, thereby placing the tripod to the side of the observer, away from any confrontation.

T&T Binocular Mounts offer a greater variety of binocular mounts than any other company, although its unfinished aluminum design gives them a homemade look. This does not detract from its performance, however. T&T's wooden mounts, including the Four-Arm Mount and the Kid's Mount, show very nice woodworking skills.

T&T's most popular mount, the ArtiMount (short for "Articulated Mount"), swings around four different axes of motion: altitude, azimuth, and binocular elevation, as well as a pivoting head. This latter capability, also found on the Universal Astronomics UniMount discussed next, lets the user swing the binoculars left or right perpendicularly to the mounting and tripod, making it much easier to enjoy binocular stargazing while sitting down. Balance is easy to achieve, thanks to the counterweight and the mount's long counterweight shaft, the longest of the group. The shaft slides into the bottom parallelogram tube, then is held in place with a captive pin.

One of the nicest features of most T&T mounts is how binoculars can be aimed at any angle and then released without having to lock them in place manually. By adjusting a large side knob before use, just enough pressure can be applied to let them stay put when released. On the downside, the aluminum side plates that hold the parallel beams in place seem a little thin, possibly leading to some flexure. But all in all, these mounts are excellent for the price.

If money is no object and you are looking for the best binocular mounts, then look no further than Universal Astronomics. These are the Cadillacs of the binocular world. The company's most popular mount, the UniMount seen in Figure 7.9, comes in three variations: UniMount Light, Millennium Uni-Mount, and Sirius UniMount. With the optional Ultra Swing Mounting System, binoculars can move in several different ways: left to right in two axes, up and down in two axes, and rotating about the central pivot or hinge.

Universal Astronomics' T-Mount and T-Mount Light are similar in design to the UniMount but are more compact. That makes them easier to transport but at the same time limits their height range. Both are more than adequate for nearly all binoculars and even some telescopes. The T-Mount Light is designed to support up to 15 pounds, while the beefier T-Mount claims a maximum weight of 30 pounds. Both feature welded aluminum frames and are completely waterproof.

Adaptability is certainly a word that comes to mind with all of Universal Astronomics' mounts. The mounting systems typically consist of several plate assemblies that slide together and can be adjusted in any direction to achieve a perfect balance. Because of this flexible design, no locking knobs are required to hold the binoculars in place; just mount the glasses, slide the plates back and forth, and adjust the counterweight until balanced. Very nice!

Table 7.9 gives some vital statistics for each of the binocular mounts mentioned here. The weight limits are based on manufacturer's data.

If the Universal Astronomics mounts are the Cadillacs of binocular mounts, then the Starchair 3000 is the Rolls-Royce. The Starchair 3000 is part

Figure 7.9 *Universal Astronomics' UniMount binocular mount.*

Table 7.9 **Binocular Mounts**

Manufacturer/Model	Height Range	Weight Limit (pounds)	Degrees of Freedom
Blaho Stedi-Vu Jr.	22″	3	2
Blaho Stedi-Vu	24″	5	2
Orion Telescopes Paragon-Plus T&T	28″	10	2
Kid's Mount	17″	4	3
Spring Mount	18″	3	3
Sidesaddle Mount	31″	8	3
ArtiMount	31″	8	4
Three-Arm Mount Mark II	25″	12	4
Four-Arm Mount	27″	20	4
Garden Mount	30″	20	4
Wheelchair Mount	31″	8	4
Universal Astronomics			
UniMount Light	32″	12	4 to 6, depending on options
Millennium UniMount	32″	30	4 to 6, depending on options
Sirius UniMount	32″	50	4 to 6, depending on options
T-Mount Light	22″	15	4 to 6, depending on options
T-Mount	22″	30	4 to 6, depending on options

CONSUMER CAVEAT: Mirror-Based Binocular Mounts

Some binocular mounts aim the glasses downward toward a mirror, which reflects starlight into the objective lenses. I am sure that there are many readers out there who own and love them. Perhaps you are one of them. If so, great! I encourage anyone to use anything that enhances his or her enjoyment of the night sky. To my way of thinking, however, these binocular mounts are compromises at best. For openers, all commercial mirror-style binocular mounts I know of use only standard aluminizing, which reflects between 85% and 88% of the light striking them, effectively reducing the binoculars' aperture. Another problem is that the mirrors used in these contraptions are usually made of float glass, which may appear flat to the eye but is far from good optical quality. Although distortions may not be apparent through low-power binoculars, image quality is likely to suffer as magnification grows. These accessories also lack the freedom of movement of parallelogram mounts. Finally, these also flip the images around, negating one of the greatest appeals of binocular astronomy, which is for the binoculars to act as direct extensions of our eyes.

binocular mount and part amusement-park ride. While most commercial binocular mounts use a counterweighted parallelogram frame, the Starchair 3000 is completely different. The name says it all. The Starchair is a motorized chair with a cantilevered binocular-support arm that is strong enough to hold even Fujinon's heavy 150-mm binoculars. Using a variable-speed joystick, an observer controls the chair's up-down, left-right motions to aim the binoculars at any point in the sky from horizon to zenith. Power for the Starchair can come from common 12-volt car battery. Two words sum it up: pretty cool.

Vibration Dampers

These make great stocking stuffers for any amateur who owns a tripod-mounted telescope. One of the biggest problems facing amateurs using such telescopes is how to combat vibration. Vibration is caused by a wide range of sources, anything from nearby automobile traffic to another person moving about or the wind.

Although nothing can guarantee shake-free viewing, Celestron, Meade, and Orion Telescopes sell a simple little gadget that can make a big difference. Vibration dampers are made of a hard outer resin and soft inner rubber polymer, isolated from each other by an aluminum ring. By simply placing one under each tripod leg, any vibration will be absorbed by the inner polymer before it can be transmitted to the telescope. It's such a simple yet effective idea, why didn't someone think of this before? Or for that matter, why not make your own? Plans can be found in chapter 8.

Observatories

Most amateurs dream of one day owning their own observatory. Apart from a backyard observatory making a statement that you are a serious astronomer, they offer many practical advantages. A permanent structure makes setting up and breaking down a telescope a snap, as well as providing shelter from earthly distractions, including light pollution and wind.

CONSUMER CAVEAT: Observatories

There is no doubt in my mind that the convenience of having a backyard observatory will dramatically increase your time at the eyepiece, as it has mine. But before you begin construction, whether using a commercial product or devising one of your own, check with your town's building department. Consider setback distances from your property line, how deep any footings must be, the depth of electrical lines running underground, and other parameters. A little research and planning beforehand will make it much easier for you in the long run. And since most building departments don't have forms for backyard observatories, you are building a tool shed with a funny roof.

Observatories are usually broken into two broad categories: domes and roll-off roofs. Most professional observatories use large domes that open a thin slit to the night sky. Some of today's most popular domed observatories for the amateur are the Home Domes and Pro Domes (Figure 7.10) from Technical Innovations. Both domes are white on the outside and deep blue on the inside, to dampen stray light from damaging the observer's dark adaptation. Shutters and slot openings overlap to prevent leaks. Home Domes' smallest models are the 6-foot-diameter HD-6S and HD-6T, for "short" and "tall," respectively. The former is best suited for attaching to an existing building, while the latter is a freestanding structure with 45-inch walls. The larger HD-10 (10 feet in diameter) is also meant to be added on to existing walls. Pro Domes come in 10- and 15-foot-diameter versions. Both have the unique ability to stack fiberglass wall rings, which allow the customer make as tall an observatory as desired. Finally, there is the Robo-Dome, a small, robotic observatory that, like the Boyd 5000C, is only large enough for a remotely controlled telescope, not an observer.

The Robo-Dome, and all Technical Innovations domes, can be controlled automatically with two optional software/hardware systems. The most sophisticated, called Digital Dome Works, controls both the dome's rotation as well as the opening and closing of the slot. Two, preferably three, people are needed for assembly of any of these products, although the manufacturer states that

Figure 7.10 *The Technical Innovations Pro Dome is a popular option for backyard amateurs who long for a domed observatory. Photo by Kevin Dixon.*

no special hoisting equipment is needed (though, obviously, it may be necessary if the dome is being placed on very high walls or on a roof). Motors to rotate the domes are sold separately. Although not necessarily required for the 6-foot models, they are recommended for the larger units.

Astro Haven makes fiberglass clamshell-style domes that combine attributes of conventional domes as well as some of the benefits of a roll-off roof observatory. Available in 7-, 12-, and 16-foot diameters, all are split at the zenith and open up fully in sections. This has the immediate benefit of not needing to rotate the dome to see the entire sky. All are normally made with white gel coat exteriors and black interiors, but other light colors are available. The 7-foot model comes preassembled, so it's a simple matter (simple if you have a forklift or a lot of friends, since it weighs 600 pounds) to take the dome, bolt it to a concrete foundation, and it's ready to go. The larger domes require a fair amount of assembly. Entry into an Astro Haven dome is either over the edge for those on the ground or, if the dome is atop a foundation, from underneath.

Other companies that offer a wide variety of domes made in various materials include one of the oldest companies in the field, Ash Domes. Observa-Dome, Astrodomes, Sirius Observatories, and Sky Domes Ltd. UK. Domes from Ash and Observa-Dome are very popular among college and public facilities, while the latter two are geared more toward the backyard amateur.

Most domed observatories have relatively high walls, which mean their telescopes must be on tall-standing mounts. Refractors and Cassegrain-style telescopes are fine for these, but Newtonians, especially those on low-riding Dobsonian mounts, are better housed in roll-off roof observatories, the second category. Unfortunately, this latter style does not offer the same protection against wind and light pollution as the domes. They do, however, give the observer a much more panoramic view of the entire sky.

Most roll-off roof observatories are constructed by amateur astronomers. Many have successfully modified a wooden garden shed by slicing off the roof and placing it on four or more wheels, which ride on a pair of tracks that extend behind the building. Another approach, Star Watcher Observatory, is shown in chapter 8.

Recently, SkyShed from Staffa, Ontario, introduced a series of well-designed roll-off roof observatories that have attracted a great deal of attention among amateurs. A SkyShed looks like a beautifully crafted tool shed, with pine clapboard walls, a door, two windows, and even a pair of flowerboxes. Sizes range from a cramped 6 feet by 6 feet to a spacious 10 feet by 14 feet. All stand about 8.5 feet high. Owners continually sing the praises of their SkySheds.

SkySheds are available in several different ways. For those who have the skills, plans are available on CD-ROM that detail complete construction from start to finish. Materials, tools, and sweat equity are up to you. In addition, three SkyShed kits are also available from the company as well as a growing string of distributors. The most basic kit includes the plans as well as a pair of adjustable support jacks for leveling the roll-off rails. An intermediate kit throws in a telescope pier that can be used to support most equatorial mounts. Finally, the deluxe kit includes all this plus corrugated steel roofing panels, wall panels, rollers and tracks, and hinges. Contact your local SkyShed dealer, listed on the SkyShed Web site (www.SkyShed.com), for delivery options.

If you are looking for a temporary shelter that can come and go as you do, consider a portable observatory. Modeled after or adapted from tents, portable observatories can give observers the best of both worlds, affording both shelter during and after a long night under the stars as well as the ability to remain nomadic.

The Kendrick Observatory Tent was the first of this kind on the market and still shows an ingenious design. It is easy for one person to set up quickly, and took this veteran camper only eight minutes the first time, without instructions. Once erected, the Observatory Tent has two rooms separated by a polyester wall. The smaller area (9 feet by 6 feet) is for sleeping and storing supplies, while the larger area (9 feet by 9 feet) is devoted to telescope space. The roof peaks at 6.5 feet but slopes off toward either end.

Actually, floor space is not the limiting factor for the size of the telescope that the Observatory Tent will accommodate. That is set by the size of the D-shaped door into the telescope area, which measures only 30 inches wide by 50 inches high. I would guess that the largest telescope you could squeeze through there comfortably would be either a 10-inch reflector or possibly a 12-inch Schmidt-Cassegrain. The observatory tent's floor, which runs across both the telescope area as well as the sleeping section, is made from a plasticized rip-stop polyester for extra strength. The floor also gives extra sleeping room, should a few extra observers decide to spend the night but could also cause some lighter telescopes to shake as people move about. A pad should be placed under any tripods with pointed feet to protect the floor from being punctured. Recent improvements to the Kendrick tent include stronger tent poles and a more opaque, matte-silver rain fly, to keep the tent cooler and darker during the day than the original fly.

With the help of the Royal Astronomical Society of Canada, North Peace International has modified two square dome tents to serve as portable observatories: the Galileo (measuring 96 inches by 96 inches by 76 inches high in the center) and the Orion (84 inches by 84 inches by 60 inches high in the center). North Peace has certainly done its homework when designing the tent frame, as it is just about the simplest that this camper of more than twenty-five years has ever seen. Unlike the Kendrick tent, which zips open across the tent's top, the single-room North Peace tents are designed so that all four sides zip open fully. Unfortunately, even with all four doors open, the tent's crisscrossing frame blocks the zenith as well as four equally spaced vertical slices of sky. Larger instruments will find this a greater hindrance than smaller telescopes, since the latter can be repositioned more readily.

Finally, for those who can't justify a full-fledged observatory, but who still need to block neighborhood lights, Dark Sky, Inc., offers a possible solution. Its Dark Sky Portable Panel Observatory consists of a framework of PVC piping and sheets of opaque canvas that shield your telescope from intrusive building and street lighting. The panels, each measuring 63 inches tall and 48 inches wide, attach to one another using large ratcheting clamps. The standard kit includes four panels, while the star party set comes with six. Riser kits are also available for extending panel height by either 12 inches or 22 inches, and windy night kits include stakes and tie-downs.

Table 7.10 **Checklist of Must-Have Accessories for the Well-Groomed Astronomer**

Commercial
☐ Red flashlight
☐ Star atlas (*Sky Atlas 2000.0* or *Herald-Bobroff AstroAtlas* recommended)
☐ Narrowband nebula filter
☐ Straight-through 8 × 50 finderscope
☐ Adjustable observer's chair
☐ Warm clothes
☐ Hand and foot warmers
☐ Bug repellent (Cutter, Deep Woods Off!)
☐ Thermos with a hot drink or soup
Homemade
☐ Eyepiece and accessory tray or table
☐ Two old briefcases, one for charts, maps, and flashlights, the other lined with high-density foam for eyepieces, filters, etc.
☐ Observation record forms

The Well-Groomed Astronomer

The list of available accessories for today's amateur astronomer could run on for page after page, but I must cut it off somewhere. To help put all of this in perspective, Table 7.10 lists what the well-groomed amateur astronomer is wearing these days. Some of the items may be readily purchased from any of a number of different suppliers; others can be made at home. A few may even be lying in your basement, attic, or garage right now.

The most important accessory that an amateur astronomer can have, however, is someone else with whom to share the universe. Although many observers prefer to go it alone, there is something special about observing with friends; even though you may be looking at completely different things, it is always nice to share the experience with someone else. If you have not already done so, seek out and join a local astronomy club. For the names and addresses of clubs near you, contact a local museum or planetarium, the Astronomical League, or, if you have access to the Internet, check out the World Wide Web; you will find addresses in Appendix E. The same list of links, as well as discussions about additional accessories and a complete, hyperlinked list of manufacturers and other references, can be found on this book's Web site at www.philharrington.net. If you have an astronomy buddy, then you have the most important accessory of all, and one that money cannot buy!

8

The Homemade Astronomer

Amateur astronomers are an innovative lot. Although manufacturers offer a tremendous variety of telescopes and accessories for sale, many hobbyists prefer to build most of their equipment themselves. Indeed, some of the finest and most useful equipment is not even available commercially, making it necessary for the amateur to go it alone.

Many books and magazines have published plans for building complete telescopes. Rather than reinvent that wheel here, I thought it might be fun to include plans for a variety of useful astronomical accessories. For this fourth edition, I have included ten new projects that range from the very simple to the very sophisticated. Their common thread is that they were created by amateurs to enhance their enjoyment of the universe. These projects are just a few samples of the genius of the amateur telescope maker.

A Swinging Chart Table

Sometimes, setting up for a night of observing can be a hassle. By the time you bring out the telescope and mount, eyepieces, charts, and a table, you might start to wonder if it's worth the effort, especially after a long day at work or school. Anything that can lessen that burden is always welcome.

To help do just that, Craig Colvin from San Jose, California, devised the neat wooden swing-arm table shown in Figure 8.1 that attaches directly to the base of his Orion SkyQuest XT8 reflector. The table is sized to fit his laptop computer perfectly, but it is just as adept at holding charts, a clipboard, or any other lightweight accessory that you might need at an arm's length. Best of all, the table pivots at the end of a rotating arm so that everything can be placed in just the right spot for use while observing.

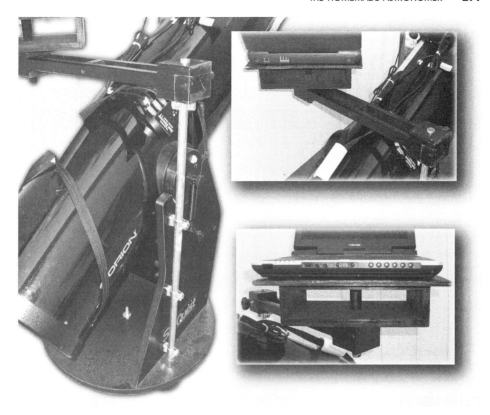

Figure 8.1 *Craig Colvin's swinging table. Photos by Craig Colvin.*

Table 8.1 lists the parts you'll need to make your own chart table. Don't be afraid to alter the dimensions in the table to suit your needs and telescope.

Begin assembling the table by cutting, gluing, and then nailing or screwing together its frame from the four table frame plywood boards, as shown in Figure 8.2. Assemble the frame so that the top and bottom are flush with the side boards. Once assembled, drill centered 1-inch holes through the top and bottom so that the holes are in alignment (use the dowel to make sure both line up correctly). Attach the 13-inch × 10-inch table to the frame using countersunk screws. Finally, insert the 6-inch-long dowel so that it slides through both mounting boards. Ideally, the fit should be tight enough that glue will not be required to hold it in place.

The arm is constructed of two equal-length 1 × 2s capped with two pairs of 3-inch × 3-inch squares at either end, all made from a hardwood of your choice. The pair of squares that will be at the table end should be drilled with aligned 1-inch holes, while the pair at the telescope end should have concentric 0.75-inch holes (or sized so that the metal conduit will slip through loosely—but not too loosely). Again, screw everything together after the holes are drilled and lined up.

Table 8.1 **Parts List: Chart Table**

Quantity	Item
1	Plywood, 0.5" thick, 13" × 10" inches (table)
2	Plywood, 0.5" thick, 3" × 7.5" long (table frame sides)
2	Plywood, 0.5" thick, 10" × 7.5" long (table frame top and bottom)
2	Hardwood, 1 × 2, 22" long (arms)
4	Hardwood, 1 × 4, 3" long (squares for arm ends)
1	Hardwood, 1 × 4, cut to wedge between arms (mid-arm spacer)
1	Hardwood dowel, 1" diameter, 6" long
1	Metal conduit, 0.5" diameter, 36" long
3	Conduit clamps, 0.5" diameter
—	Assorted fasteners
—	Wood glue
—	Outdoor spar urethane

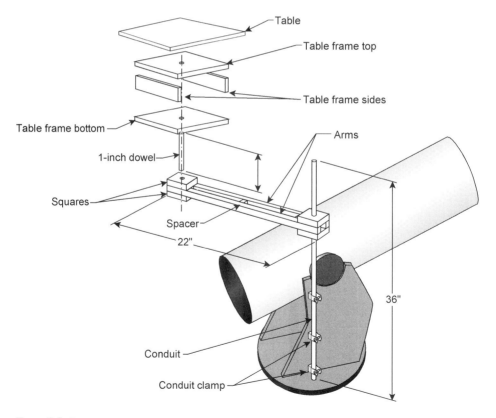

Figure 8.2 *An expanded view of the Colvin swinging table.*

Attach the 36-inch-long conduit to your telescope's side board using three conduit clamps, making sure they are also aligned so that the conduit will be perfectly vertical, and you're done.

Slide the table's dowel into the matching end of the arm, and then the entire assembly onto the conduit. Colvin notes that the weight of his laptop puts enough pressure on the conduit to keep the arm at whatever height it is adjusted to without slipping. This also lets him slide the arm up and down at will. If, however, you find your table slides too easily on the conduit, simply put a small hose clamp on the conduit where you want the table to stay. The clamp will still let the table swing to the side easily.

If you want to make the table larger for a full-size star atlas, keep in mind that heavier objects could cause the conduit to flex. If that happens, the telescope will likely shake as well. If you want a larger table surface, you should consider upgrading to a larger conduit as well.

Lazy Laser Collimator Collimator

As discussed in chapter 9, checking that the optics are properly collimated in any telescope, especially Newtonian reflectors with fast focal ratios, is critical to image quality. Many people incorrectly assume that means they need several expensive tools, including a laser collimator. The idea behind a laser collimator is that it sends out a perfectly straight, narrow beam. When one of these gadgets is placed in a reflector's focuser, the beam should bounce off the secondary mirror to the primary, back to the secondary, and into the focuser. If the optics are properly collimated, the beam's reflected image should overlap the beam itself exactly.

The merits of laser collimators were discussed in chapter 7, but here, let me ask those readers who have already purchased one a simple question: Is your laser collimator collimated? Wouldn't it stand to reason that if the collimator itself is off, then the results it produces will be as well? How could you

Figure 8.3 *Craig Stark's Lazy Laser Collimator Collimator.*

Table 8.2 **Parts List: Lazy Laser Collimator Collimator**

Quantity	Item
1	2 × 4 lumber, about 8 or 9 inches long (give or take depending on the collimator)
5	Nails (2.5-inch common nails)

tell? The easiest test is to rotate the collimator slowly while holding its aim absolutely steady. A properly collimated collimator will show only a point of light. But if the laser's aim is off inside the collimator's housing, the beam will trace out a tiny circle.

Most collimators come with instructions for adjusting the aim (typically by removing an access plate and turning three screws), but unless you can check it accurately, you will probably do much more damage than good. I suppose that you could buy an expensive optical bench with the appropriate testing apparatus. That's what the manufacturers hopefully do. Or, you can do what Craig Stark, a professor at Johns Hopkins University, did and make a "lazy laser collimator collimator" on the cheap.

Stark's collimator rig, shown in Figure 8.3, takes about fifteen minutes to make using the items listed in Table 8.2. Embarrassingly simple, isn't it? Stark just cautions to make sure that there are no burrs or ridges on the nails that will scratch the collimator's barrel.

To make the collimator, simply drive in four of the nails at approximately 45-degree angles to create a pair of Vs. Space the Vs so that they will support the collimator at both ends and be sure to take into account where the on-off switch or any adjustment screws are, since you don't want the collimator to hit them when it's rotated in the Vs. Finally, pound in the fifth nail to act as sort of a backstop and you're done. Does it matter that the Vs aren't exactly at 45 degrees? No. Or that they aren't perfectly symmetrical? No. Or that the nails you used are a little longer or a little shorter? No again. All that matters is that the nails don't wobble in the wood, and that the wood itself is reasonably flat so that it doesn't rock when placed on a bench or a table.

To test your collimator, simply lay it into the Vs, press it against the backstop, and turn it on. Aim toward a wall 10 or more feet away and rotate the collimator slowly. Does the red dot stay perfectly still as you rotate the collimator, or does it trace out a small circle? If the latter, your collimator has a problem and must be adjusted. Check the instructions that came with it to find out how. Take your time, never rush, and you'll be ready to head back to your telescope in no time, collimated collimator in hand.

Tom's Chair

Comfort at the eyepiece is critical to anyone's success as an observer. Nothing will drain the fun out of astronomy quicker than a bad back. If you've looked through any of the three previous editions of *Star Ware*, then you will have seen plans for an observing chair in each. This edition is no different, with a nicely crafted chair designed and built by Ron Boe of Phoenix, Arizona. He built the chair originally for Tom Watson, a friend in Tucson; hence, the name "Tom's Chair."

From Figure 8.4, it's clear that Boe is an artist when it comes to woodworking. One look at the laminated butcher-block beams used for the chair's two vertical supports speaks to that loud and clear. Although you can certainly make your own chair out of any material you choose, Table 8.3 lists the materials used in the original design.

With the materials gathered, it's time to begin cutting wood using the dimensions in Figure 8.5 as a guide. Boe prefers the more finished look that gluing, mortising, and keying pieces together gives rather than using nails and screws to hold things together. If this is your preferred woodworking method as well, be sure to take that into account when measuring and cutting the wood.

First, assemble the front, box-shaped frame. If you are cutting from single pieces of wood, glue and nail or screw the four pieces together (two side pieces and two spacers, top and bottom). If assembling butcher-block style, cut the frame side strips from six 45-inch-long 1×4s spaced with three frame spacers of 8-inch-long 1×4s at the top and bottom. Alternately, purchase a premade butcher block and use a router to cut out the central opening. Once everything is glued and pinned, set the pieces aside to dry.

The rear legs can be fashioned from single 2×4 beams or again from laminated 1×4 beams, as in the original chair. When assembly of each is complete, put the legs down to set.

Figure 8.4 *Tom's Chair is an elegant observing chair. Photo by Ron Boe.*

Table 8.3 **Parts list: Tom's Chair**

Quantity	Item
6	Hardwood, 1 × 4, 43" long (front frame side strips, 3 per side)
3	Hardwood, 1 × 4, 3" long (front frame spacers, top)
3	Hardwood, 1 × 4, 7" inches long (front frame spacers, bottom)
2	Hardwood, 1 × 4, 44.5" inches long (rear legs)
2	Hardwood dowel, 2" diameter, 3" long (leg pivots)
1	Hardwood dowel, 2" diameter, 15" long (rear support dowel)
2	Hardwood, 1 × 4, 15" long (rear support brace)
2	Hardwood, 1 × 4, 11" long × 2.5" wide at bottom, tapered to 0.75" at top (front brace)
3	Hardwood, 1 × 4, 28" long (front foot)
1	Hardwood, 1" thick, 10" × 10" (seat)
9	Hardwood, 1 × 4, long (seat wedge)
—	Assorted screws, nails, wood glue
—	Outdoor spar urethane

The front of the chair is supported by a 28-inch-wide foot, also cut from laminated 1 × 4s. Center the foot on the frame and secure in place, either with wood screws or another method. In Boe's original chair, the frame is mortised into the foot. "But not a through mortise," Boe advises, "since I didn't want end grain exposed to the ground."

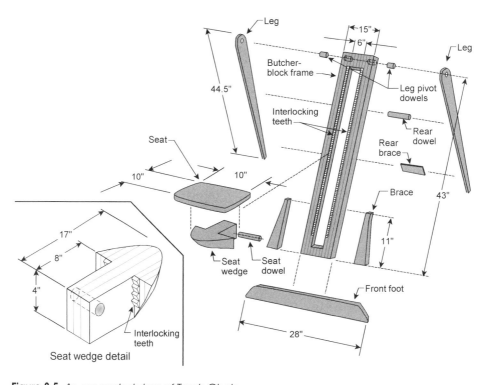

Figure 8.5 *An expanded view of Tom's Chair.*

To assemble the chair frame, drill two 2-inch holes, each about 1.5 to 2 inches deep, into the top end of the front frame as shown. Make sure both holes lie along the same axis or the rear legs may pinch when folded. Slip one of the two 2-inch dowels into each hole so that about 1 inch protrudes. Drill a mating hole through the top of each rear leg, and slip the legs over the dowel ends. Finally, put a 2-inch dowel section as well as a support plate between the rear legs approximately where shown, to stiffen the legs and ensure they won't slip off the front frame.

To strengthen the front frame further, Boe added a pair of braces on either side, between the frame and horizontal foot. Once again, the original plan called for butcher-block laminated beams mortised into place on two sides.

Cut the 10-inch × 10-inch seat from a piece of 1-inch-thick hardwood. Boe tapered the underside so that the full thickness of the wood coincided with the seat wedge. The wedge, which is shaped something like a rounded arrowhead, is also carved from a butcher block assembled from nine 1 × 4s. Approximate dimensions are shown on the drawing.

To make sure the chair seat has a good grip as it is moved up and down to match eyepiece height, Boe incorporated sets of interlocking teeth on either side of the chair frame's vertical opening as well as on either side of the seat block. He explained, "I cut the teeth using a 0.5-inch V bit in my router." A V bit is probably the best way to do this with any kind of good finish and accuracy. "I first made gage blocks by thickness planing some stock down to 0.5 inch, then chopping it up." Boe further advises, "Do one cut, add a gage block to move the router over, cut, and repeat until you run out of blocks. Cut that section out so you have something that will index on the teeth as you work across the work. Since this was time consuming, I did a fairly wide piece and then cut it up for parts. A real pain and my wrists hurt for weeks, but the results came out well." The teeth were fashioned out of mahogany for its strength, but the rest of the chair could be made from pine or a less-expensive hardwood.

LYBAR Chair

There is no doubt that Tom's Chair is as beautiful to look at as it is functional. That combination is what drew me to it in the first place, as I am a great admirer of woodworking artisans. But at the same time, I recognized that it was well beyond my abilities. So, when it came time to consider making an observing chair, I began to look around for a design that was within my limited skills but also functional. It had to be adjustable, but, at the same time, simple. The Denver Chair, detailed in the second edition of this book, is certainly an easy project, but in the course of my research, one popped up that was even simpler to make.

Professor Craig Stark, who also conceived the Lazy Laser Collimator Collimator described earlier in this chapter, has also come up with an adjustable observing chair that is the ultimate in simplicity. He calls it the LYBAR Chair (see Figure 8.6), which is an acronym for "Lift Your Butt and Rotate." To use the chair, simply flip it around to one of its three height positions until the seat matches your needs at the time. Stupidly brilliant, isn't it?

Figure 8.6 *The LYBAR Chair is an ideal observing chair for astronomers whose woodworking skills are limited.*

To appreciate the beauty of the LYBAR Chair, consider Stark's design goals. He wanted a chair that (a) could give him 8 inches of travel from low position to high, matching the travel distance of his telescope's eyepiece from horizon to zenith; (b) was 20 inches off the ground at its highest point; and (c) in his words, "I'm lazy; I'd like to say that I'm busy, but it really comes down to being lazy." You have to admire someone like that!

The parts list in Table 8.4 includes everything you need to make your own chair in about 15 minutes! Hard to believe that's all there is to it, but there isn't anything else.

The dimensions shown in Figure 8.7 are for my version of the chair, designed to match me (I'm 6 foot 4 inches tall) and my 18-inch telescope. I also added the 13-inch intermediate support to better steady my larger creation. For Stark's original chair, he used two pieces 18 inches long and another

Table 8.4 **LYBAR Chair**

Quantity	Item
2	1 × 6 hardwood (or pressure-treated wood), 31″ long
2	1 × 6 hardwood (or pressure-treated wood), 23″ long
1	1 × 6 hardwood (or pressure-treated wood), 13″ long
12	#10 decking screws, 2″ long
—	Carpenter's glue or other exterior-use wood adhesive
—	Outdoor spar urethane

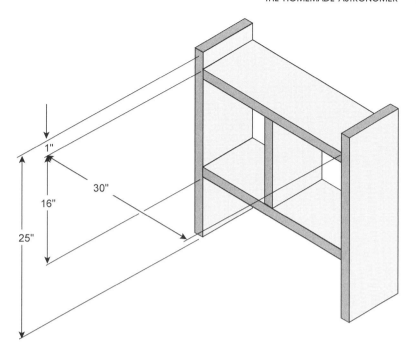

Figure 8.7 *Plans for the LYBAR Chair.*

two 17 inches long. That gave him the three seating heights he needed for his 8-inch Schmidt-Cassegrain and 4-inch refractor—12 inches, 16 inches, and 20 inches. Size yours to suit your particular needs. However you lay out the dimensions, be sure to allow 1-inch "legs" when the chair is in its intermediate position, as shown in the photograph. That way, when you flip the chair around to its lowest position, the "upper seat board," which is now on the bottom, won't be lying directly on the grass, or whatever happens to be on the ground.

When I asked Stark about the sturdiness of the chair, he remarked that it's held his 185 pounds without a problem. If it makes you feel more secure, add corner brackets or blocks to the joint, or use dado joints. But regardless, always heed Stark's advice to "never use the chair as a stepstool or with your feet off the ground."

Focuser Handle

As magnification increases, getting an image in sharp focus becomes even more important, but it also becomes more difficult. Part of the problem is due to the minute focuser adjustments needed at about 150× or more. It seems that just the slightest tweak will move the image through sharpest focus and out the other side. To make fine focusing easier, many amateurs spend big money on

Figure 8.8 *Florian Boyd's focuser handle.*

dual-speed focusers in the hope of getting a smoother adjustment. Although most of these fancy focusers work fine, their cost is often high.

A focuser's tuning resolution can be improved in a couple ways. One is to use finer internal gearing, which is what most two-speed focusers do when they incorporate an internal speed reducer. But a second simpler and cheaper way is to increase the radius of the focusing knob(s). Tele Vue, for instance, makes a nice focus lever (Tele Vue part number FCL-0001) that plugs right into matching holes in its focuser knobs.

To enjoy the same benefit at a fraction of the cost, Florian Boyd from Palm Springs, California, came up with a clever lever for his Tele Vue TV-76 refractor. Best of all, it costs just a couple of dollars and can be made in about thirty minutes. Figure 8.8 shows the result and Table 8.5 itemizes the modest parts list.

Cut the dowel to length, making sure that it isn't so long that it hits part of the telescope or mounting. Drill two holes at one end of the dowel, spaced to match the U-bolt. For a more finished look, use a rotary tool to round out a portion of the dowel halfway between the holes to match the circumference of the focusing knob. Paint the dowel with a color of your choice and set it aside to dry. After, insert the ends of the U-bolt through the matching holes in the dowel and slip everything over the knob. Use a pair of washers and wing nuts to hold the U-bolt, tightening down just enough to grip focuser knob securely.

Pushing the end of the dowel up and down will give you a much finer touch for crucial focusing, while also lessening vibrations. Owners of re-flectors can also make something similar for their telescopes, although the

Table 8.5 **Parts List: Focus Handle**

Quantity	Item
1	Hardwood dowel, 0.75" diameter × 12" long
1	U-bolt, sized to match the inside diameter of a focusing knob
	Paint, as needed

dowel will have to be significantly shorter so that it misses the telescope when turned.

A Simple Dew Heater for Unity Finders

"Beware falling dew points!" A sign with that warning should be posted next to almost every observing site. As the temperature drops at night, the relative humidity will usually rise, increasing the air's dampness. If the temperature drops low enough, it will hit the dew point, the temperature when water vapor in the air begins to condense on the ground, cars, and optics. It's a common problem that nearly all of us face. To combat the problem, many amateurs resort to heater wraps, dew shields, and even hair dryers. Some are pretty elaborate, drawing so much power that they need a controller to vary their heat output. If dew is severe enough, there isn't much choice.

Often, one of the first things to dew over is a unity finder, especially the windows common to the Telrad and the Rigel QuikFinder. Even if humidity isn't that serious, the fully exposed window will often fog over and wipe out the view. Since the windows aren't critical optical components, they can certainly be cleaned with a shirt sleeve, but it's much more elegant to keep them clean in the first place rather than having to constantly wipe them dry. For that, Jack Kellythorne of Felton, California, cooked up the simple series circuit using a 9-volt transistor battery, a battery clip, a single 390-ohm resistor,

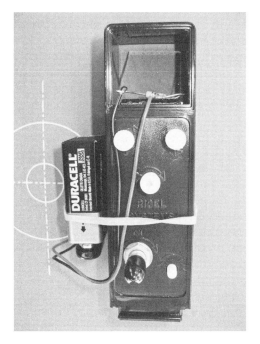

Figure 8.9 *Jack Kellythorne's antidew heater. Photo by Jack Kellythorne.*

Table 8.6 *Parts List: Unity Finder Dew Heater*

Quantity	Item
1	Resistor, 390-ohm, 0.5-watt (Radio Shack catalog number 271-1114 or equivalent)
1	Battery clip (Radio Shack catalog number 270-326 or equivalent)
1	Battery snap connector (Radio Shack number 270-325 or equivalent)
1	9-volt battery
1	Velcro, sized to suit application
—	Silicon adhesive

and a piece of Velcro shown in Figure 8.9. Kellythorne writes, "My whole plan was to keep it simple; no adjustments, just the absolute minimum power required at the precise point of application, namely inside the cover at the bottom of the window."

Table 8.6 lists the few parts needed to make Kellythorne's heater.

The heater is simply made by soldering each end of the resistor to the ends of the battery snap connector. When powered, the small resistor will give off just enough heat to keep the window clear. To work correctly, however, the resistor must be placed directly under the window, where its heat will be trapped. If it's located on top, the resistor's heat will simply radiate skyward with little or no effect. To hold the resistor in place, Kellythorne suggests drilling a small hole in each side of the finder housing next to the bottom of the window pane and threading the resistor wires out each side before soldering on the battery holder. When the resistor is right where you want it, dab small blobs of silicone adhesive over its ends to hold it place. Attach the battery holder onto the finder's housing using the Velcro and/or thick rubber band, and you're done.

To turn the power on, simply connect the battery to the snap connector. To turn it off, disconnect it. It's a simple project, but one that will let you keep your sanity on nights when others are closing up their scopes because of fogged finders.

Vibration Suppression Pads

I can still recall the shaky view I had through my first telescope, a 4-inch reflector back in 1969. The slightest breeze seemed to be magnified to at least a 6 on the Richter Earthquake Scale. This common gripe, echoed by new stargazers everywhere, is due to a combination of underdesigned mountings and poorly built tripods. Although the only way to fix a mount is to upgrade to a beefier model, we can help quell a tripod's jitters by placing a vibration suppression pad under the tip of each tripod leg. Each suppression pad has two parts: a soft, inner pad that "floats" in a hard, outer cup. The soft inner pad serves as a shock absorber by isolating the tripod from the ground.

Although several companies make vibration suppression pads, Jim Dixon from Little Rock, Arkansas, suggests that we can each make our own in a mat-

Figure 8.10 *Jim Dixon's homemade vibration suppression pads.*

ter of minutes. His pads, shown in Figure 8.10, are made from matched pairs of rubber and plastic cups normally used under the legs of a table or a chair. Table 8.7 gives the short parts list. Like Dixon and me, you can probably find everything you'll need at Home Depot, Lowe's Home Center, or a local hardware store.

As it turns out, the inside diameter of the larger, hard rubber cup measures 1.875 inches, while the outside diameter of the soft rubber cup measures about 1.8 inches. As a result, one will fit inside the other with a little room to spare.

To turn these parts into a vibration suppression pad, all you need to do is squirt enough silicone sealant into the larger cup to cover its bottom and lay the smaller cup on top. That's it. Just make sure that there is a complete layer of silicone under the soft cup, since the two furniture cups must not touch

Table 8.7 **Parts List: Vibration Suppression Pads**

Quantity	Item
1 pack	Hard plastic furniture leg cups, 1.875" inside diameter (such as Shepherd Hardware #89082 for a spiked bottom, which is good on grass, or #89088 for a smooth bottom, best for pavement), sold four per pack
1 pack	Soft rubber furniture leg cups, 1.5" inside diameter (Shepherd Hardware #89075), sold four per pack
1 tube	Silicone sealant (GE Silicone II, or equivalent)

each other directly. Set each pad aside and let the silicone dry. Although the instructions on the sealant say that it dries completely in twenty-four hours, I found that mine took at least twice that to solidify completely.

Simple Accessory Tray

One of the nice things about tripod-mounted telescopes is that they often come with small accessory trays built right into the tripod's center support arms. These trays are very convenient for holding small items, such as eyepieces, filters, or even small rechargeable batteries used to power dew-suppression devices. Dobsonian-mounted telescopes, however, rarely have trays, apart from small metal holders for three or four eyepieces. Recognizing that need, here are plans for an accessory tray that can be hung off the front of the base of any wooden Dobsonian mount. As you can see in Figure 8.11, it's a simple idea that will prove very handy time and again.

The tray itself is made from 1 × 10 oak, although any hardwood will do. Cut the length of the tray to suit your needs. In my case, my tray measures 12 inches long, which is perfect for holding a rechargeable 12-volt battery. Depending on what will be placed on the tray, surround the outside edges with 0.25 × 2-inch oak trim strips, to keep things from rolling off. The two precut shelf brackets (sometimes marketed as *corbels*) are available from Home Depot, Lowe's Home Center, and many hardware stores. Table 8.8 lists everything you need to make your own accessory tray.

Begin by cutting the tray and trim to size, using Figure 8.12 as a guide. After sanding the tray, trim, and brackets smooth, use wood glue and two #8 wood screws to attach each shelf bracket to the bottom of the tray. Attach the trim strips to the top side of the tray using wood glue and small finishing nails for the trim strips. Set everything aside to dry.

While the glue is drying, measure the spacing between the flush metal hangers that are built into the shelf brackets, and transfer those measurements onto the front of your telescope's Dobsonian rocker box. Double check that the

Figure 8.11 *A simple accessory tray attached to a Dobsonian mount. The inset shows one type of metal mounting clip that comes built into each of the shelf's brackets.*

Table 8.8 **Parts List: Simple Accessory Tray**

Quantity	Item
1	Tray, 1 × 10 oak, length to match needs
3	Trim strips, 12″ × 0.25″ oak
2	Shelf brackets, precut
4	Wood screws, flat head, #8 × 1″ long
4	Wood screw, flat head, #6 × 0.5″ long
1	Mouse pad or adhesive-backed foam rubber
10	Small finishing nails
—	Wood glue
—	Outdoor spar urethane

mounting points are square and level, screw a #6 × 0.5″ long wood screw into each point on the rocker box. Note that the size and length of the screws used for mounting the tray will depend on the size of the brackets' flush metal hangers as well as the thickness of the rocker box, so adapt the hardware to your needs accordingly. Screw each wood screw into the rocker box so that only the flat head and short length of the screw's shank protrudes.

Alternatively, if you would prefer not to mar your telescope base, fashion a pair of hangers from strips of metal that hang over the top edge of the rocker box. That way, the tray can be set in place and removed easily when needed. Attach the hanger straps to the back of the shelf brackets using short flathead wood screws. Line the inside of each metal strap with adhesive-back felt to prevent scratching.

Once the wood glue has dried, stain or paint the tray to match your telescope base and set everything aside to dry. Afterward, you may want to line the tray surface with some adhesive-backed foam rubber or possibly a computer mouse pad as cushioning. Finally, line up the four screws in the rocker box with the flush metal hangers on the back of the brackets and hang the tray in place.

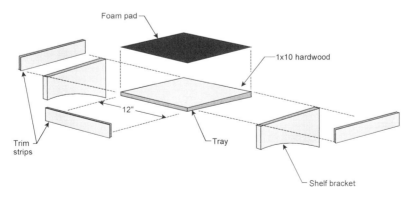

Figure 8.12 *The plans for the accessory tray.*

Figure 8.13 *Ed Hitchcock's equatorial table. Photo by Ed Hitchcock.*

Equatorial Table

Dobsonian mounts have their pros and their cons. On the plus side, they are exceptionally simple to make and use, and very inexpensive to manufacture. On the negative side, however, they force us to keep up with the sky manually by nudging our telescopes left to right and up or down. Unless you observe at the North or South Pole, or exactly on the equator, it's impossible to add a single motor drive to a Dobsonian mount and expect it to track the sky precisely. Or is it?

One way around this problem is to place the telescope, mount and all, on what is called an equatorial table. Some commercial models are discussed in chapter 7, but here we'll look at one constructed at home by Ed Hitchcock, a creative amateur from Toronto, Ontario. Using some common materials and basic tools, Hitchcock designed and built the equatorial table shown in Figure 8.13 for his 8-inch Sky-Watcher reflector. There are several designs out there for equatorial tables, but Hitchcock's impressed me for its simplicity of construction.

An equatorial table consists of a flat base (the table itself) riding on top of a pair of circle segments, or arcs, one large and the other small (see Figure 8.14). When these arcs are tilted at the proper angle and polar aligned, the table—and everything on it—will track the sky in a single, fluid motion. In effect, the table is a truncated horseshoe mount.

Table 8.9 lists some of the material needed for the equatorial table.

Before getting into specific construction details, the first step in designing an equatorial table is to make sure it fits your telescope. If you own an 8-inch reflector like Hitchcock's, then you can use the numbers here. Otherwise, you

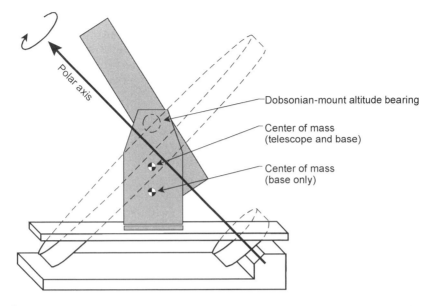

Figure 8.14 *A generic equatorial tracking table showing components. Notice how the polar axis passes through the centers of the two bearing disks.*

will need to determine the center of mass of your scope and base. The center of mass must lie below the mount's axis of rotation, or you risk the chance of the scope tipping over. To do this, first weigh the base and tube separately on a bathroom scale. Next, tip the mount's rocker box (without the telescope) on its side, on top of a 2 × 4 or other support so that it rocks up and down. Move

Table 8.9 **Parts List: Equatorial Table**

Quantity	Item
1	Lazy Suzan bearing, 4" diameter
1	0.5" threaded rod
1	$1/2$" threaded rod connector
1	$1/2$" threaded rod coupler
4	$1/2$" locknuts
2	0.5" bearings
4	Rollerblade wheels (cheapest available)
4	0.25" bolts and T-nuts for wheels
3	0.38" bolts, wing nuts and T-nuts for feet
2' × 2'	Plywood, $3/4$" thick for the north bearing
3' × 4'	Plywood, $1/2$" thick for the ground board and table
2'	Hardwood, 2 × 2 (for the platform's reinforcing beam)
3'	Hardwood, 1 × 6 (for the south bearing and support)
1	Bracket (attached between threaded rod connector and bearing; sized to suit)
—	Outdoor spar urethane

it back and forth along the 2 × 4 slowly until it balances. Mark that location (the mount's center of mass) in the center of one of the side supports. The center of mass for the telescope and mount will lie along the line between that point and the center of the telescope's altitude bearing. To find out where, first measure the length of the line between the center of mass of the base and the center of the tube bearings, then apply the formula: Center of mass = (length of line × weight of tube) ÷ (weight of base + weight of tube). Mark this location on the side support, as well.

Now, sketch your scope on a sheet of graph paper (see Figure 8.14). Draw a line just above the combined center of mass, tilted at an angle that's approximately the same as your latitude (round it off to an even 5°, since you'll need to cut wood to this angle later). This line is the axis of rotation of your platform. Draw the platform top under your scope, and then figure out where your bearings need to be, keeping in mind that the bearings must be perpendicular to the axis of rotation.

The north (larger) bearing is cut from two pieces of 0.75-inch plywood that have been glued together. For Hitchcock's 8-inch reflector, the bearing's radius equaled 17.5 inches. Use a jigsaw to rough cut the radius into a 90° pie-shaped piece of plywood (which works out to be about a 2-foot arc in this case), and then use a router and a sander to smooth the curve. To do this, Hitchcock fashioned a version of the jig shown in Figure 8.15 using a hinge attached to two perpendicular pieces of scrap wood, one per side of hinge. Screw one piece of scrap to the pie-shaped arc so that the hinge's pivot is located exactly on the centerline of the arc. Hold the second piece of scrap in a vice so that it is flush with your workbench. To smooth the arc's surface, swing the assembly across your router, which must also be anchored to your workbench. Replace the router with a sander and repeat the process to make sure there are no bumps or irregularities that would stop the table from moving smoothly. Finally, cut off the arc from the pie-shaped wedge itself. Be sure to cut it off at your latitude's complementary angle. (In other words, if you are at 40° latitude, cut the arc at a 50° angle.) This way, the arc's edge will fit flush against the underside of the table when assembled. The telescope platform itself is constructed of 0.5-inch-thick plywood. In order to brace it and bearings, as well as to keep everything from flexing under the telescope's weight, a central backbone made from 2 × 2 hardwood stock is attached to the platform's underside.

For the smaller south bearing, mount a 4-inch lazy-Susan bearing on a wooden block as shown in Figure 8.16. The block's face must be angled to match the complement of your latitude angle. Most home centers and hardware stores sell these, or you can order one online from Lee Valley Tools (www.leevalley.com), which is where Hitchcock got his. Afterward, attach a 6-inch wooden disk to the exposed side of the lazy Susan, drilled with a 0.5-inch hole in the center and a second hole, offset by 2 inches. These will be used to engage the table once it is set in place. Of course, they also allow the table to be removed easily for transport and setup.

The trickiest part of the project is locating the two pairs of rollers in the proper locations. One roller in each pair carries the weight of the table and

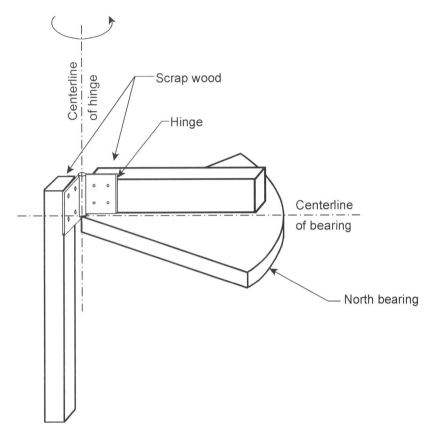

Centerline of hinge

Scrap wood

Hinge

Centerline of bearing

North bearing

Figure 8.15 *Both truncated bearings must be both smooth and accurate. As explained in the text, Hitchcock fashioned a jig out of two perpendicular pieces of scrap wood and a hinge to cut the larger bearing segment.*

telescope, while the second keeps the table from falling backward. The two pairs must be located far enough apart so the table moves smoothly, but also close enough so the arc doesn't run off one after a short time tracking the sky. Figure 8.16 shows the final numbers from Hitchcock's table, but yours may need to be changed. Draw everything to scale on a piece of paper before cutting and drilling wood.

The DC motor used to move the table in time with Earth's rotation couples to a captured threaded rod, which in turn is attached to the underside of the table with a bracket that can be fashioned either from scrap metal or from an angle bracket sold at hardware stores. The bracket is attached to a threaded rod coupler, which moves along the threaded rod as the rod turns. A pair of U-bolts holds the coupler to the bracket, which is also bolted to the face of the north bearing. A slot cut in the bracket allows the bolt to move vertically, but not laterally. As the rod is turned by the motor, the plate pushes on the bolt, pivoting the platform in time with the sky.

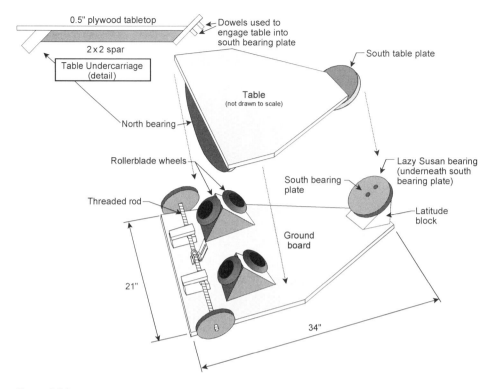

0.5" plywood tabletop

Dowels used to engage table into south bearing plate

2 x 2 spar

Table Undercarriage (detail)

South table plate

Table
(not drawn to scale)

North bearing

Rollerblade wheels

Lazy Susan bearing (underneath south bearing plate)

South bearing plate

Threaded rod

Latitude block

Ground board

21"

34"

Figure 8.16 *An expanded view of Ed Hitchcock's equatorial table.*

With the dimensions listed here, the threaded rod must be rotated at about 1 rpm for the platform to counter Earth's rotation. There are many ways to motorize the platform. Hitchcock chose to remount the motor and gearbox from a child's toy crane using little more than a hinge and an elastic band. If you want to drive the platform with great precision and are handy with a soldering iron, you can buy a surplus stepper motor and speed regulator kit from www.electronickits.com in the United States or www.qkits.com in Canada.

The Hitchcock equatorial table is the most complex homemade accessory described in this edition, or, for that matter, any previous edition of *Star Ware*. Space constraints, however, make it impossible to give precise step-by-step assembly instructions and photographs. Recognizing that readers will need additional instruction, photographs, and feedback on making their own version of this very useful project, a special section devoted to constructing equatorial tables and other tracking platforms can be found in the *Star Ware* section of www.philharrington.net.

Although an equatorial table takes extra effort and close attention to detail to construct correctly, those who own them often ponder aloud how they ever observed without one. And the best part of Hitchcock's version detailed here is that it's constructed of materials that are both readily available and inexpensive.

Star Watcher Observatory

Ever since I became interested in astronomy back in 1968, I have wanted my own observatory. That's not an unusual dream for any hobbyist with more than a passing interest in studying the night sky. The plans and schemes that I have seen published in books and magazines and later on the Internet only served to whet my appetite further. That boyhood dream finally turned into a reality in the summer of 2004. But before I lifted a hammer, I had to decide what type of observatory would best serve my needs as well as budget and construction skills. Domed observatories are traditionally the first that come to mind, and while there is no denying their usefulness, I quickly realized that building such a structure was beyond my abilities. Instead, I considered a roll-off roof observatory, where the roof retracts on a track to give a full view of the sky. Once again, however, I didn't feel I possessed the carpentry skills to make such an elaborate design.

Tempering those visions of domes and roll-off roofs, I designed a structure that would serve my needs as an observer, match my shoestring budget, and be within the grasp of my limited talent as a carpenter. The end result is the roll-away observatory shown in Figure 8.17. My 18-inch f/4.5 reflector sits on a concrete pad, which is surrounded by a wooden deck. When not in use, the telescope is covered by a minimalist shelter mounted on casters. It's a simple matter to roll the shelter off the telescope for viewing, then roll it back over the telescope when the night is done.

Table 8.10 lists the items used to make the shelter illustrated in Figure 8.18, but keep in mind that all were sized to house an 18-inch Newtonian reflector.

Figure 8.17 *The author's roll-away Star Watcher Observatory.*

Table 8.10 **Parts List: Star Watcher Observatory Shelter**

Quantity	Item
2	2 × 4 lumber, 66″ long (front corners)
5	2 × 4 lumber, 60″ long (rear wall)
3	2 × 4 lumber, 96″ long (long wall frame)
2	2 × 4 lumber, 40″ long (long diagonal braces, wall)
2	2 × 4 lumber, 16″ long (short diagonal braces)
2	2 × 4 lumber, 36″ long (short wall frame, bottom, wall)
5	2 × 4 lumber, 36.5″ long (roof rafters)
4	2 × 4 lumber, 66″ long (door frames)
4	2 × 4 lumber, 48″ long (door frames)
8	2 × 4 lumber, 17″ long (door frame diagonal braces)
1	Plywood, exterior, 36.5″ × 96″ × 0.38″ thick (roof)
2	Plywood, exterior, 36.5″ × 66″ × 0.25″ thick (side walls)
1	Plywood, exterior, 62″ × 96″ × 0.25″ thick (back wall)
2	Plywood, exterior, 48″ × 66″ × 0.25″ thick (door panels)
3	Plywood, exterior, 0.25″ thick (trim strips, sized to suit)
—	PVC flashing or corrugated plastic roofing panels
—	Rubber, anti-skid stair treads (skirt around bottom of shelter)
4	Casters, phenolic/hard rubber, 8″ diameter, nonpivoting
—	Door hardware
4	Screened vents, approximately 6″ × 12″
—	Decking screws, nails, white paint

Be sure to make your observatory fit your expectations and needs, and not the other way around. Take into account your equipment and observing site. Think it through on paper before committing it to wood.

Step one is to lay out the deck's footprint and level the land. To determine the size of the deck, take into account the length of your telescope, the likely size of the shelter, and how many people are likely to join you at the eyepiece. Remember, you will need enough room to move the shelter off the telescope. That direction is liable to be blocked from viewing, so choose the layout accordingly. My best advice is to make the deck larger than you think you need. Mine ended up being 16 feet square, which is just about perfect for my 80-inch-long telescope.

Next, decide where the concrete pad will lie. Measure the footprint of your telescope mount, and make the pad a couple inches wider. The base of my 18-inch's Dobsonian mount measures about 30 inches square, so I ended up with a 31-inch × 31-inch pad. Because the telescope simply sits on top, it is unnecessary to sink the pad more than 8 inches into the ground (unless you live in an especially cold area, where frost could cause upheaval). If you need a permanent pier, however, you must sink the base below your local frost line. And don't forget to add some rebar to prevent the pad from cracking in cold weather.

As the concrete cures, begin to lay out the deck and supports. Use 4-inch × 4-inch or 6-inch × 6-inch pressure-treated beams for the deck sleepers and 2-inch × 6-inch or 2-inch × 8-inch boards for the decking itself. In my design,

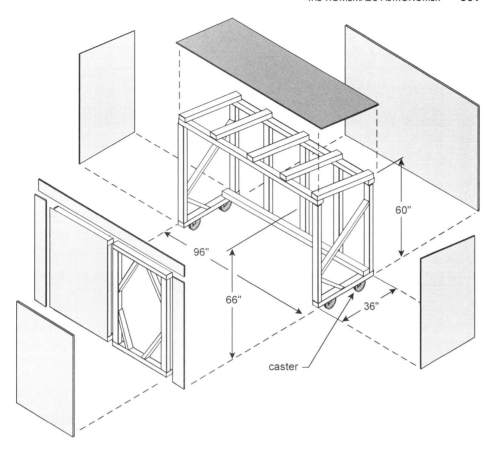

Figure 8.18 *An expanded view of the Star Watcher Observatory shelter.*

the deck is supported by 8-inch × 16-inch concrete blocks dug into the ground and laid on beds of gravel, rather than be supported by fixed footings. The weight of the pressure-treated wood prevents shifting, even in the winter when the ground freezes. Again, if you live in an especially cold climate, you may need more permanent support. Be sure to leave about a half-inch gap between the deck and the concrete pad to eliminate telescope vibration when walking about.

To block wind as well as any local light pollution, the deck should be surrounded by a stockade fence or a similar opaque wall. Estimate how high the fence needs to be to block light by measuring the height of your telescope's eyepiece at its highest point off the ground, and add a few inches for good measure. The fence will probably have to be higher for Newtonians than for refractors and Cassegrain-design telescopes.

The shelter is constructed from 2-inch × 4-inch framing and should be just large enough so that the telescope can be stored at a tilt of 45° or less. This way, any dew or moisture on the optics will drip off rather than puddle. In my case, the shelter is 96 inches wide by 36 inches deep. The front wall measures 66 inches high, while the back is 60 inches high. That difference lets the sloped roof shed any water or snow away from the doors. Frame the shed so that adjacent 2×4 beams are no more than 16 inches apart, which is probably the same spacing as in the walls of your house.

The shelter walls are constructed from 0.25-inch exterior-grade plywood, while the roof is made from 0.38-inch exterior plywood. Although these are quite thin, anything thicker will needlessly add to the weight of the shed, making it more difficult to roll on and off the telescope. To save further on weight, cover the roof with overlapping PVC flashing or corrugated plastic roofing panels rather than traditional asphalt roofing shingles.

The barn-style doors of the shed are also made of 2×4 framing and 0.25-inch plywood. Make the doors large enough so that they overlap the top and sides of the shelter opening. Hang both to the shed's frame using 4-inch strap hinges. To help maintain weather-tightness, line the edge of the door opening with adhesive-backed, foam-rubber insulation.

The shed rides on four casters, two per side as shown. After experimenting with different sizes and types of casters, I ended up using four 8-inch wheels made out of phenolic (plastic). Pneumatic wheels proved problematic because of their greater rolling resistance, while smaller, solid casters couldn't support the shed's weight without flattening. The casters move along a track made from 1-inch $\times$ 6-inch pressure-treated wood; 2-inch $\times$ 2-inch side beams prevent the wheels from derailing.

To keep the shed in place when not in use, drill two oversized holes through the shed frame, and two matching holes directly below into the tracks. Install 0.38-16 threaded inserts into the track holes. Later, when closing the observatory after an observing session, run two long, threaded rods through the holes in the frame and into the inserts below to keep wind from moving the shed.

Paint the shed with exterior semigloss white paint to reflect as much of the Sun's light and heat as possible. In especially hot climates, you might consider lining the inside of the shed with foam-board rigid insulation, although I haven't found it necessary. Install at least four vents along the back wall and/or sides of the shed to promote circulation. I also added a small, solar-powered vent fan to one door, just to keep the air moving during the day. Finally, add a hinged shelf along the back wall for eyepieces, charts, and other accessories. Use strap hinges and chains to support the shelf when open, and a small hook to lock it closed when not in use.

To make the bottom of the shed as weatherproof as possible, install strips of rubber along the bottom. For these I used rubber, antiskid stair treads purchased at the local home-improvement center. Put the rubber treads so that they are just above the deck, in effect shrinking the gap between the two to about a pencil diameter. I would also recommend lining the inside of the shed along the bottom of the frame, either with additional rubber treads or thin plywood, to act as a double baffle.

Some may need to bring power to their observatories, while others will want to dress it up with trim, paint, and surrounding landscaping. Regardless of how you personalize your observatory, you will find that the convenience of having your telescope ready for action at a moment's notice greatly increases your time and enjoyment at the eyepiece.

9

Till Death Do You Part

"A telescope alone does not an astronomer make." Sure, telescopes, binoculars, eyepieces, and other assorted contraptions are important ingredients for the successful amateur astronomer, but there is so much more to it than that. This chapter contains lots of little tidbits to help you get the most out of your telescope, ranging from setup and maintenance to traveling with a telescope.

Before proceeding, you should look through the earlier chapters to first learn about how your new telescope works, both optically and mechanically. It will also help you to become familiar with the terminology, such as declination and right ascension axes, collimation, astigmatism, and spherical aberration, before going on.

It's a Setup

Few things in the life of a budding stargazer are more exciting than the prospect of getting that first telescope. It seems almost a rite of passage, as if to make the statement "Look at me, I'm an amateur astronomer." That's why it's so important to choose that first telescope wisely. Making the right choice can fuel a fire that will lead to a lifelong interest. But buying the wrong telescope can just as quickly extinguish that flame.

One complaint that new amateurs are often heard saying is that their new telescopes are so complicated. They are difficult to set up and even more difficult to use. Although some companies include good instruction manuals and setup instructions with even their most basic telescopes, others include scant instructions at best. After all, what good is a telescope if the owner can't even set it up correctly?

To help end that frustration, let's examine three of the most common types of beginner telescopes: an equatorially mounted refractor, a Dobsonian-mounted Newtonian reflector, and a small instrument on a computerized GoTo automated mount. Although the instructions here are designed to be generic rather than specific to a particular brand or model, they should give you a good idea of how to set up and use that new telescope.

If you don't have instructions for your telescope, and are still having trouble after you have read through these steps, check the *Star Ware* section of my Web site, www.philharrington.net. There you will find a comprehensive list of links for a large number of telescope manuals.

Setting up a Dobsonian Mount

Let's begin with the simplest type of amateur telescope to assemble. As discussed in chapters 3 and 5, Dobsonian mounts are offshoots of altitude-azimuth mounts. Most are easy to set up, since they often arrive nearly fully assembled. Here are a few steps to make sure that everything is as it should be.

1. Begin by unpacking the telescope and mount from their boxes, but only if you have enough time to put everything together slowly. Never rush. Check all parts for damage, laying everything out on the floor as you go.

2. Identify the telescope tube assembly, rocker box, base, accessory tray (if supplied), and any miscellaneous hardware that might have been supplied. Some telescopes come with a small combination wrench and screwdriver for assembly, while others require that you use your own tools, including Phillips and straight-blade screwdrivers, pliers, and possibly an adjustable wrench.

3. Depending on the telescope model, you may have to put together the rocker box and base, as shown in Figure 9.1a. In most cases, these are made from precut pieces of laminated particle board, predrilled for easy assembly. But take heed. If you have ever assembled a prefab bookcase or cabinet, you'll know not to tighten down on the screws too much for fear of damaging the boards.

4. Insert and tighten the bolt that passes from the ground board through the bottom of the rocker box (see Figure 9.1b). Most use either a jam nut, a lock washer, or, more typically, a nut with a nylon insert to keep the bolt from loosening or tightening too much during use. If your telescope came with only a standard hex nut, take a trip to your local hardware store for a nut with a nylon insert. This will help keep the ground board and rocker box from loosening during use.

5. Place the telescope tube assembly's two side bearings into the rocker box's two side cutouts as shown in Figure 9.1c. Many imported reflectors use springs that loop between the round bearings and anchor points on the box's two side boards. If your reflector came with these, hook them onto the anchor points by pulling the tab attached to each spring.

6. Finally, attach and align the finder to the telescope, as detailed next.

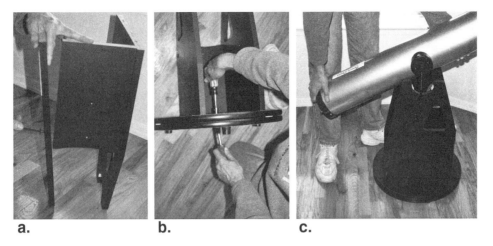

a.   b.   c.

Figure 9.1 *Three of the steps in assembling a typical Dobsonian mount. See the text for details.*

Setting Up an Equatorial Mount

Outwardly, a German equatorial mount appears much more complicated than a Dobsonian mount, therefore, many beginners immediately feel a sense of angst when setting one up for the first time. But beyond the extra parts, like a counterweight, possibly a pair of flexible slow-motion arms or knobs, and a pair of setting circles, they are little more than an alt-azimuth mount tilted so that the azimuth (horizontal) axis matches the observer's latitude. Indeed, if an alt-azimuth mount were placed on either of the Earth's two poles, it would be instantly transformed into an equatorial mount (since its azimuth axis would be pointing directly at the celestial pole).

1. As with the Dobsonian, unpack the telescope and mount from its boxes and inspect the parts for damage, laying everything out on the floor as you go. Identify the telescope tube assembly, equatorial mount, tripod legs, accessory tray (if supplied), and any knobs and miscellaneous hardware that might have been supplied. Again, some telescopes come with their own assembly tools, while others may require Phillips and straight-blade screwdrivers, pliers, and possibly an adjustable wrench.
2. Start assembly by attaching each tripod leg to the bottom of the equatorial mount using the supplied screws, washers, and nuts. Tighten everything finger-tight only for now, as shown in Figure 9.2a.
3. Stand the tripod up, keeping the legs fully retracted.
4. Tighten the attachment screws holding the legs to the base plate of the equatorial mount with a wrench or tool as needed. These screws often loosen up after use, so make a mental note to check them each time before using the telescope.
5. Attach the equatorial mount to the tripod base plate as shown in Figure 9.2b. This is usually done by threading a long threaded rod through the base plate and into the underside of the equatorial mount.

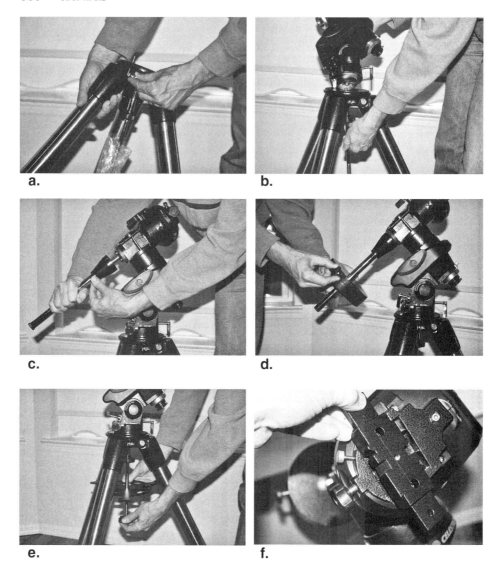

a.   b.

c.   d.

e.   f.

Figure 9.2 *Six of the steps in assembling a typical German equatorial mount. See the text for details.*

6. Swing the mount around so that it is aimed toward the north (to the south if you are in the southern hemisphere), with the connection for the counterweight shaft aimed toward the ground.
7. Insert the counterweight shaft into the open end of the mount's declination axis, tightening it fully as shown in Figure 9.2c.
8. Next, either install or tighten the latitude adjustment screws. Often shaped like a T, these screws are used to hold the mount at the desired tilt to match your latitude. Most mounts have a protractorlike scale with an arrow indicator next to this screw to help estimate the tilt.

9. Slide the counterweight(s) onto the shaft, locking them about halfway along to keep them from sliding back and forth as shown in Figure 9.2d. If there is a shaft end knob, attach it to keep the counterweight from slipping off should its lock screw loosen inadvertently.

10. Attach the accessory tray between the tripod legs. In the case of the illustrated mount, the tray is braced against the inside of each tripod leg to make the mount more rigid as shown in Figure 9.2e. Other mounts have accessory trays that attach to the inside of each leg with a hinged bracket.

11. Attach the slow-motion control arms, if equipped. Although both are interchangeable, most people prefer putting the longer one on the declination control, because it often requires a longer reach.

12. Attach the telescope mounting rings to the tube's mounting plate. In most cases, the plate will clip into the equatorial mount and is locked in place by one or two thumbscrews (see Figure 9.2f, telescope not shown for clarity). Tighten the locks on the right ascension and declination axes, if equipped.

13. Lay the telescope tube into the open rings, close the upper half of each ring over the tube, and lock the ring halves in place finger-tight with the thumbscrews. Once set, tighten the two rings to the equatorial head using a small wrench.

14. Attach the finder's base to the telescope tube. This is usually done by slipping the base over a pair of protruding screws and tightening it in place with two thumb nuts.

15. Slip the finder onto the bracket, tightening it in place by hand.

The telescope is now assembled, but it needs to be balanced to work correctly. To do this, keep one hand on the telescope tube to keep it from swinging, loosen the right ascension axis (keep the declination shaft locked), and swing the instrument around until the counterweight shaft is aiming horizontally. Perform the following steps in this order:

1. Loosen the equatorial mount's right ascension axis lock and swing the telescope to the left or right of the mount.

2. Loosen the counterweight slightly so that it can slide back and forth.

3. Lock the counterweight in place when it exactly balances the weight of the telescope. The telescope is now balanced in right ascension.

4. Now, tighten the right ascension axis and loosen the declination lock, keeping one hand on the tube.

5. Turn the telescope so that its tube is horizontal and facing either east or west (relative to the mount's axes).

6. Loosen the tube rings until the telescope can move back and forth. Slide the tube until it stays perfectly horizontal when you release it. Lock the rings in place. The telescope is now balanced in declination, and ready to use.

Setting Up a GoTo Computerized Mount

Many newcomers to astronomy are lulled into thinking that a computer-controlled GoTo telescope mount takes the guesswork out of finding objects in the sky. Not true. Although their pointing accuracy is amazingly good, these mounts still must be set up and initialized before they can work properly. The instructions here are intended to be generic enough to work with most popular, non-GPS, GoTo telescopes, although some—notably, high-end models—require different setup steps.

1. Once again, unpack the telescope and mount from its boxes and inspect the parts for damage, laying everything out on the floor as you go.
2. Follow steps 2 through 4 for "Setting up an Equatorial Mount" to assemble the telescope's tripod.
3. Place the GoTo mount on top of the tripod. The two are often connected by a large knob that threads onto the base of the mount from underneath the tripod head. Make sure to tighten the knob securely.
4. Next, slide the finder base into the small mounting shoe that is already attached to the tube. Lock the base with the small thumbscrew in the shoe.
5. If assembling a refractor or a catadioptric, place the star diagonal into the focuser.
6. Place the telescope tube assembly into the clamp ring that is attached to the mount. If the telescope came already attached to the mount, I recommend separating the two before assembly to prevent damaging the telescope in case there is a problem securing the mount to the tripod.
7. Close the clamp ring around the telescope tube and tighten the securing thumbscrew loosely.
8. Slide the telescope forward and aft in the ring until it is balanced. Be sure to loosen any axis lock screw that might be on the mount first. Be sure not to let go of the telescope tube at this step to prevent it from slipping in the clamp ring, possibly causing damage.

Aligning a Finder

The last step to setting up any telescope involves adjusting the aim of the finder. Unless the finder is exactly parallel to the telescope, it will prove of little use when trying to point at a target. Be sure to do this step in the daytime, because it's much easier to see what you're doing in the light than at night. Here's how to do it.

1. Attach the finder's mounting bracket to the telescope as shown.
2. Slide the finderscope into the mounting rings and tighten the small thumbscrews to prevent it from slipping out. Chances are there are two, three, or six thumbscrews.
3. Aim the telescope toward a distant identifiable object. Although the Moon, a bright star, or a bright planet may be used, I suggest using instead a terrestrial object such as a distant light pole or mailbox. The reason is quite simple: they don't move. Celestial objects appear to move because of the Earth's rotation, making constant realignment of the telescope necessary just to keep up.
4. Center the target in the telescope's field and lock the mounting's axes.
5. To alter the finder's aim if the mounting has six thumbscrews:

    a. Adjust the front three thumbscrews until the finder's tube is centered in the front mounting ring.

    b. Now, loosen the back three screws. Move the finder by hand until the target is centered in the finder's crosshairs. Once set, tighten all the adjustment screws and check to see that the finder or telescope did not shift in the process.

6. If the finder mount has only three adjustment screws:

    a. Aim at a terrestrial target as above and lock the telescope's axes.

    b. Look through the finder to see which direction is out of alignment.

    c. Loosen the two opposite screws and move the finder by hand until the target is centered.

    d. Tighten all thumb screws, making a final check to see that the finder is aimed correctly.

If the mount has two thumbscrews and a spring-loaded piston, alternately loosen and tighten the thumbscrews to move the finder's aim up or down and left to right until the target is centered. The piston will press against the finder as the screws are adjusted, ensuring that it never slips during the process.

Love Thy Telescope as Thyself

Unlike so many other products in our throwaway society, where built-in obsolescence seems the rule, telescopes are designed to outlast their owners. They require very little care and attention, cost nothing to keep, and eat very little. With a little common sense on the part of its master, a telescope will return a lifetime of fascination and adventure. But if neglected or abused, a telescope may not make it to the next New Moon. Are you a telescope abuser?

Storing Your Telescope

Nothing affects a telescope's life span more than how and where it is stored when not in use. Storing a telescope will be addressed further along in the discussion, but first let's consider where to keep it. The choice should be based on several different factors. A good storage place should be dry, dust-free, secure, and large enough to get the telescope in and out easily. Ideally, a telescope should always be kept at or near the outside ambient air temperature. Doing so reduces the cooldown time required whenever the telescope is first set up at night. The quicker the cooldown time, the sooner the telescope will be ready to use.

    Without a doubt, the best place to keep a telescope is in an observatory. It offers a controlled environment and easy access to the night sky. Of course, not everyone can afford to build a dedicated observatory nor is an observatory always warranted. Clearly, an observatory is pointless if the nearest good observing site is an hour's drive away. In cases such as these, a few compromises must be struck.

    If an observatory is not in the stars, other good places to store telescopes include a vented walled-off corner of an unheated garage or a wooden tool shed. I keep some of my telescopes in a corner of my garage. The corner is completely walled off to protect the optics from the inevitable dust and dirt that accumulates in a garage as well as from any automotive exhaust that could damage delicate optical surfaces. A pair of louvered vents was installed in the outside wall of the garage to let air move freely in and out of the telescope room, reducing the risk of mildew. Wooden tool or garden sheds share many of the advantages of observatories and garages for telescope storage, but again I recommend installing one or two louvered vents in the shed's walls for air circulation. Metal, vinyl, and plastic sheds are not as good because they can build up a lot of interior heat on sunny days.

Many amateurs choose to store their equipment in basements. Basements are certainly secure enough and large enough to qualify. Furthermore, they offer easy access provided there is a door leading directly to the outside. Their cool temperatures also keep the optics closer to that of the outside air. Although all these considerations weigh in their favor, most basements fail when it comes to being dry and free of dust. If a basement is your only alternative, invest in a dehumidifier. Clothes closets, another favorite place to hide smaller telescopes, also fall short since clothes act as dust magnets. Remember, unless a spot meets all of the criteria, continue the search.

Regardless of where a telescope is kept, seal the optics from dust and other pollutants when not in use. Usually, this is simply a matter of putting a dust cap over the front of the tube. Most manufacturers supply their telescopes with a custom-fit dust cap just for this purpose. Use it diligently. If the telescope did not come with a dust cap, or if it has been lost over time, then a plastic shower cap makes a great substitute. If you want a more-sophisticated look, some of the companies listed in Appendix A sell dust caps made of rubber in a wide variety of sizes. Furthermore, if the telescope or binoculars came with a case, use it. Not only will a case add a second seal against dust, it also will protect the instrument against any accidental knocks or bumps.

A dark, damp telescope tube is the perfect breeding ground for mold and mildew. To avoid the risk of turning your telescope into an expensive Petri dish, be sure that all of its parts are dry before sealing it up for the night. Tilt the tube horizontally to prevent water from puddling on the objective lens, primary mirror, or corrector plate. No matter how careful you are, optics are bound to become contaminated with dust eventually. A moderate amount of dust has, surprisingly, little effect on a telescope's performance. But if there's a great deal of it, or if the optics have become coated with a film or mildew, the observer sees dimmer, hazier views that lack clarity and "zip."

An optic should be cleaned only when stains are apparent to the eye; otherwise leave well enough alone. Never clean a telescope lens or mirror just for the sake of cleaning it, because every time an optic is touched, there is always the risk of damaging it. Remember the rule that if it ain't broke, don't fix it! The methods described here are for cleaning outer optical surfaces only. Unless you really know what you are doing, I strongly urge against dismantling sealed telescopes (such as refractors and catadioptrics), binoculars, and eyepieces. Dirt and dust will never enter a sealed tube if it is properly stored and protected. Nevertheless, if an interior lens or mirror surface in a sealed telescope becomes tainted by film or mildew, it should be disassembled and cleaned only by a qualified professional. If you attempt it yourself, you may discover that the telescope was much easier to take apart than it is to put back together! Contact the instrument's manufacturer for their recommendations. I know someone who once decided to take apart his 4-inch achromatic objective lens to give it a thorough cleaning. All was going well until it came time to put the whole thing back together. Seems he tightened a retaining ring in the lens cell just a little too much and . . . *crack*! The edges of the crown element fractured. Don't make the same mistake.

Never start to clean a telescope if time is short. For instance, it is not time to decide that your telescope is absolutely filthy as the Sun is setting on a crystal-clear night. Instead, check the optics well beforehand so there are no surprises. To help you along the way, I have divided the cleaning process into two parts: one for lenses and corrector plates, and another for mirrors.

Cleaning Lenses and Corrector Plates

Begin the cleaning process by removing all abrasive particles that have found their way onto the lens or corrector plate. This does *not* mean blowing across the lens with your mouth; you'll only spit all over it. Instead, use either a soft camel-hair brush, as shown in Figure 9.3a, or a can of compressed air, both available from photographic supply stores. Some brushes come with air bulbs to blow and sweep at the same time. If the brush is your choice, lightly whisk the surface of the lens in one direction only, flicking the brush free of any accumulated dust particles at the end of each stroke.

Many amateurs prefer to use a can of compressed air instead of a brush for dusting optical surfaces, because this way the lens is never physically touched. Hold the can perfectly upright, with the nozzle away from the lens at least as far as recommended by the manufacturer. If the can is too close or tilted, there is a chance that some of the spray propellant will strike the glass surface and stain it. Also, it is best to use several short spurts of air instead of one long gust.

 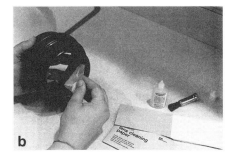

Figure 9.3 *The proper way to clean a lens or a corrector plate: (a) lightly brush the surface with a soft camel-hair brush and then (b) gently wipe the lens with lens tissue and lens-cleaning solution.*

With the dust removed, use a gentle cleaning solution for fingerprints, skin oils, stains, and other residue. Don't use a window glass cleaning spray or other household cleaners. They could damage the lens' delicate coatings beyond repair. Photographic lens-cleaning fluid can be used, but can occasionally leave a filmy residue behind.

One of the best lens-cleaning solutions can be brewed right at home. In a clean container, mix three cups of distilled water with a half cup of pure-grade isopropyl alcohol. Add two or three drops of a mild liquid dishwashing soap (you know, the kind that claims it will not chap hands after repeated use).

Dampen a piece of sterile surgical cotton (not artificial cotton balls, which can scratch optical coatings) or lens tissue with the solution. Do not use bathroom tissue or so-called facial tissue; they are usually impregnated with perfumes and dyes and tend to break down in water. Squeeze the cotton or lens tissue until it is only damp, not dripping, and gently blot the lens, as shown in Figure 9.3b. Avoid the urge to use a little elbow grease to get out a stubborn stain. The only pressure should be from the weight of the cotton wad or lens tissue. Once done, use a dry piece of lens tissue to blot up any moisture.

The steps for cleaning the corrector plate of a catadioptric telescope are pretty much the same as just detailed. The only difference is in the blotting direction. In this case, begin with the damp cotton or tissue at the secondary mirror holder in the center of the corrector and move out toward the edge. Follow a spokelike pattern around the plate, using a new piece of cotton or tissue with each pass. As you stroke the glass, turn the cotton or tissue in a backward-rolling motion to carry any grit up and away from the surface before it has a chance to be rubbed against the optical surface. Overlap the strokes until the entire surface is clean. Again, gently blot dry.

Mirror Cleaning

Cleaning a mirror is much like cleaning a lens in that special care must be exercised to prevent damaging the fine optical surface. The mirror's thin aluminized coating is extremely soft, especially when compared to hard, abrasive dirt, and it can be easily gouged. This is not meant to scare you out of cleaning your mirror if it really needs it, only to heighten your awareness.

The operation for cleaning a telescope's primary or secondary mirror typically requires removing it from the telescope and the cell that holds it in place. Consult your owner's manual for more details on mirror removal. With the mirror lying on a table, blow compressed air across its surface to rid it of any large dust and dirt particles. (Remember, keep the can of compressed air vertical.) Do not use a brush for this step to avoid any possibility of damage.

Next, inspect the mirror's coating for pinholes and scratches. A good coating can last ten years and even longer if the mirror has been well cared for. To check its condition, hold the mirror, its reflective side toward you, in front of a bright light. It is not unusual to see a faint bluish image of the light source through the mirror if the source is especially bright, but its image should appear the same across the entire mirror. If not, there may be thin, uneven spots in the coating. Any scratches or pinholes in the coating will become immediately obvious as well. A few small scratches or pinholes, while not desirable, can be lived with. But if scratches or pinholes abound, or if an uneven coating is detected, then the mirror should be sent out for recoating. Appendix A lists several companies that recoat mirrors. Consult any or all of them for prices and shipping details. It is also a good idea to have the secondary mirror recoated at the same time, since it probably suffers from the same problems as the primary. In fact, most coating companies will include the secondary at no additional cost.

If the coating is acceptable, bring the mirror to a sink. Be sure to clean the sink first and lay a folded towel in the sink as a cushion, just in case—*oops!*—the mirror slips. Run lukewarm tap water across the mirror's reflective surface. This should lift off any stubborn dirt particles that refused to dislodge themselves under the compressed air. End with a rinse of distilled water. Tilt the mirror on its side next to the sink on a soft, dry towel and let the water drain off the surface. Examine the mirror carefully. Is it clean? If so, quit.

If you want to go further, thoroughly clean the sink to remove any gritty particles that may not have washed down the drain. Next, fill the sink with enough tepid tap water to immerse the mirror fully and add a few drops of gentle liquid dish soap (the same as used in the lens-cleaning solution). As shown in Figure 9.4a, carefully lower the mirror into the soapy water and let it sit for a minute or two. With a big, clean wad of surgical cotton, sweep across the mirror's surface ever so gently with the same backward-rolling motion, being careful not to bear down. Now is not the time to act macho. After you've rolled the cotton a half-turn backward, discard it and use a new piece. If stains still exist after this step is completed, let the mirror soak in the water for five to ten minutes and repeat the sweeping with more new cotton.

With the surface cleaned to your satisfaction, drain the sink once again. Run tepid tap water on the mirror for a while to rinse away all soap. Then turn off the tap and pour room-temperature distilled water across the surface for a final rinse. Finally, rest the mirror on edge on a towel where it may be left to air dry in safety. I will usually rest it against a pillow on my bed (Figure 9.4b). Tilt

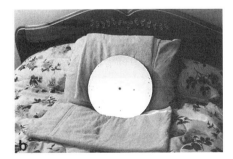

Figure 9.4 *The proper way to clean a mirror: (a) after using compressed air or a camel-hair brush to remove any loose contaminants, wash the mirror in the sink using a gentle liquid dishwashing detergent and surgical cotton; (b) after rinsing the mirror with distilled water, tilt it nearly vertically to prevent water from puddling.*

the mirror at a fairly steep angle (greater than 45°), its edge resting on the soft towel, to let any remaining water droplets roll off without leaving spots. Close the door behind you to prevent any nonastronomers from touching the mirror. When the mirror is completely dry, reassemble the telescope. Recollimate the optics using the procedure described later in this chapter, and you're done!

Other Tips

Other telescope parts also require occasional attention as well. For instance, rack-and-pinion focusing mounts will sometimes bind if not lubricated occasionally. To prevent this from happening, spray a little dry Teflon lubricant on the mount's small driving gear. To do this, remove the screws (typically two) that hold the small plate onto the side of the focuser housing, taking care not to lose anything along the way. With the plate out of the way, look inside at the small pinion gear that meshes with and drives the drawtube in and out. Squirt a little (very little) lubricant on the pinion teeth, replace the cover plate, and wipe off any drips as required.

Readers who own Chinese and Taiwanese telescopes already know that manufacturers feel obliged to coat nearly every moving part of the telescope with thick grease that turns very sticky in actual use. This "glubrication" can really gum things up at times, since not only can it impede a focuser from turning smoothly or slow a clock drive to a crawl, it also attracts dust and dirt that can do great damage to the gearing. My best advice is to use a solvent or degreaser to remove the glue-grease, and then relube everything with a dry Teflon spray. Just be very careful when using the solvent, because you do not want it to wind up on any optical surfaces. If possible, remove the piece to be cleaned from the telescope (especially, the focuser) and clean it away from the optics. Replace everything when done.

If a metal telescope mounting begins to move roughly or starts to bind, spray a small amount of dry Teflon lube on the axes' bearing points. This will keep the telescope moving freely and evenly, rather than binding and grabbing. Some manufacturers recommend this be done at specific intervals, while others make no mention of it at all. If nothing is said in the owner's manual, do it once a year or so.

The typical wood-Formica-Teflon construction of Dobsonian mounts requires little in the way of maintenance. If your Dobsonian does not move freely in altitude or azimuth, however, take the mount apart and apply a little furniture polish or car wax on the contact surfaces. Buff, put the mounting back together, and take it for a test spin. The difference should be immediately noticeable. If not, consider replacing the pads with furniture slides, available at most hardware stores.

Some clock drives also need an occasional check to keep them happy. Carefully remove the drive's protective cover plate or housing and apply a bit of thin grease between the two meshing gears. While the drive is open, put a drop or two of thin oil on the motor's shaft as well. Finally, reassemble the drive and turn it on. Listen for any noises. Most clock drives hum as they slowly turn. If unusually loud, angry, or grinding noises are coming from it, turn the drive off immediately and contact the manufacturer for recommendations.

Get It Straight! (Collimation)

Is that hapless stargazer in Figure 9.5 you?

There is nothing more frustrating or disappointing to an amateur astronomer than to own a telescope that doesn't work as expected. "That lousy company," you think to yourself. "I spent over a thousand dollars on a telescope, and what do I get? A big, expensive piece of junk!" All too often we are quick to blame the poor images produced by our telescopes on faulty optics and poor workmanship. Yet this is not necessarily the case. Although some telescopes are truly lacking in quality, most work just fine if they are in proper tune. Even with the finest optics, a telescope will show nothing but blurry, ill-formed images if those optics are not in proper alignment. Correct optical alignment, or collimation, is necessary if we expect our telescopes to work to the best of their ability.

What exactly is meant by *collimation*? A telescope is said to be collimated if all of its optics are properly aligned to one another. Refractors, for instance, are collimated when the objective lens and eyepiece are perpendicular to a line connecting their centers (of course, this is assuming the lack of a star diagonal). A Newtonian reflector is collimated when the optical axis of the primary mirror passes through the centers of the secondary mirror and the eyepiece. Furthermore, the eyepiece focusing mount must be at a right angle to the optical axis. Finally, for the discussion here, all Cassegrain reflectors and all Cassegrain-based catadioptric telescopes will be lumped together. To be precisely collimated, their primary and secondary mirrors must be parallel to

Figure 9.5 *From* Russell W. Porter *by Berton Willard. Reprinted with permission of the author.*

each other, with the center of the secondary lying on the optical axis of the primary. The eyepiece holder must be perpendicular to the mirrors, with the telescope's optical axis passing through its center. Refer to Figure 2.4 in chapter 2 for clarification if these telescope descriptions are unfamiliar.

The optics in some types of telescopes are apt to go out of alignment more easily than others. For example, it is rare to find a refractor that is not properly collimated unless its tube is bent or warped, usually the result of abuse and mishandling. On the other hand, most Schmidt-Cassegrain telescopes will experience some collimation difficulties in their lives, while Newtonian and Cassegrain reflectors are notoriously easy to knock out of alignment.

It's easy to tell if a telescope is collimated properly. On the next clear night, take your telescope outside and center it on a bright star using high magnification. Make sure the star is in the center of the field. Place the star slightly out of focus and take note of the diffraction rings that result. If the telescope is properly collimated, then the rings should appear concentric around the star's center. If, however, the rings appear oval or lopsided—and remain oriented in the same direction when you turn the focuser both inside and outside of best focus—then the instrument is in need of adjustment.

Collimating a Newtonian

Here's how to collimate a Newtonian reflector. Remember that in general, the faster the telescope—that is, the lower the f-number—the more critical its collimation. By following these steps today, you will have a better telescope tonight.

1. Begin by purchasing or making a *sight tube*. A sight tube is nothing more than a long, empty tube with a pair of crosshairs at one end and a precisely centered pinhole at the other. The outer diameter of the sight tube must match the inside diameter of the telescope's eyepiece holder (typically, 1.25 inches). Several companies sell excellent sight tubes, as you have seen in chapter 7. Or, you can make one yourself from plans available in the *Star Ware* section of www.philharrington.net.

2. Check that the center of your telescope's primary is marked. If your telescope's mirror has a centered dot, then skip to step 5; otherwise, you will need to put one there. To do this, the mirror and its mounting cell must be removed from the bottom of the tube. Most cells are held in the tube by three or four screws protruding through the tube's side. Lay the telescope on a bed or sofa and slowly unscrew each of these screws, making certain that you are holding onto the back of the mirror cell to prevent the cell and mirror from falling or rolling into the tube. With all screws removed, slowly extract the mirror and cell from the tube, being especially careful not to touch the mirror's surface. Mark each hole in the cell and on the tube so that you can line each up again in the same orientation afterward.

3. Place the mirror on a large piece of tissue paper and trace its diameter with a pencil. Cut out the tracing precisely along its edge. Fold the paper circle in half, then in half again so that you end up with a 90° wedge. Snip

off the tip of the wedge with a pair of scissors and unfold the paper. The resulting hole marks the center of the mirror tracing.

4. Using the tracing as a template, make a tiny mark at the center of the mirror with a permanent felt-tip marker. Carefully check the spot's accuracy with the ruler, and if acceptable, stick an adhesive-backed hole reinforcer (the kind used to strengthen loose-leaf paper) on the mirror around the black dot. Don't worry about this central spot blocking part of the primary. It's located in the shadow of the secondary and never sees light anyway. Reinstall the primary in the reverse order that you removed it.

5. Aim the telescope toward a bright scene, such as the daytime sky (but not at or near the Sun!), turn the focuser in as far as it will go, insert the sight tube, and take a look. Identify all of the parts that are pictured in Figure 9.6a. Ideally, you should see the secondary mirror centered in the sight tube's crosshairs. If it is, skip to the next step; if not, move to the front of the telescope tube and look at the back of the secondary mirror holder. Most Newtonians use a four-vane spider mount to hold and position the secondary in place. Spider mounts typically grasp the secondary in a holder supported on a central bolt. Loosen the nut holding the bolt in place and move the

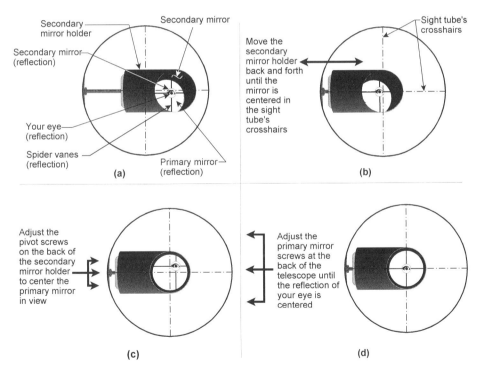

Figure 9.6 *Collimating a Newtonian reflector is as easy as 1-2-3. (a) The view through an uncollimated telescope. (b) Adjust the secondary mirror's central post until the mirror is centered under the focuser tube. (c) Turn the secondary mirror's three (possibly four) adjustment screws until the reflection of the primary mirror is centered. (d) Finally, adjust the primary mirror until the reflected image of your eye is centered in view.*

secondary mirror holder in and out along the optical axis until the secondary's outer diameter is centered in the sight tube, as shown in Figure 9.6b. Before tightening the nut, check to make sure that the secondary is not turned away from the eyepiece, since it can also rotate as it is moved.

6. For the secondary mirror to be adjusted correctly, the primary mirror must appear centered. To do this, most spider mounts have three equally spaced screws that, when turned, pivot the secondary's tilt. By alternately loosening and tightening the screws, pivot the secondary mirror until the primary appears centered in the sight tube, as shown in Figure 9.6c.

7. Aligning the primary mirror will go much faster if you have an assistant. First, look at the back of the primary mirror's cell. You should see three (sometimes six) screws facing out. These adjust the tilt of the primary. Each works a little differently to accomplish the same thing. Cells with three adjustment screws rely on stiff springs to press outward against the cell's plates to keep the mirror in position. Those that use six screws, called *push-pull* cells, are set up in pairs. One screw in each pair (usually the inner one) adjusts the mirror, while the other prevents the mirror from rocking. If your telescope uses this type of mirror mount, the three outer screws must be loosened slightly before an adjustment can be made.

8. With your helper at the adjustment screws, look back through the sight tube. You should see the reflection of the primary centered in the crosshairs, as shown in Figure 9.6d. In bright light, the primary's central spot should also be seen. Following your instructions, have your assistant turn one or more of the screws—slightly. Did the primary's reflection move a little? If so, did the reflection of the center dot move close or farther from the sight tube's crosshairs? If closer, then have your assistant continue turning the screw in the same direction; if not, then turn the screw back to where it was and try another one. Continue turning the screws in and out until the reflected image of the center dot and your eye in the secondary mirror appear centered. Again, do this step very slowly and deliberately, no more than a quarter-turn at a time.

If you own an f/6 or slower Newtonian, the mirrors should now be collimated adequately. If, however, your Newtonian is below f/6, then some fine-tuning will be needed. For this, another tool, called a *Cheshire eyepiece,* is strongly recommended. A Cheshire eyepiece, really a misnomer since no lenses are involved, has a large hole in its side and, inside, a shiny 45° face. By shining a flashlight through the hole, light will bounce from the Cheshire eyepiece to the secondary mirror, then to the primary mirror, back to the secondary, and out through the small peephole in the Cheshire.

1. For the final collimation adjustment, shine a flashlight beam into the Cheshire's side opening and look through the peephole. Centered in the dark silhouette of the secondary mirror will be a bright donut of light—the reflection of the Cheshire's mirrored surface. The dark center is actually the peephole in that surface. Adjust the primary mirror until the mirror's black reference spot is centered in the Cheshire eyepiece's donut.

2. With the telescope collimated, take it back out on the next clear night and let the optics fully adjust to the outside air temperature. Using a medium-

power eyepiece, place a moderately bright star in the center of the field. Defocus the image until it is transformed from a point into a tiny disk surrounded by bright and dark rings. If the mirrors are properly collimated, then the rings should look concentric, like a bull's-eye. If not, one or both of the mirrors may need a little fine tweaking. Be sure to recenter the star in the field every time an adjustment is made.

Some collimating tools combine a sight tube and a Cheshire eyepiece, giving you both tools in one. Although this sounds handy, these can be difficult to use because the crosshairs tend to block the Cheshire's alignment cues.

Collimating a Schmidt-Cassegrain

Unlike the Newtonian, where both the primary and secondary mirrors can be readily accessed for collimation, commercially made Schmidt-Cassegrain telescopes (SCTs) have their primary mirrors set and sealed at the factory. As such, an owner cannot adjust the primary mirror if misalignment ever occurs. Fortunately, SCTs are rugged enough to put up with the minor bumps that might occur during setup without affecting collimation.

This leaves only the secondary mirror to adjust, as shown in Figure 9.7. The following checklist should help walk you through the process.

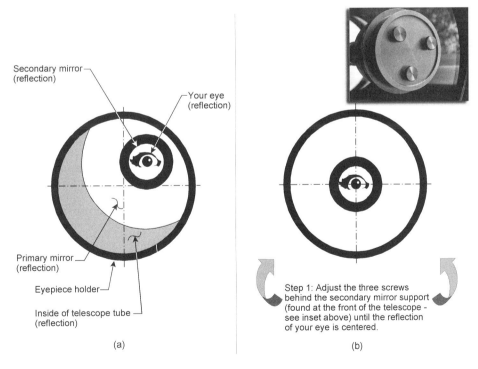

Secondary mirror
(reflection)

Your eye
(reflection)

Primary mirror
(reflection)

Eyepiece holder

Inside of telescope tube
(reflection)

(a)

Step 1: Adjust the three screws behind the secondary mirror support (found at the front of the telescope - see inset above) until the reflection of your eye is centered.

(b)

Figure 9.7 *Collimating a Schmidt-Cassegrain involves adjusting the secondary mirror (a) until its reflection appears centered in the primary as you look through the empty focuser (b). The inset at upper right shows an example of the secondary mirror's three adjustment knobs. Inset photo courtesy of Bob's Knobs.*

1. Take a look at the secondary mirror mount centered in the front corrector plate. There, you will see the heads of three adjustment screws spaced 120° apart. (Note that on some models, a plastic disk covers the screw heads. If so, it must be removed—carefully—to expose the adjustment screws.) In addition, some telescopes have a fourth large screw or nut in the center of the secondary holder. If so, *do not touch it*! Loosening that central screw will release the secondary mirror from its cell and drop it into the tube. Talk about a quick way to ruin your day!

2. Remove the star diagonal and insert a sight tube into the eyepiece holder. Take a look. Ideally, you should see your eye centered in the secondary mirror. If the view is more like Figure 9.7a, turn one or more of the adjusting screws until everything lines up as in Figure 9.7b.

3. Now, set the telescope up as you would for an observing session, giving it adequate time to cool to the night air.

4. With the telescope acclimated to the outdoor temperature, remove the star diagonal if so equipped, and insert a medium-power eyepiece into the telescope.

5. Center the instrument on a very bright star. Defocus the star until it fills about a third of the eyepiece field.

6. Take a look at the dark spot on the out-of-focus disk. That's the dark silhouette of the secondary mirror. For the telescope to be properly collimated, the secondary's image must be centered on the star, creating a doughnut-like illusion. If the doughnut is asymmetric, then the secondary mirror needs further adjustment (but again, make sure the star is perfectly centered in the field of view).

7. Make a mental note of which direction the silhouette favors, go to the front of the telescope, and turn the adjustment screw that most closely coincides to that direction ever-so-slightly. (Since the collimation screws on most SCTs can only be turned using either a screwdriver or an Allen wrench to turn them, many owners have replaced them with no-tool-required knobs. Bob's Knobs are especially popular and will make this chore much easier.)

8. Return to the eyepiece, recenter the star in view, and look at the dark spot again. Is it better or worse? If it's worse, turn the same screw the opposite way; if it's off in a different direction, turn one of the other screws and see what happens. Continue going back and forth between eyepiece and adjustment screws until the dark spot is perfectly centered in the star "blob."

9. To double-check the adjustment, switch to a high-power eyepiece. Place the star back in the center of the field, and defocus its image slightly. If the secondary's dark outline still appears uniformly centered, then collimation was a success and the telescope is set to perform at its best. If not, repeat the procedure but with finer adjustments.

10. If the image is still not correctly aligned even after repeated attempts, then there is a distinct possibility that the primary mirror is not square to the secondary mirror. Focus on a rich star field. If any coma (ellipticity) is evident around the stars at the center of view, chances are good that the pri-

mary is angled incorrectly. In this case, your only alternative is to contact either the dealer where the telescope was purchased or the manufacturer.

To learn more about the fine art of telescope collimation, consult the book *Perspectives on Collimation* by Vic Menard and Tippy D'Auria.

Put Your Telescope to the Test

There is no way to guarantee that every telescope made, even by the finest manufacturers, will work equally well. Although companies have quality control measures in place to weed out the bad from the good, a lemon is bound to slip through every now and then. That is why it is important to check an instrument soon after it is purchased. Examine the telescope for any overt signs of damage. Next, look at the optics. They should be free of obvious dirt and scratches. The mounting should be solid and move smoothly. If any problems are detected, contact the outlet that the telescope was purchased from immediately so that the problem can be rectified.

Although it is easy to tell clean, scratch-free optics from those that are not, a poor-quality lens or mirror may not be immediately obvious. Fortunately, an elaborate fully equipped optical laboratory is not required to check the accuracy of a telescope's optics. Instead, all that you need are a well-collimated telescope, a moderate-power eyepiece, and a clear sky. With these ingredients, any amateur astronomer can perform one of the most sensitive and telling optical tests available: the star test. A fourth ingredient (this book) will help interpret the star test's results.

Before you run a valid star test, make certain that the optics are properly collimated. It is impossible to get accurate results from the star test if a telescope's optics are misaligned. Now might be a good time to review that procedure (found earlier in this chapter) if you are uncertain.

Once everything is in alignment, select a star with which to run the star test. Which star? That depends on your telescope. First, if your scope does not have a clock drive, an equatorial mount, or is not polar aligned, it is best to aim toward a star near the celestial pole, since these move more slowly in the sky than, say, stars near the celestial equator. Which individual star depends on the telescope's aperture. For 4-inch and smaller instruments, select a 1st- or 2nd-magnitude star; 6- to 10-inch telescopes are best checked with a 3rd- or 4th-magnitude star, while larger instruments may be best tested with a 5th-magnitude star. Use a moderately high-power eyepiece for the test. Harold Suiter (*Star Testing Astronomical Telescopes.* Richmond, VA: Willmann-Bell, 1994) recommends a magnification equal to 25× per inch (1× per millimeter) of aperture. So, for a 6-inch telescope, magnification should equal 150×, and so on.

Focus the star precisely, and examine its image closely. Recall from chapter 1 that, at higher magnifications, a star will look like a bull's-eye—that is, a bright central disk (the Airy disk) surrounded by a couple of faint concentric rings (diffraction rings). By definition, any telescope that claims to have *diffraction-limited* optics must show this pattern. Is that what you see?

Probably not, or at least not at first glance. Many factors affect the visibility of diffraction rings, such as telescope collimation, warm-air currents inside the telescope tube, the steadiness of the atmosphere, and the focal ratio of the telescope. Poor atmospheric steadiness, or *seeing* as it commonly referred to, causes star images to *boil*, making it impossible to detect fine detail. Diffraction rings can be seen only under steady seeing conditions (see the section in chapter 9 titled "Evaluating Sky Conditions" for more on this). The aperture of the telescope also plays a big role in seeing diffraction rings. The larger the aperture, the smaller the diffraction pattern and thus the more difficult it is to see.

Slowly rack the image slightly out of focus. Just how far out is a matter of focal ratio. Most people void the star test simply by being too aggressive when defocusing the star's image. Again referring to Suiter, he recommends for an f/10 instrument, the eyepiece should be moved no more than 0.20 inch off focus, while an f/4.5 telescope requires the eyepiece be moved a mere 0.06 inch! In either case, the star's light should enlarge evenly in all directions, like ripples expanding after a small stone is tossed into a calm body of water.

Begin by moving the eyepiece inside of focus. Examine the out-of-focus star image. It should look like one of the patterns illustrated in Figure 9.8. Then, reverse the focusing knob, bringing the star slightly outside of focus. Examine the ring pattern again. If the telescope has first-rate optics (that is, substantially better than merely "diffraction-limited"), both extra-focal images should appear identical.

What if they don't? Compare the exact shapes of the inside-focus and outside-focus patterns with those shown in Figure 9.9. Are the patterns at least circular? No? What geometric pattern do they resemble? Oval? If the oval shapes looks the same both inside and outside of best focus, the optics are doubtlessly out of collimation. If the images are dancing wildly, then either the telescope optics are not yet acclimated to the outside air temperature or the atmosphere is too turbulent to perform the test. What if the images are either triangular or hexagonal? If the patterns have one or more sharp corners, then the optics are probably being pinched, or distorted, by their mounts. Pinched optics are especially common in Newtonian reflectors when the clips holding the primary in the mirror cell are too tight.

"My telescope never focuses stars sharply. I just did the star test, but the out-of-focus patterns appear oval. When the eyepiece is turned from one side of focus to the other, the ovals flip orientation 90°." If that is your telescope, then it suffers from *astigmatism*. Astigmatism may be caused by poorly figured optics, but it may also be caused by pinched optics, uncollimated optics, or by the cooling-down process after the scope is taken outdoors. In Newtonian reflectors, a misaligned secondary mirror is the most frequent cause of astigmatism. If the axis of the star-oval is parallel to the telescope tube, suspect the secondary mirror.

The most common optical defect found in amateur telescopes is *spherical aberration*. Spherical aberration becomes evident when a mirror or lens has not been ground and polished to its required curvature. As a result, light from around the edge of the optic comes to a focus at a different distance than light from the center. Spherical aberration comes in two varieties: one caused by undercorrected optics and one caused by overcorrected optics. Both produce

Here's how a star should look . . .

Inside Focus   In Focus   Outside Focus

(a) through an unobstructed telescope (e.g., a refractor)

Inside Focus   In Focus   Outside Focus

(b) through an obstructed telescope (e.g., a reflector or a catadioptric)

Figure 9.8 *What should a star look like through the perfect telescope? Ideally, all light should come to a common focus, causing a star to appear as a tiny disk (the Airy disk) surrounded by concentric diffraction rings. The star should expand to look like one of the illustrations here when seen out of focus, the image being identical on either side of focus: (a) through a refractor (unobstructed telescope) and (b) through a reflector or a catadioptric (obstructed telescope).*

similar effects—on one side of focus, the outermost part of the bull's-eye pattern is brighter or sharper than on the other side of focus.

What if, even when a star is brought out of focus, rings cannot be seen at all? Instead, all that is seen is a round, mottled blob. This condition, more common in reflectors and catadioptrics than in refractors, indicates a rough optical surface. Mirror makers have an especially appropriate nickname for this problem: *dog biscuit*.

Although not as accurate as the star test, another simple way to check optical quality of reflecting telescopes is the Ronchi test. The Ronchi test uses

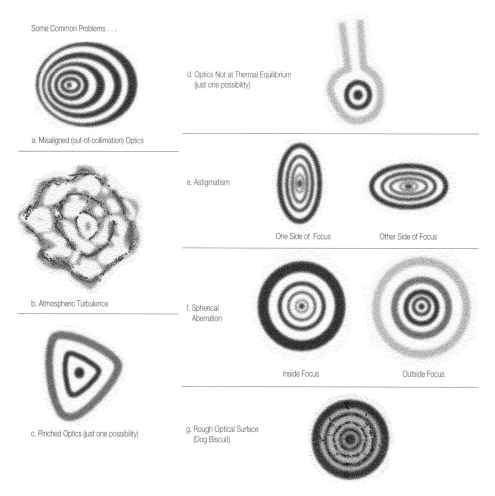

Some Common Problems . . .

a. Misaligned (out-of-collimation) Optics

b. Atmospheric Turbulence

c. Pinched Optics (just one possibility)

d. Optics Not at Thermal Equilibrium
(just one possibility)

e. Astigmatism

One Side of Focus   Other Side of Focus

f. Spherical
Aberration

Inside Focus   Outside Focus

g. Rough Optical Surface
(Dog Biscuit)

Figure 9.9 *Something is amiss if the star test yields any of these results: (a) optics are out of collimation; (b) turbulent atmospheric conditions; (c) optics are being pinched or squeezed by their mounting; (d) optics have not acclimated to the outside air temperature; (e) optics suffer from astigmatism; (f) optics suffer from spherical aberration; or (g) optics suffer from rough optical surface.*

a diffraction grating made of many thin parallel lines printed on clear glass or plastic. The idea is to hold the diffraction grating up to the reflector's eyepiece holder, with the eyepiece removed, and examine a star's blurred image through the grating.

To perform a Ronchi test, you will need a 100-line-per-inch diffraction grating, available from Schmidling Productions, Orion Optics (UK), or Edmund Industrial Optics, among others. With the grating in hand, set up your telescope outside and allow it to cool to the outside air temperature. While you are waiting for the optics to acclimate, check and adjust the instrument's collimation. The Ronchi test will not work properly with poorly aligned optics.

Center the telescope on a bright star, remove the eyepiece (and star diagonal, if used), and hold the diffraction grating up to the empty eyepiece holder. Turn the focusing knob slowly in one direction until you see an enlarged disk with four superimposed black lines. Are the lines straight or are they curved or crooked? If the latter, in what direction are they bent? Repeat the test by turning the focusing knob in the opposite direction. You'll see the star shrink in size, then enlarge again. Continue defocusing until the star's disk expands and four dark lines are visible again. As before, examine their curvature. Note how, this time, the lines curve in the opposite direction.

Depending on how much and in what manner the lines curve, your optics may have one of several problems, or no problem at all. Figure 9.10 interprets some of the common results. If the lines are straight, your telescope probably enjoys a properly corrected mirror or lens. (Even if it passes the Ronchi test, be sure to double-check the instrument with the more-telling star test. Some instruments will pass the Ronchi test but fail the star test miserably.) If, however, the lines are bent or crooked, the telescope suffers from one of the imperfections shown in Figure 9.10. Be forewarned that, like the star test, an improperly collimated or uncooled telescope may show the same result. Before you accuse your instrument of poor optics, double-check the alignment, let the optics cool, and then repeat the test.

Both the star test and the Ronchi test examine all the optical components collectively; they can ascertain if there is a problem, but they cannot immediately distinguish which optical component is at fault. To point the finger of blame in a reflector, try rotating the primary mirror 90° and retesting. If the defect in the pattern turns with the mirror, the primary is at fault. If it does not, the secondary mirror is the culprit.

The moral to all this is simple: there is no substitute for good optics. Pay close attention to manufacturers' claims and challenge those that appear too good to be true. If you are considering a telescope made by the XYZ Company,

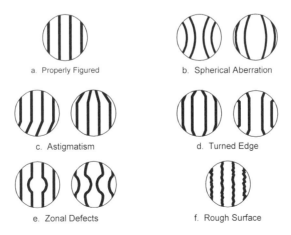

Figure 9.10 *Some possible results from the Ronchi test and their causes. As explained in the text, ideally you should see straight lines during the test.*

write or call to inquire about its optics. Who makes them? What tests are performed to evaluate optical quality? And finally, what kind of guarantee is offered? Remember, if you want to approach perfection, you will probably have to pay a premium for it.

Have Telescope, Will Travel

Not long ago, the most traveling a telescope would ever do was the trip from the house to the backyard. Not anymore. With the increased popularity of large regional or national star parties, as well as the ever-worsening problem of light pollution, many amateur telescopes routinely travel tens or hundreds of miles in search of dark skies.

Whenever a telescope is picked up and moved, the risk of damage is present. That is why some hobbyists tend to shy away from driving around with such delicate instruments. With a little forethought and common sense, though, this risk can be minimized.

Let's first examine traveling with a telescope by car. If you own, say, a subcompact sedan and an 18-inch Newtonian, you have a big problem with only one solution: buy another car. If your telescope and automobile are already compatibly sized, however, it is just a matter of storing the instrument safely for transport. Take apart as little of the telescope as necessary. If there is room in the car to lay down the entire instrument, mounting and all, by all means do so. Take advantage of anything that can minimize the time required to set up and tear down.

If the telescope tube must be separated from its mounting, it is usually best to place the tube into the car first. Be sure to seal the optics from possible dust and dirt contamination. Usually this means simply leaving the dust caps in place, just like when the telescope is stored at home. If the telescope does not come with a storage case, protect the tube from bumps by wrapping it in a clean blanket, a quilt, or a cheap sleeping bag. Strategically placed pillows and pieces of foam rubber can also help to minimize screw-loosening vibrations. If possible, strap the telescope in place using the car's seat belt.

Next comes the mounting. Carefully place it into the car, making sure that it does not rub against anything that may damage it, or that it may damage. Wrap everything with a clean blanket for added protection. Again, use either the car's seat belts or elastic bungee cords to keep things from moving around during a sharp turn. Be sure to secure counterweights, which can become dangerous airborne projectiles if the car's brakes are hit hard.

Transporting telescopes by air presents many additional problems. Their large dimensions usually make it impossible to carry them on a plane and store them under the seat. Therefore, we must be especially careful when packing these delicate instruments.

Some owners of 8-inch and smaller Schmidt-Cassegrain telescopes prefer to wrap their instruments in foam rubber and place them in large canvas duffle bags and carry them onto the plane, placing the scopes into the overhead compartments. Although this method usually works (note that some overhead compartments are not large enough to do this), the chance of damaging the telescope is great. This practice, therefore, is not recommended.

Scott Roberts of Meade Instruments recommends placing the telescope in its carrying case, and then placing the case in a double-walled cardboard shipping carton. Use the original padding that came with the telescope from the factory, if available. He also recommends, for Meade 10-inch and larger instruments, reinstalling the mirror lock shipping bolt that originally came with the telescope, and that the RA and Dec locks should be released to prevent gear damage should the telescope be dropped in shipment.

Doskocil makes a series of equipment cases that are designed to handle the stress of airline travel, while Celestron says that its deluxe hard case (Celestron part number 302070) is strong enough to take the jostling that air cargo can go through. These cases are packed with cubed high-density foam rubber, so that they may be customized to hold any comparably sized instrument. Celestron's other cases, as well as Meade's soft cases, are not appropriate for airline travel.

Even if you will be using one of these high-impact cases, take a few additional precautions before sending your baby off onto the loading ramp. Begin by completely enclosing the telescope in protective bubble wrap, available in larger post offices and stationery stores. Bubble wrap is a two-layer plastic sheet impregnated with air to form a multitude of cushioning bubbles. The bubbles are available in many different sizes—select the larger variety. Another measure of protection is to place the telescope (case and all) into a heavy-duty multiwalled cardboard box. These boxes may be purchased from independent companies that specialize in wrapping packages for shipment.

Owners of other types of telescopes face even greater challenges. First and foremost, the optics must be removed and placed into a suitable case to be carried on the plane. The empty tube may then be bubble-wrapped, surrounded by Styrofoam popcorn (both inside and outside the tube), and placed into a strong wooden crate. Make certain that the shipping carton can be used again on the return trip, and bring along a roll of packing tape or duct tape just in case an emergency repair is needed.

Because of their weight, tripods and mountings pose special problems. I have traveled by air with a large tripod by first wrapping it in two thick sleeping bags and then packing everything in a large tent carrying case. Heavy equatorial mountings, on the other hand, must be packed professionally. Once again, seek a local crating company for help.

There is no industry-wide policy regarding the transport of telescopes by air. Some airlines permit telescopes to be checked as luggage provided they do not exceed size and weight restrictions. (The purchase of optional luggage insurance is strongly recommended.) Other airlines will not accept telescopes as check-in items at all. In these instances, you must ship the instrument separately ahead of time via an air cargo carrier. Due to the amount of paperwork involved, especially on international shipments, air cargo services usually require that advance arrangements be made. Contact your airline well ahead of departure to find out exact details and damage insurance options.

Make sure that each piece of luggage has both a destination ticket and an identification tag, and that both are clearly visible on the outside. Information on the ID tag should include your name, complete address, and telephone number. Although permanent plastic-faced identification tags are preferred, most check-in points provide paper tags that may be filled out on the spot. I always make it a habit to include a second identification label inside my luggage as well, just in case the outside tag is torn off.

Whatever you do, don't forget to bring along all the tools needed to reassemble the telescope once you arrive, but always keep them in a piece of checked baggage.

Finally, compile an inventory of all equipment that you plan to bring. Include a complete description of each item, such as dimensions, color, serial number, manufacturer, and approximate value. United States Customs requires owners to register cameras and accessories with them on a Certificate of Registration for Personal Effects Taken Abroad form before departure. Contact your nearest Customs office for further information. Keep a copy of the list with you at all times while traveling, just in case any item is lost or stolen. Carriers will be able to find the missing piece quicker if they know what to look for.

10

A Few Tricks of the Trade

Up to this point, we have examined how to choose a telescope and accessories, as well as how to set up, care for, and maintain them. The only item left to discuss is how to use the stuff. Clearly, if a stargazer lacks the knowledge and skills to use his or her equipment, it is doomed to spend more time indoors gathering dust than outdoors gathering starlight. Here is a look at some techniques and tricks used by amateur astronomers when viewing the night sky.

Evaluating Sky Conditions

Clearly, nothing affects our viewing pleasure more than the clarity of the night sky. Just because the weather forecast calls for clear skies does not necessarily mean that the sky is going to be clear. As amateur astronomers are quick to discover, "clear" is in the eye of the beholder. To most people, a clear sky simply means an absence of obvious clouds. But to a stargazer, a truly clear sky is much more.

To an astronomer, sky conditions may be broken down into two separate categories: *transparency* and *seeing*. Transparency is the measure of how clear the sky is, or in other words, how faint a star can be seen. Many different factors, such as clouds, haze, and humidity, contribute to the sky's transparency. The presence of air pollutants, both natural and otherwise, also adversely affect sky transparency. Artificial pollutants include smog and other particulate exhaust, while volcanic aerosols and smoke from large fires (for instance, forest fires) are forms of natural air pollution. Still, the greatest threat to sky transparency comes not from nature but from ourselves. We are the enemy, and the weapon is uncontrolled, badly designed nighttime lighting.

Today's amateur astronomers live in a paradoxical world. On one hand, we are truly fortunate to live in a time when modern technology makes it possible for hobbyists to own advanced equipment once in the realm of the professional only. On the other, we hardly find ourselves in an astronomical Garden of Eden. Although technology continues to serve the astronomer, it is also proving to be a powerful adversary. The night sky is under attack by a force so powerful that unless drastic action is taken soon, our children may never know the joy and beauty of a supremely dark sky. No matter where you look, lights are everywhere: buildings, gigantic billboards, highway signs, roadways, parking lots, houses, shopping centers, and malls. Most of these lights are supposed to cast their light downward to illuminate their earthly surroundings. Unfortunately, many fixtures are so poorly designed that much of their light is directed horizontally and skyward. The result is light pollution, the bane of the modern astronomer.

Have you ever driven down a dark country road toward a big city? Long before you get to the city line, a distinctive glow emerges from over the horizon. Growing brighter and brighter with each passing mile, this monster slowly but surely devours the stars; first the faint ones surrender, but eventually nearly all succumb. Several miles from the city itself, the sky has metamorphosized from a jewel-bespangled wonderland to a milky, orange-gray barren desert. Although I know of no astronomer advocating the total and complete annihilation of all nighttime illumination, we must take a critical look at how it can be made less obtrusive.

Responsible lighting must take the place of haphazard lighting. But let's face it, not many people will be interested in light conservation if the point is debated from an astronomical perspective only. To win them over, they must be convinced that more efficient lighting is good for them. The public must be educated on how a well-designed fixture can provide the same amount of illumination over the target area as a poorly designed one, but without extraneous light scattered toward the sky and at a lower operating cost. That latter phrase, *lower operating cost,* is the key to the argument. The cost of operating the light will be lower because all of its potential is specifically directed where it will do the most good. The wattage of the bulb may now be lowered for the same effect, resulting in a lower cost. Everybody wins—taxpayers, consumers, and, yes, even the astronomers!

Can one person affect a change? That was the dream of David Crawford, the person behind the International Dark-Sky Association (IDA). The nonprofit IDA has successfully spearheaded anti-light-pollution campaigns in Tucson, San Diego, and many other towns and cities. It provides essential facts, strategies, and resources to light-pollution activists worldwide. For more information on how you can join the fight against light pollution, contact the IDA at 3545 North Stewart, Tucson, Arizona, 85716, or on the Internet at www .darksky.org.

From mid-northern latitudes, the clearest nights usually take place immediately after the crossing of an arctic cold front. After the front rushes through, cool, dry air usually dominates the weather for twenty-four to forty-eight hours, wiping the atmosphere clean of smog, haze, and pollutants. Such nights are characterized by crisp temperatures, high barometric pressure, and low relative humidity.

To help judge how clear the night sky actually is, many amateurs living in the Northern Hemisphere use the stars of Ursa Minor (the Little Dipper) as a reference, because they are visible every hour of every clear night in the year. Observers located south of the equator can use the stars of Crux (the Southern Cross) for the same purpose. Figure 10.1 shows the major stars of Ursa Minor and Crux with their corresponding visual magnitudes. Note that in each case the decimal point has been eliminated to avoid confusing it with another star. Therefore, the "20" next to Polaris indicates it to be magnitude 2.0, and so on. You may find that all of the Dipper stars are visible only on nights of good clarity, while other readers may be able to make them out on nearly every night. Still others, observing from light-polluted environs, may never see them all.

The night sky is also judged in terms of *seeing*, which refers not to the clarity but rather the sharpness and steadiness of telescopic images. Frequently, on nights of exceptional transparency (for instance, clear and dark), the twinkling of the stars almost seems to make the sky come alive in dance. To many, twinkling

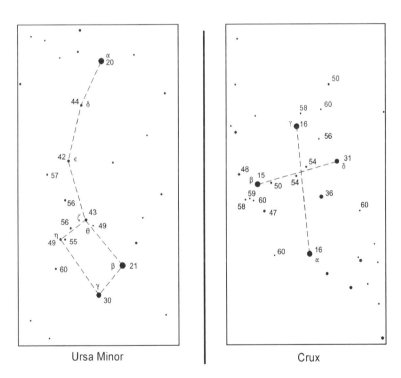

Ursa Minor   Crux

Figure 10.1 *How clear is the sky tonight? Many amateurs in the Northern Hemisphere use the visibility of the Little Dipper (left, Ursa Minor) as a gauge, while amateurs south of the equator use the Southern Cross (right, Crux). In both figures, decimals have been omitted to avoid confusing them with stars. Therefore, a 2.0-magnitude star is shown as "20" on the chart, and so on.*

adds a certain romantic feeling to the heavens, but to astronomers, it only detracts from the resolving power of telescopes and binoculars.

The twinkling effect, called *scintillation,* is caused by turbulence in our atmosphere. Density differences between warm and cold layers refract or bend the light passing through, causing the stars to flicker. Ironically, the air seems steadiest when a slight haze is present. Although any cloudiness may make faint objects invisible, the presence of *thin* clouds can enhance subtle details in brighter celestial sights such as double stars and the planets.

To help observers in North America evaluate their local sky conditions, Canadian amateur astronomer Attilla Danko offers a wonderful free Internet service called Clear Sky Clocks. Each of the hundreds of observing sites in Danko's directory uses meteorological data from the Canadian Meteorological Centre to predict if and when it will be clear in the next forty-eight hours. Danko's Web site graphically predicts cloud cover, transparency, and darkness. These factors appear as three horizontal rows of squares, each square representing an hour and varying in color from dark blue to white. Check www .cleardarksky.com for more information.

Although the Earth's atmosphere greatly influences seeing conditions, image steadiness also can be adversely affected by conditions inside and immediately surrounding the telescope. If you are like most amateur astronomers, you probably store your telescope somewhere inside your house or apartment—your warm house or apartment. Moving the telescope from a heated room out into the cool night air immediately sets up swirling heat currents as the instrument and its optics begin the cooling process toward thermal equilibrium. Peering through the eyepiece of a warm telescope on a cool night is like looking through a kaleidoscope, with the stars writhing in strange ritualistic dances. As the telescope tube and its optics reach equilibrium with the outside air, the images will begin to settle down. The night's observing may then begin in earnest.

How long it takes for these heat currents to subside depends on telescope size and type as well as the local weather. Newtonian and Cassegrain reflectors seem to acclimate the fastest. Still, these telescopes can require at least one hour in the spring, summer, and fall, and up to two hours in the winter (the greater the temperature change, the longer it will take). Refractors and catadioptric instruments, especially Maksutovs because of their thick corrector plates, need up to twice as long!

Several steps can be taken to minimize the time required for instrument cooling. First, find a cool and dry place to store the telescope when not in use. If a dedicated observatory is impractical, then good alternatives include a vented wooden garden shed or a sealed-off corner of an unheated garage. This way, the telescope's temperature will always be close to the outside temperature even before it is set up. If you must drive with your telescope to an observing site, try to travel with the heater off on the way there.

Tilting the open end of a reflector vertically like a smokestack can help speed air-temperature equalization because warm air rises. In the case of sealed tubes, where a lens or a corrector prevents the "smokestack" effect, it is best to turn the instrument broadside to the wind. The cooling breeze will help wick away heat as it is radiated by the telescope tube.

If the telescope is of your own making, select a material other than metal for the tube. Wood, cardboard, and fiberglass have a lower heat capacity than metal, and therefore will have less heat to radiate once set outside. Unfortunately, nonmetal tubes can also retain the warm air inside the tube, but in practice, the benefit outweighs the disadvantage.

A variation on this theme is not to use a solid tube at all, but rather an open truss framework to support the optics. Such an idea has been popular among observatory-installed telescopes for a century or more, as it speeds thermal equalization while cutting down the instrument's weight. Many Dobsonian-style reflectors use this principal for the same reasons, as well as the ability to break a large instrument down into a relatively small package for transport. Unfortunately, while the speed of temperature adaptation is increased, so are the chances of stray light intruding into the final image. Open tubes are also more susceptible to interference from the observer's body heat, turbulence caused by crosswinds, and possible dewing of the optics. A sleeve made from black cloth wrapped around the truss acts to slow these interferences but will not eliminate them.

As noted in chapter 5, some manufacturers build cooling vents into the design of their instruments—a good idea. Some even offer small, flat "muffin" fans to help rid their telescopes of trapped pockets of warm air. Amateur telescope makers should consider including some type of venting system in their instruments as well, for these additions go a long way in speeding up a telescope reaching temperature equilibrium. Just be sure to seal the vents against dust infiltration when the telescope is not in use.

A telescope is not the only piece of equipment that needs to adjust to the change in temperature after being brought outdoors for the first time at night. The Earth must as well. The ground you place the instrument on, having been exposed to the comparatively warm temperatures of daytime, also has to adapt to the cool of night. Different surfaces absorb heat better than others. Concrete and blacktop are the worst offenders because they readily absorb and retain heat. As the heat radiates back out at night, turbulent microclimates can affect seeing. Grass, although also requiring a cooldown period, is better because it does not retain as much heat.

Your Observing Site

This brings up another hot topic among amateur astronomers today, which is where to view the sky from. Choosing a good observing site is becoming increasingly difficult. The ideal location should be far from all sources of light pollution and civilization in general. In addition, it should be as high above sea level as possible to avoid low-lying haze and fog, it must be safe from social ne'er-do-wells and other possibly harmful trespassers, and have an obstacle-free view of the horizon in all directions. Wouldn't it be nice if this was a description of your backyard? While we can all dream of finding such a Shangri-la, a few compromises usually must be made.

Over the years I have used several different observing sites with varying degrees of success. Many national, state, county, and local parks and beaches offer excellent areas, but their accessibility may be limited to daytime hours only or by residency. Ask your local park office if special access is available. The local authority that oversees the state parks near my home offers a star-gazing permit. This permit allows after-hours access to more than half a dozen parks for a small annual fee. The parks not only have much better horizons than most observing sites but also offer the added benefit of around-the-clock security patrols, an important consideration in these times! (Unfortunately, the patrol cars are outfitted with more lights than a small city, but I guess you have to take the good with the bad.)

Other good alternatives include both private and public golf courses. They have wide-open expanses but may also suffer from restrictions and from excessive security lighting around the clubhouse. If the owner of the course is apprehensive at first, why not offer to run a free observing session for club members in return for nighttime access? Although they may not be bona fide amateur astronomers, most people jump at the chance to see celestial wonders such as the rings of Saturn and the Moon. Flat farmland can also provide a secluded view, but unless the land is your own, be sure to secure permission from the owner beforehand. The last thing you want is to be chased by a gun-wielding farmer at two o'clock in the morning!

Many things must be considered when judging an observing site, especially safety and accessibility. But no matter where it is, a good observing site must be easy to reach. I know many urban and suburban amateurs who never see starlight during the week. Instead, they restrict their observing time to weekends only, because the closest dark-sky site is more than an hour's drive away. Isn't that a pity? First, these amateurs may spend more time commuting to the stars than they do actually looking at them. Second, odds are they are forsaking many clear nights each month just because they believe the sky conditions closer to home are unusable.

The late Janet Mattei, director of the American Association of Variable Star Observers, once said to me, "Even if your observing site isn't the best, make the best of your observing site." Ask yourself honestly, "Are the local sky conditions really that bad?" Remember, a telescope will show the Moon, the Sun, and the five naked-eye planets as well from the center of a large city as it will from the darkest spots on Earth. Hundreds, if not thousands, of double and variable stars are also observable through even the most dismal of sky conditions. True, there is something extra special about observing under a star-filled sky, but never forgo a clear night just because the ambience is less than ideal.

Star Parties and Astronomy Conventions

Perhaps as a reflection of the increasing interference of light pollution, the past few decades have seen a tremendous growth in regional, national, and

Figure 10.2 *The granddaddy of all amateur astronomy conventions, Stellafane hosts thousands of hobbyists each year atop Breezy Hill in Springfield, Vermont.*

international star parties and astronomy conventions (Figure 10.2). At these events, hundreds, even thousands, of amateur astronomers travel to remote spots to hear top-notch speakers, set up their homemade and commercially purchased telescopes, and enjoy dark-sky conditions far superior to the skies back home.

More than a dozen major conventions and hundreds of smaller get-togethers are held across the country and around the world throughout the year. To help spread the word about when and where they will occur, astronomy magazines contain monthly listings giving information and addresses for further information, while the *Star Ware* section of this book's Web site, www.philharrington.net, offers links to listings on the Internet. Check to see if there is an upcoming event near you. If so, try to attend. Astronomy conventions and star parties are great ways to meet new friends, learn a lot about your hobby and science, see many different telescopes, and get a terrific view of the night sky.

Finding Your Way

Once a site is selected, it is time to depart on your personal tour of the sky. Although most novice enthusiasts begin with the Moon and the brighter planets, most objects of interest in the sky are not visible to the unaided eye or even through a side-mounted finderscope. How can a telescope be aimed their way if the observer cannot see the target to point at in the first place? That is where observing technique comes into play. To locate these heavenly bodies, one of three different methods must be used. Before any of these systems can be discussed, it is best to become fluent in the way astronomers specify the location of objects in the sky.

Celestial Coordinates

Like the Earth's spherical surface, the celestial sphere has been divided up by a coordinate system. On Earth, the location of every spot can be pinpointed by unique longitude and latitude coordinates. Likewise, the position of every star in the sky may be defined by *right-ascension* and *declination* coordinates.

Let's look at declination first. Just as latitude is the measure of angular distance north or south of the Earth's equator, declination (abbreviated *Dec.*) specifies the angular distance north or south of the celestial equator. The celestial equator is the projection of the Earth's equator up into the sky. If we were positioned at 0° latitude on Earth, we would see 0° declination pass directly through the zenith, while 90° north declination (the North Celestial Pole) would be overhead from the Earth's North Pole. From our South Pole, 90° south declination (the South Celestial Pole) is at the zenith.

As with any angular measurement, the accuracy of a star's declination position may be increased by expressing it to within a small fraction of a degree. We know there are 360° in a circle. Each of those degrees may be broken into sixty equal parts called *minutes of arc*. Furthermore, every minute of arc may be broken up to sixty equal *seconds of arc*. In other words:

one degree (1°) = 60 minutes of arc (60′)

= 3,600 seconds of arc (3,600″)

When minutes of arc and seconds of arc are spoken of, an angular measurement is being referred to, not the passage of time.

Right ascension (abbreviated *R.A.*) is the sky's equivalent of longitude. The big difference is that while longitude is expressed in degrees, right ascension divides the sky into 24 equal east-west slices called *hours*. Quite arbitrarily, astronomers chose as the beginning or zero-mark of right ascension the point in the sky where the Sun crosses the celestial equator on the first day of the Northern Hemisphere's spring. A line drawn from the North Celestial Pole through this point (the Vernal Equinox) to the South Celestial Pole represents 0 hours right ascension. Therefore, any star that falls exactly on that line has a right ascension coordinate of 0 hours. Values of right ascension increase toward the east by 1 hour for every 15° of sky crossed at the celestial equator.

To increase precision, each hour of right ascension may be subdivided into 60 minutes, and each minute into 60 seconds. A second equality statement summarizes:

1 hour R.A. (1 h) = 60 minutes R.A. (60 m)

= 3,600 seconds R.A. (3,600 s)

Unlike declination, where a minute of arc does not equal a minute of time, a minute of R.A. does.

The stars' coordinates do not remain fixed. Because of a 26,000-year wobble of the Earth's axis called *precession,* the celestial poles actually trace circles on the sky. Right now, the North Celestial Pole happens to be aimed almost exactly at Polaris, the North Star. In 13,000 years, however, it will have shifted away from Polaris, and instead point toward Vega, in Lyra. The passage of another 13,000 years will find the pole aligned with Polaris once again.

Throughout the cycle, the entire sky shifts behind the celestial coordinate grid. Although this shifting is insignificant from one year to the next, astronomers find it necessary to update the stars' positions every fifty years or so. That is why you will notice that the right ascension and declination coordinates are referred to as "epoch 2000.0" in this book and most other contemporary volumes. These indicate their exact locations at the beginning of the year 2000, but are accurate enough for most purposes for several decades on either side.

Star-Hopping

The simplest method for finding faint sky objects, and also the one preferred by most amateur astronomers, is called *star-hopping*. Star-hopping is a great way to learn your way around the sky while developing your skills as an observer. Before a telescope can be used to star-hop to a desired target, its finderscope must be aligned with the main instrument. If you are unfamiliar with that process, pause here and return to page 308, where the process is detailed.

With the finder correctly aligned to the telescope, the fun can begin. After deciding on which object you want to find, pinpoint its position on a detailed star atlas. Scan the atlas page for a star near the target that is bright enough to be seen with the naked eye. Once a suitable star is located on the atlas, turn your telescope (or binoculars) toward it in the night sky. Looking back at the atlas, try to find little geometric patterns among the fainter suns that lie between the naked-eye star and the target. You might see a small triangle, an arc, or perhaps a parallelogram. Move the telescope to this pattern, center it in the finderscope, and return to the atlas. By switching back and forth between the finderscope and the atlas, hop from one star (or star pattern) to the next across the gap toward the intended target. Repeat this process as many times as it takes to get to the area of your destination. Finally, make a geometric pattern among the stars and the object itself. For instance, you might say to yourself, "My object lies halfway between and just south of a line connecting star A and star B." Then locate star A and star B in the finder, shift the view to the point between and south of those two stars, and your target should be in (or at least near) the telescope's field of view. Don't worry if you get lost along the way; breathe a deep sigh and return to the starting point.

Let's imagine that we want to find M31, the Andromeda Galaxy. (M31 is this galaxy's catalog number in the famous Messier listing of deep-sky objects.) Begin by finding it on Figure 10.3. M31's celestial address is right ascension 00 hours 42.7 minutes (often written 00h 42.7m), declination +41° 16'. Start at Alpheratz, the star that appears to be shared by the Great Square and Andromeda, then find the group of stars that arcs from Alpheratz through Delta Andromedae and Mirach to Almach. Center your finderscope on Mirach, then take a turn to the northwest, to the faint naked-eye star Mu

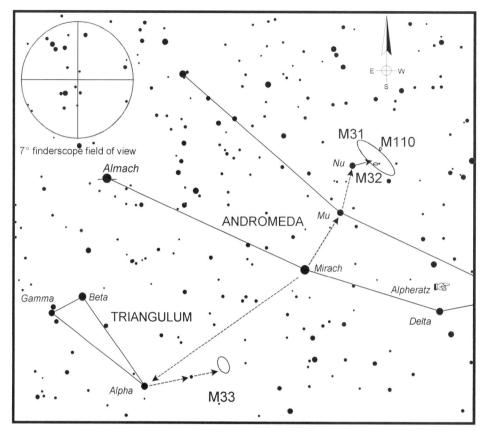

Figure 10.3 *Star-hopping is the only way to find sky objects, according to some observers. Here are suggested plans of attack for locating M31, the Andromeda Galaxy, and M33, the Great Spiral in Triangulum. Both are in the northern autumn sky.*

Andromedae. Continue farther north to Nu Andromedae, which is fainter still but readily visible through finders and binoculars. Look to the west of Nu for a large smudge of light. That's M31. In fact, from even moderately dark skies, M31 is visible to the naked eye, holding the distinction of being the most distant sight visible without telescopic aid (technically, two more distant galaxies—M33 in Triangulum and M81 in Ursa Major—can be glimpsed with the naked eye from extremely dark locations).

Easy, right? Now let's tackle something a little more challenging. Many observers consider M33, the Great Spiral Galaxy in Triangulum, to be one of the most difficult objects in the sky to find. Most references note it as 6th magnitude, which means that if the galaxy could somehow be squeezed down to a point, that would be its magnitude. However, M33 measures a full degree across, so its *surface brightness,* or brightness per unit area, is very low.

M33 and its home constellation are also plotted on Figure 10.3. How can we star-hop to this point? One way is to locate the naked-eye stars Alpha Trianguli and our old friend Mirach. Aim at the former with the finderscope and move slightly toward the latter. About a quarter of the way between Alpha Trianguli and Mirach, and a little to the south, lies a 6th-magnitude star. Spot it in your finderscope. Connect an imaginary line between it and Alpha, and extend that line an equal and opposite distance from the 6th-magnitude star. There, you should see the galaxy's large, hazy glow. Nevertheless, when you peer through your telescope at that point, nothing is there! Recheck the map. No, that's not it; the telescope is aimed in the right direction. It's got to be there! What could be wrong?

At frustrating times such as this, it is best to pull back from your search and take a breather. Then go back, but this time, look a little to one side of where the galaxy should be. This trick is called *averted vision*. By glancing at it with peripheral vision, its feeble light will fall on a more sensitive area of the eye's retina. If that doesn't do it, try gently tapping the side of your telescope with a finger or two. Sometimes, very slight motion will render an invisible object visible. Suddenly, there it is! You will wonder how you ever missed it before!

Setting Circles

This brings us to the second method that amateur astronomers use to find faint objects in the sky: setting circles. Many equatorial mounts still come with a pair of small mechanical setting circles, one on each axis. The circle on the polar axis is divided into 24 equal segments, with each segment equal to 1 hour of right ascension. The setting circle attached to the declination axis is divided into degrees of declination, from 90° north to 90° south. In theory, once the equatorial mounting has been accurately aligned to the celestial pole, a sky object can be located by dialing in its pair of celestial coordinates.

Is that all there is to it? Unfortunately, no. In this day of digital everything, it's a rare mounting that comes with old-style setting circles that are more functional than decorative. If your telescope has these, save yourself hours of frustration and learn how to star-hop or retrofit the telescope with either digital setting circles or computerized GoTo control, if available.

One look through chapter 5 shows that computerized telescopes rule the astronomical world these days. All of these marvels of modern technology fall into one of three categories based on their design: PushTo, equatorial GoTo, and alt-azimuth GoTo.

PushTo telescopes, such as Orion's SkyQuest Intelliscopes as well as any equipped with digital setting circles (Figure 10.4), rely on the observer to move the telescope toward a preselected target until indicators on the small setting circle control box's readout reaches zero. This is much faster and quieter than a GoTo telescope, draws much less battery power, and is every bit as accurate.

Equatorial GoTo telescopes sit on German equatorial mounts outfitted with a DC-powered motor on each axis. These motors are, in turn, connected to a computer that aims the telescope at a selected target automatically, provided the mount is aligned to the celestial pole beforehand. Popular examples include the Celestron Advanced Series CG5-GT, the Meade LXD75, and the Vixen Sphinx.

Figure 10.4 *Digital setting circles are handy for finding objects that pose too difficult a challenge to locate by star-hopping. But, please, don't use them as an excuse not to learn how to star-hop.*

Finally, the mounts that come with Celestron NexStar and Meade LX200 telescopes, among others, are examples of alt-azimuth GoTo mounts. These also computer-driven, but they are based on an altitude-azimuth configuration that does not require polar alignment to work properly. Instead, once the user initializes the mount by inputting date, time, and location (versions that incorporate GPS technology do this automatically), the onboard computer figures out where it is and where everything is in the sky. All PushTos and GoTos require the user to align the mounts to two target stars to complete initialization.

Polar Alignment

GoTo or not, in order for an equatorial mount to track the sky correctly, its polar axis must first be aligned to the celestial pole. Here's how to go about it. Begin by leveling the telescope. This may be done by adjusting the length of each tripod leg (if so equipped) or by placing some sort of a block (for instance, a piece of wood or a brick) under one or more corner footpads of the mount. Don't spend a lot of time trying to make the mount perfectly level; close is good enough. In fact, to be completely correct, it is not necessary to level the mount at all. The only thing that matters is that the polar axis be aimed at the celestial pole. In practice, however, it is easier to polar align a mount that is level than one that is not, so take the time to do so. As an aid, many instruments come outfitted with bubble levels on their mountings.

Next, check to make sure that your finderscope is aligned to the telescope and that the entire instrument is parallel to its polar axis. This latter step is usually accomplished simply by swinging the telescope around until the declination circle reads 90°. Most declination circles are preset at the factory, though some are adjustable; others may have slipped over the years. If you believe that the telescope is not parallel to the polar axis when the declination circle reads 90°, consult your telescope manual for advice on correcting the reading. If your manual says nothing, or worse yet, you cannot find it, try this test. Align the telescope to the polar axis as best you can by eye and lock the declination axis. Using only the horizontal and vertical motion (azimuth and altitude respectively) of the mounting, center some distant object, such as a treetop or a star, in your finderscope's view. Now, rotate the telescope about the polar axis only. If the telescope/finderscope combination is parallel to the polar axis, then the object will remain fixed in the center of the view; in fact the entire field will appear to pivot around it.

If, however, the object moved, then the finder and the polar axis are not parallel to one another. Try it again, but this time pay close attention to the direction that the object shifted. If it moved side to side, shift the entire mounting in azimuth (horizontally) exactly half the horizontal distance it moved by. If it moved up and down, then the mounting's altitude (vertical) pivot is not set at the correct angle. Loosen the pivot and move the entire instrument one-half the vertical distance that the object moved. Since, in all likelihood, it shifted diagonally, this will turn into a two-step procedure. Take it one step at a time, first eliminating its horizontal motion, then the vertical.

With the polar axis and telescope now parallel, it is time to set the polar axis parallel to the Earth's axis. Some equatorial mounts come with a polar-alignment finderscope built right into the polar axis. These come with special clear reticles surrounding the celestial pole. Others can be outfitted with optional polar scopes that thread into the polar axis. If you are polar aligning the telescope simply for visual use or to use a motorized GoTo system, the alignment need only be approximate to get satisfactory results. To do this, simply sight through the mount's polar scope opening, turning the mounting left and right and tilting its angle up and down until the celestial pole (in the case of Northern Hemisphere operation, Polaris) appears in the opening. Be sure to move the entire mount, tripod and all as necessary, but do not move the telescope.

Polar-alignment finderscopes and computer-driven mounts are certainly handy, but for those of us without such luxuries, the following alignment method should work well. With the telescope level and the declination axis locked at +90°, point the right ascension axis by eye approximately toward Polaris. Release the locks on both axes and swing the instrument toward a star near the celestial equator. Once aimed at this star, spin the right ascension circle (taking care not to touch the declination circle) until it reads the star's right ascension. Table 10.1 suggests several suitable stars for this activity. Note that the stars' positions are given at five-year intervals. Choose the pair closest to your actual date. (Although this slight shift is not of much concern when aligning a telescope to use setting circles, it is of great consequence for long-exposure through-the-telescope astrophotography.)

Table 10.1 **Suitable Stars for Setting-Circle Calibration**

Star	Epoch	Right Ascension	Declination
Alpheratz (Alpha Andromedae)			
	(2005.0)	00h 08.6m	+29° 07′
	(2010.0)	00h 08.9m	+29° 09′
	(2015.0)	00h 09.2m	+29° 10′
Hamal (Alpha Arietis)			
	(2005.0)	02h 07.5m	+23° 29′
	(2010.0)	02h 07.8m	+23° 31′
	(2015.0)	02h 08.0m	+23° 32′
Aldebaran (Alpha Tauri)			
	(2005.0)	04h 36.2m	+16° 31′
	(2010.0)	04h 36.5m	+16° 32′
	(2015.0)	04h 36.8m	+16° 32′
Procyon (Alpha Canis Minoris)			
	(2005.0)	07h 39.6m	+05° 13′
	(2010.0)	07h 39.9m	+05° 12′
	(2015.0)	07h 40.1m	+05° 11′
Regulus (Alpha Leonis)			
	(2005.0)	10h 08.6m	+11° 57′
	(2010.0)	10h 08.9m	+11° 55′
	(2015.0)	10h 09.2m	+11° 53′
Arcturus (Alpha Boötis)			
	(2005.0)	14h 15.9m	+19° 10′
	(2010.0)	14h 16.1m	+19° 08′
	(2015.0)	14h 16.4m	+19° 07′
Altair (Alpha Aquilae)			
	(2005.0)	19h 51.0m	+08° 53′
	(2010.0)	19h 51.3m	+08° 54′
	(2015.0)	19h 51.5m	+08° 55′
Polaris (Alpha Ursae Minoris)			
	(2005.0)	02h 38.5m	+89° 17′
	(2010.0)	02h 44.7m	+89° 19′
	(2015.0)	02h 51.2m	+89° 20′
South Pole Star (Sigma Octantis)			
	(2005.0)	21h 12.1m	−88° 56′
	(2010.0)	21h 17.0m	−88° 55′
	(2015.0)	21h 20.3m	−88° 54′

Swing the telescope back toward the celestial pole, stopping when the setting circles read the position of Polaris (also given in Table 10.1). If the telescope is properly aligned with the pole, Polaris should be centered in view. If not, lock the axes and shift the entire mounting horizontally and vertically until Polaris is in view. Repeat the procedure again, until Polaris is in view when the circles are set at its coordinates. Due to the coarse scale of most circles supplied on amateur telescopes, a polar alignment within about 0.5° to 1° of the celestial pole is usually the best you can get. Digital setting circles are much more accurate, but still the alignment need not be overly precise to be useful.

Incidentally, although the previous instructions outline the procedure for aligning to the North Celestial Pole, the actions are the same for observers in the Southern Hemisphere, except the mounting must be aligned to the South Celestial Pole. Everywhere you see "Polaris" or "North Star," substitute "Sigma Octantis" or "South Star." Sigma Octantis, a 5.5-magnitude sun, is located about 1° from the South Celestial Pole. Its celestial coordinates are also given in Table 10.1. Although not as obvious as its northern counterpart, Sigma works well for the task at hand.

With the mounting now polar aligned, the setting circles calibrated, and the clock drive switched on, it is now a relatively simple matter to swing the telescope around until the circles read the coordinates of the desired target. If all was done correctly beforehand, the target should appear in, or at least very near, the eyepiece's field of view. Be sure to use your eyepiece with the widest field when first looking for a target's field before moving up to a higher-power ocular.

Alt-Azimuth GoTo and PushTo Mounts

PushTo telescopes with digital setting circles and GoTo computer-driven telescopes are much easier to get up and running than equatorially mounted instruments. In fact, neither requires any polar alignment at all, except for needing to aim the telescope in the general direction of north (or south, in the Southern Hemisphere) to get going. You don't even need to be able to see, or for that matter, even know Polaris. Instead, a built-in electronic compass will aim the telescope automatically.

Begin by turning on the power to the setting circles or onboard computer. Although both are likely to be battery-powered, the latter draws quite a bit of power, draining a set very quickly. If possible, plug the GoTo control into a rechargeable battery or an AC wall outlet, or use an extension cord. Let the computer initialize. PushTo digital setting circles require that you tilt the telescope tube vertically, and aim the telescope at two of the stars in the onboard database.

GoTo telescopes are more complex but still fairly straightforward. Some will prompt you to enter a variety of information, such as the date, the time, your location (which may be selected from the unit's built-in database), and possibly the telescope mode. GPS models gather this information automatically from GPS satellites.

Once everything is initialized, both work pretty much in the same way. Select a target from the object database. In the case of GoTo telescopes, press "go" and the telescope does the rest. For PushTo scopes, move the telescope by hand, watching the indicator arrows on the hand control box as you go until the location reads "00." Some fine adjustment is inevitable, but if you set everything up correctly, the accuracy of either system should be sufficient to get a target in the field of a low-power eyepiece.

Now that the techniques for using PushTo and GoTo telescopes are familiar, here is why they should not be used, especially by beginners who are unfamiliar with the sky. Using setting circles or a computer to help aim a telescope reduces the observer to little more than a couch potato sports spectator flipping television channels between football games on a Sunday afternoon. Where is the challenge in that? Observational astronomy is not meant to be a spectator's sport; it is an activity that is best appreciated by doing. There is something very satisfying in knowing the sky well enough to be able to pick out an object such as a faint galaxy or an attractive double star using just a telescope, a finder, and a star chart. Stalking an elusive sky object is much like searching for buried treasure: you never know what else you are going to uncover along the way.

Maybe I have been hard on computerized telescopes, maybe even a bit unjustifiably. Make no mistake; they are very useful tools when used for the right reasons. There is no question that they serve a very real purpose for advanced amateurs who are involved in sophisticated research programs, such as searching for supernovae in distant galaxies or estimating the brightness of variable stars. Both of these activities involve rapid, repetitive checking of a specific list of objects. Setting circles can also come in handy when light pollution makes star-hopping difficult or during a public star party. But they should never be used as a crutch. Most amateur astronomers will do much better by looking up and learning how to read the night sky rather than looking down at the setting circles.

Eye Spy

Although we may think of our binoculars and telescopes as wonderful optical instruments, there is no optical device as marvelous or versatile as the human eye. Experts estimate that 90% of the information processed by our brains is received by our eyes. There is no denying it; we live in a visual cosmos.

From an astronomical point of view, we are most interested in the eye's performance under dimly lit conditions, our so-called *night vision.* Try this test on the next clear, moonless night. Go outdoors from a well-lit room and look toward the sky. Chances are you will be able to pick out a few of the brightest stars shining in what looks like an ink-black background. Face away from the sky, wait three or four minutes, and then look up again. You will immediately notice that many more stars seem to have dawned in that short time, an indication that the eye has begun to adapt to its new, darker environment.

Complete the exercise by turning away from the sky a second time, making certain to block your eyes from any stray light. Wait another fifteen to twenty minutes, then look skyward once again. This time, there are even more stars than before. Over the ensuing time, the eye has become fully dark adapted. Not only has the pupil dilated but a shift in the eye's chemical balance has occurred. The buildup of a chemical substance called *rhodopsin* (also known as *visual purple*) has increased the sensitivity of the eyes' low-level light receptors, known as *rods*. Most people's eyes become adjusted to the dark in twenty to thirty minutes, although some require as little as ten minutes or as long as one hour. This is enough to begin a night's observing, but complete dark adaptation takes up to another hour to occur.

Although the eye's sensitivity to dim lighting increases dramatically during the dark adaptation process, it loses most of its sensitivity to color. As a result, visual observers can never hope to see the wide range of hues and tints in sky objects that appear so vividly in astronomical photographs. Nebulae are good examples of this occurrence. Unlike stars, which shine across a broad spectrum, nebulae are only visible at specific narrow wavelengths. Different types of nebulae shine at different wavelengths. For instance, emission nebulae, those clouds that are excited into fluorescence by the energy from young stars buried deep within, shine with a characteristic reddish color. Unfortunately for us, red—which is toward the long end of the visible spectrum—is all but invisible to the human eye under dim illumination. As a result, emission nebulae are among the toughest objects for the visual astronomer to spy.

The eye is best at perceiving color among the brighter planets, such as Mars, Jupiter, and Saturn, as well as some double stars where the color contrast between the suns can be striking. Most extended deep-sky objects display little color, apart from the greenish and bluish tints of some brighter planetary nebulae, and the blue, yellow, and reddish-orange tinges of some star clusters.

Although the eye's blind spot does not adversely affect our night vision, the fact that the fovea centralis is populated only by cones does. What this means, quite simply, is that the center of our view is not the most sensitive area of the eye to dim light, especially when it comes to diffuse objects such as comets and most deep-sky objects. Instead, to aid in the detection of targets at the threshold of visibility, astronomers use a technique called *averted vision*. Rather than staring directly toward a faint object, look a little to one side or the other. By averting your vision in this manner, you direct the target's dim light away from the cone-rich fovea centralis and onto the peripheral area of the retina, where the light-sensitive rods stand the best chance of revealing the subject of your search.

Another way to detect difficult objects is to tap the side of the telescope tube very lightly. Your peripheral vision is very sensitive to motion, so a slight back-and-forth motion to the field of view will frequently cause faint objects to reveal themselves. I know it sounds a bit strange, but try it. It works, but be gentle.

Here is a tip for all readers who wear eyeglasses. If you suffer from either nearsightedness or farsightedness, it is best to remove your glasses before looking through the eyepiece of a telescope or binoculars, refocusing the image until everything is sharp and clear. Of course, if you use contact lenses, leave them in place when observing as you would when doing any other activity. If you suffer from astigmatism, it is best to leave the glasses on. Alternatively, Tele Vue sells DIOPTRX corrective lenses for more than twenty of its eyepieces, including some in its Plössl, Nagler Type 4 and 5, Radian, and Panoptic lines. Attaching to the top of an eyepiece in place of its eyecup, DIOP-TRX lenses can make a big difference for those who would otherwise be forced to view through telescopes wearing glasses because of astigmatism. Test results by those with astigmatism show positive results. DIOPTRX models are available in 0.25-diopter steps, from 0.25 up to 2.5 diopters, and must be matched to your eyeglass prescription. All include anodized aluminum housings and rubber eyecups.

Frequently, localized light pollution will also mask a faint object. The distraction caused by glare seen out of the corners of your eyes from nearby porch lights, streetlights, and so on, can be enough to cause a faint celestial object to be missed. To help shield our eyes from extraneous light, many eyepieces and binoculars sold today come with built-in rubber eyecups. Although they prove adequate under most conditions, eyecups may not block out all peripheral light. Here are a couple of tricks to try if eyecups alone prove inadequate.

This first idea was already mentioned in my book *Touring the Universe through Binoculars,* but I think it merits repeating here. Buy a cheap pair of ordinary rubber underwater goggles, the kind you can find at just about any toy or sporting goods store. Cut out half of the goggles' front window. If you prefer to use one eye over the other for looking through your telescope, make certain to cut out the correct side. Of course, both windows would need to be cut out if used with binoculars. Spray the goggles with flat black paint and they're done. The blackened goggles should provide enough added baffling to keep stray light from creeping around the eyepiece's edge and into your eyes.

Here's another approach to the same situation that works well for viewing through telescopes, but it is not really applicable to binocular use. This involves wearing a dark turtleneck shirt or sweater, but not in the way you are used to. Instead of slipping it on from bottom to top, stick your head into the shirt through the neck opening. Let the shirt rest on your shoulders. Whenever you look through the eyepiece, simply pull the shirt up and over your head to act as a cloak against surrounding lights. This is a variation on the idea first used by photographers more than a century ago, and it still works today. The only drawback, apart from looking a little strange to civilians, is that the eyepiece may dew over more quickly because of trapped body heat. Still, I can sometimes see up to a half-magnitude fainter just by using this makeshift cloaking device. Give it a try.

Record Keeping and Sketching

One of the best habits an amateur astronomer can develop is keeping a logbook of everything he or she sees in the sky. Recording observations serves the dual purpose of both chronicling what you have seen as well as how you have developed as an observer. It's also a great way to relive past triumphs on cloudy nights.

Although you are free to develop your own system, I prefer to record each object on a separate sheet of paper. On it, I include the date and location, comments on sky conditions, a description of the equipment used for the observation, and a few descriptive notes and a drawing. You will find a generic observation form from the *Star Ware* section of www.philharrington.net, which you can feel free to customize as you wish. Most of the entries on the form should be self-explanatory. *Transparency* rates sky clarity on a scale of 0 (complete overcast) to 10 (perfect). *Seeing* refers to the steadiness of the atmosphere, from 0 (rampant scintillation) to 10 (very steady, with no twinkling even at high power). Since both of these are subjective judgments, as previously noted, always include the magnitude of the faintest star visible to the naked eye. This helps put the other two values in perspective.

It is difficult, if not impossible, to convey the visual impact of subtle heavenly sights with words alone. That is why including a drawing with all written observations is so important. The drawing should convey the perceived image as accurately as possible. Right now, you might be thinking, "I can't even draw a straight line." That's all right, since few objects in space are straight. Actually, sketching celestial objects is not as difficult as one might think at first. It just takes a little practice.

Although astrophotographers require elaborate, expensive equipment to practice their trade, the astro-artist can enjoy his or her craft with minimal apparatus. Besides a telescope or tripod-mounted binoculars, all one needs to begin is a clipboard, a pad of paper, and a few pencils. The delicate textures of celestial bodies are best rendered using an artist's pencil with soft lead such as H or HB or with sketching charcoal. (Just be careful with the charcoal, as it tends to get away from those unfamiliar with it.) As for surface media, most astro-artists prefer smooth white paper as opposed to rag bond typing paper. The grain of rag paper tends to overwhelm the fine shading of most astronomical sketches. Most sketching pads have a fine surface grain ideal for the activity, but even computer printer paper will suffice.

Sketching a sky object is a multistep process that requires patience and close attention to detail. First, examine the target at a wide range of magnifications. Select the one that gives the best overall view as the basis for the sketch. Begin the drawing by lightly marking the positions of the brighter field stars surrounding the target object. Next, go back to each star and change its appearance to match its perceived brightness. Convention has it that brighter stars are drawn larger than fainter ones. Once the field is drawn accurately, lightly depict the location and shape of the target itself. The last step is to shade in the target to match its visual impression. Using your finger, a smudging tool, or a soft eraser to smudge the lead, recreate the delicate shadows and brighter regions. Remember, the drawing is, in effect, a negative image of the object. As such, the brightest areas should appear the darkest on the sketch, and vice versa. Finally, examine the target again with several different eyepieces, penciling in any detail that previously went unseen.

Here are some final tips for beginning astro-artists. First, never try to hurry a drawing, even if it is bitterly cold and your hands are turning blue. Better to put the pencil down, take a five- or ten-minute warm-up break, and then go back. Second, do not try to create a Rembrandt at the telescope. Instead, make only a rough (but accurate) sketch at the telescope itself; the final drawing may be made later indoors at your leisure. Last, avoid the urge to add a little stylistic license, such as putting spikes around the brighter stars or drawing in detail that isn't quite visible, but that you think is there from looking at photographs. Remember, a good astro-artist is an impartial reporter.

Once filled out, the observation record may be filed in a large loose-leaf notebook. It is handiest to separate observations by category. Individual headings might include the Moon, planets, variable stars, deep-sky objects, and so on. Interstellar objects may be further broken down first by type and then by increasing right ascension beginning at 0 hours. Members of the solar system might be separated by object and then filed chronologically.

The eye of the experienced observer can detect much more tenuous detail in sky objects than a beginner can spot. Does this mean that the veteran has better eyesight? Probably not. Like most things in life, talent is not inborn. It has to be nurtured and developed with time. That's why it pays to keep notes. You will be amazed at how far your observing skills have come when you look back at your early entries a few years later.

Observing vs. Peeking (a Commentary)

Are you an astronomical observer or an astronomical peeker? There is a difference—a big one! A peeker flits from one object to the next, barely looking at each before . . . *swish* . . . it's off to another. He or she never writes down what was seen, let alone makes a simple drawing. Whenever asked what he or she saw during an observing session, the peeker will only say, "Oh, I don't know; just some stuff." The fact of the matter is, peekers usually cannot remember what they have seen and what they have not.

An observer takes a slow, methodical approach to the study of the night sky. Most compile long lists of objects they want to see before going outside in an effort to use each moment under the stars as effectively as possible. Unlike our friend the peeker, the true observer is not out to break any land-speed records. He or she prefers to take it a little more slowly, savoring each photon that reaches the eye.

Perhaps it is a sign of the hectic times in which we live, but more and more amateur astronomers seem to be peekers. If you are one of them, I have to ask, "What's your rush?" Take a deep breath and relax. Resist the urge to race impulsively across the sky. By taking a slower, more deliberate tour, you will see the heavens in a new and exciting light. Become an astronomical observer, a connoisseur of the universe.

Epilogue

The universe holds a lifetime of interesting sights for each of us to enjoy. In fact, there are so many sights that many observers simply do not know where to begin. My advice? Start with the easy stuff. Take a tour of the Moon, and learn its more striking features in the same way that you have learned your own neighborhood. Many amateurs tire of the Moon, probably because they consider it too easy, yet getting to know its many features intimately can take years of careful study. Next, if there is a bright planet visible in the sky, be sure to pay it a visit. Once again, avoid the peeker syndrome mentioned earlier. Instead, savor the view slowly. Drink it in like a fine wine, rather than guzzling it like cheap Ripple. Wait for those fleeting moments when our atmosphere steadies and you see your target in exquisite detail. Finally, step beyond our solar system into the deep sky, where thousands of star clusters, nebulae, and galaxies all lie in wait. There are many guidebooks available that will take you there, although naturally I favor *Star Watch,* the companion to this book.

Yes, indeed, we live at a remarkable time. In many ways, this is the golden age of amateur astronomy. Never before have we enjoyed such an amazing selection of equipment. Huge Newtonian reflectors, exquisite refractors, and astonishingly sophisticated catadioptric instruments are all at our beck and call. The personal computer has also served to revolutionize the hobby, just as it has influenced nearly every other aspect of our daily lives.

I must reiterate one final thought. Please don't end this book by thinking that astronomy is only for the rich. That may be an easy impression to leave with, given some of the huge price tags that some telescopes and accessories carry. But I must impress upon you that it is not important how much money you have to spend or how much equipment you own. Of the following two telescopes, which would you say is the better? The first is a humble 3.1-inch refractor that cost less than $200. The owner uses it every clear night. The second is a beautifully crafted 20-inch reflector than cost close to $10,000. Because of its size, however, its owner only uses it three or four times a year. So, which would you say is better? To my mind, the answer is clear. The small refractor, modest as it is, takes its owner into the universe far more often than the huge reflector.

What matters most is that be it a small, inexpensive instrument or a pricey, advanced scientific device, the best telescope in the world is the one that *you* use often and enjoy to its fullest. See you under the stars.

Appendix A
Specs at a Glance

Binoculars

Brand/Model	Magnifi-cation	Aper-ture (mm)	Field of View (degrees)	Type of Focusing	Exit Pupil (mm)	Eye Relief (mm)	Weight (ounces)	Tripod Adaptable?	Type of Prisms	Coat-ings[2]
Price Range: Under $50										
Barska X-Trail	7	35	8	C	5	11	18	Yes	BK-7	FC
Barska X-Trail	10	50	7	C	5	13	29	Yes	BK-7	FC
Barska X-Trail	20	50	3.2	C	2.5	7	29	Yes	BK-7	MC
Celestron Upclose	7	35	8	C	5	11	24	Yes	BK-7	FC
Celestron Upclose	10	50	7	C	5	11	25	Yes	BK-7	FC
Celestron Upclose	12	50	5.2	C	4.2	12	35	Yes	BK-7	FC
Meade TravelView[3]	7	35	7.2	C	5	11	20	No	BK-7	FC
Meade TravelView[3]	7	50	6.8	C	7.1	7	31	No	BK-7	FC
Meade TravelView[3]	10	50	7	C	5	13	30	No	BK-7	FC
Meade TravelView[3]	12	50	4.3	C	4.2	12	28	No	BK-7	FC
Meade TravelView[3]	16	50	4.2	C	3.1	8	28	No	BK-7	FC
Orion WorldView	8	40	8	C	5	15	23	Yes	BK-7	FC
Orion WorldView	10	50	6.5	C	5	15	29	Yes	BK-7	FC
Orion WorldView	12	50	5.5	C	4.2	12	26	Yes	BK-7	FC
Price Range: $50 to $100										
Apogee Astro-Vue	10	60	5.2	C	6	15	40	Yes	BaK-4	MC
Apogee Astro-Vue	12	60	5.7	C	5	15	40	Yes	BaK-4	MC
Barska X-Trail	12	60	5.7	C	5	19	41	Yes	BK-7	MC
Barska X-Trail	15	70	4.4	C	4.7	20	45	Yes	BaK-4	FC
Bushnell H2O[3]	8	42	8.2	C	5.3	17	27	Yes	BK-7	MC
Bushnell H2O[3]	10	42	6.6	C	4.2	16	27	Yes	BK-7	MC
Bushnell H2O[3]	12	42	5.5	C	3.5	15	27	Yes	BK-7	MC
Bushnell Legacy	8	40	8.5	C	5	14	24	Yes	BaK-4	MC

Brand/Model	Magnification	Aperture (mm)	Field of View (degrees)	Type of Focusing	Exit Pupil (mm)	Eye Relief (mm)	Weight (ounces)	Tripod Adaptable?	Type of Prisms	Coatings[2]
Bushnell Legacy	10	50	7.2	C	5	9	28	Yes	BaK-4	MC
Bushnell Natureview[3]	8	42	7.8	C	5.3	19	26	Yes	BaK-4	FMC
Celestron OptiView LPR	10	50	7	C	5	13	30	Yes	BK-7	FC
Celestron Outland LX	8	40	6.8	C	5	18	26	Yes	BaK-4	FMC
Celestron Outland LX	10	50	6	C	5	18	29	Yes	BaK-4	FMC
Celestron SkyMaster	12	60	5.7	C	5	18	44	Yes	BaK-4	MC
Celestron SkyMaster	15	70	4.4	C	4.7	18	48	Yes	BaK-4	MC
Eagle Optics Triumph[3]	7	35	9.2	C	5	12	20	Yes	BK-7	MC
Eagle Optics Triumph[3]	8	42	8.2	C	5.3	14	23	Yes	BK-7	MC
Eagle Optics Triumph[3]	10	50	6.5	C	5	13	30	Yes	BK-7	MC
Nikon Action	7	35	9.3	C	5	12	24	Yes	BaK-4	MC
Nikon Action	8	40	8.2	C	5	12	26	Yes	BaK-4	MC
Oberwerk 8 × 56	8	56	6	C	7	24	36	Yes	BaK-4	FMC
Oberwerk 11 × 56	11	56	6	C	5.1	19	36	Yes	BaK-4	FMC
Orion Outsider[3]	8	40	9.4	C	5	18	29	Yes	BaK-4	FC
Orion Outsider[3]	10	50	7.6	C	5	18	33	Yes	BaK-4	FC
Orion Scenix	7	50	7.1	C	7.1	20	28	Yes	BaK-4	FC+MCobj
Orion Scenix	8	40	9	C	5	12	24	Yes	BaK-4	FC+MCobj
Orion Scenix	10	50	7	C	5	12	28	Yes	BaK-4	FC+MCobj
Pentax XCF	10	50	6.5	C	5	13	31	Yes	BaK-4	MC
Pentax XCF	12	50	5.6	C	4.2	11	31	Yes	BaK-4	MC
Pentax XCF	16	50	3.5	C	3.1	13	32	Yes	BaK-4	MC
Pentax XCF	8	40	8.2	C	5	13	27	Yes	BaK-4	MC
Swift Aerolite 738[3]	8	40	9	C	5	10	22	No	BK-7	FC
Swift Aerolite 736[3]	10	50	7	C	5	10	26	No	BK-7	FC

Price Range: $101 to $150

Brand/Model	Magnification	Aperture (mm)	Field of View (degrees)	Type of Focusing	Exit Pupil (mm)	Eye Relief (mm)	Weight (ounces)	Tripod Adaptable?	Type of Prisms	Coatings[2]
Barska X-Trail	20	80	3.6	C	4	15	58	Yes	BaK-4	MC
Barska X-Trail	30	80	2.1	C	4	10	69	Yes	BaK-4	MC
Bushnell Legend	10	50	6.5	C	5	15	28	Yes	BaK-4	FMC
Bushnell Legend	12	50	5.5	C	4.2	15	28	Yes	BaK-4	FMC
Bushnell Natureview[3]	10	42	6	C	4.2	15	25	Yes	BaK-4	FMC
Nikon Action	10	50	6.5	C	5	12	34	Yes	BaK-4	MC
Nikon Action	12	50	5.5	C	4.2	9	34	Yes	BaK-4	MC
Nikon Action	16	50	4.1	C	3.1	12	32	Yes	BaK-4	MC
Nikon Action EX ATB	7	35	9.3	C	5	17	28	Yes	BaK-4	MC
Nikon Action EX ATB	8	40	8.2	C	5	17	30	Yes	BaK-4	MC
Oberwerk FMC Mini-Giant	9	60	5.5	C	6.6	16	42	Yes	BaK-4	FMC
Oberwerk FMC Mini-Giant	12	60	5.7	C	5	14	42	Yes	BK-7	FMC
Oberwerk FMC Mini-Giant	15	60	4.1	C	4	12	42	Yes	BaK-4	FMC
Oberwerk FMC Mini-Giant	20	60	3	C	3	10	42	Yes	BaK-4	FMC
Oberwerk Giant	11	70	4.5	C	6.3	23	50	Yes	BaK-4	MC
Oberwerk Giant	15	70	4.3	C	4.6	16	50	Yes	BaK-4	MC
Oberwerk Mariner[3]	7	50	7	C	7.1	24	45	Yes	BaK-4	FMC

Brand/Model	Magnifi-cation	Aper-ture (mm)	Field of View (degrees)	Type of Focusing	Exit Pupil (mm)	Eye Relief (mm)	Weight (ounces)	Tripod Adaptable?	Type of Prisms	Coat-ings[2]
Oberwerk Mariner[3]	8	40	8.4	C	5	18	35	Yes	BaK-4	FMC
Orion UltraView	8	42	8.2	C	5.3	22	27	Yes	BaK-4	FMC
Swift Armored Aerolite 737H[3]	7	50	7	C	7.1	16	25	Yes	BK-7	FC
Swift Cougar 704B[3]	10	50	7	C	5	14	29	Yes	BaK-4	FC

Price Range: $151 to $200

Brand/Model	Magnifi-cation	Aper-ture (mm)	Field of View (degrees)	Type of Focusing	Exit Pupil (mm)	Eye Relief (mm)	Weight (ounces)	Tripod Adaptable?	Type of Prisms	Coat-ings[2]
Leupold Wind River Mesa[3]	8	42	6.5	C	5.3	18	27	Yes	BaK-4	MC
Meade Astronomy	9	63	5.8	C	7	22	44	Yes	BaK-4 roof	FC
Nikon Action EX ATB	10	50	6.5	C	5	17	36	Yes	BaK-4	MC
Nikon Action EXATB	12	50	5.5	C	4.2	16	37	Yes	BaK-4	MC
Oberwerk 20 × 80 LW	20	80	3.5	C	4	13	58	Yes	BaK-4	FMC
Oberwerk Mariner	10	60	5.3	C	6	23	51	Yes	BaK-4	FMC
Orion Mini Giant	8	56	5.8	C	7	18	32	Yes	BaK-4	FMC
Orion Mini Giant	9	63	5	C	7	26	36	Yes	BaK-4	FMC
Orion Mini Giant	12	63	4.7	C	5.3	18	40	Yes	BaK-4	FMC
Orion UltraView	7	50	6.5	C	7.1	22	32	Yes	BaK-4	FMC
Orion UltraView	10	50	6.5	C	5	22	32	Yes	BaK-4	FMC
Parks GR[3]	7	50	6.8	C	7.1	20	25	Yes	BaK-4	FMC
Parks GR[3]	10	50	5.3	C	5	18	26	Yes	BaK-4	FMC
Parks Wide Angle[3]	8	42	8.2	C	5.25	18	25	Yes	BaK-4	FMC
Parks ZWCF[3]	12	50	6	C	4.2	6	26	Yes	BK-7	FC
Pentax PCF WP II	8	40	6.3	C	5	20	28	Yes	BaK-4	FMC
Pentax PCF WP II	10	50	5	C	5	20	34	Yes	BaK-4	FMC
Pentax PCF WP II	12	50	4.2	C	4.2	20	35	Yes	BaK-4	FMC

Price Range: $201 to $300

Brand/Model	Magnifi-cation	Aper-ture (mm)	Field of View (degrees)	Type of Focusing	Exit Pupil (mm)	Eye Relief (mm)	Weight (ounces)	Tripod Adaptable?	Type of Prisms	Coat-ings[2]
Apogee Astro-Vue	20	100	3	C	5	18	136	Yes	BaK-4	MC
Barska Cosmos	20	80	3	I	4	15	88	Yes	BaK-4	MC
Barska Cosmos	25	100	3	I	4	15	130	Yes	BaK-4	MC
Carton Adlerblick F591	7	50	6	C	7.1	23	28	Yes	BaK-4	FMC
Carton Adlerblick F594	8	42	6.5	C	5.3	16	22	Yes	BaK-4	FMC
Carton Adlerblick F595	10	42	6.5	C	4.2	10	21	Yes	BaK-4	FMC
Celestron SkyMaster	20	80	3.2	C	4	17	94	Yes	BaK-4	MC
Celestron Ultima DX	10	50	6.5	C	5	19	30	Yes	BaK-4	FMC
Leupold Wind River Mesa[3]	10	50	5.5	C	5	18	32	Yes	BaK-4	MC
Oberwerk 20 × 80	20	80	3.5	C	4	18	72	Yes	BaK-4	MC
Oberwerk 20 × 80 Deluxe II	20	80	3.2	C	4	18	112	Yes	BaK-4	FMC
Oberwerk 20 × 90	20	90	3.2	C	4.5	17	138	Yes	BaK-4	FMC
Orion GiantView	15	70	4	I	4.7	18	48	Yes	BaK-4	FMC
Orion GiantView	20	80	3.2	I	4	17	112	Yes	BaK-4	FMC
Orion Little Giant II	11	70	4.5	C	6.4	18	47	Yes	BaK-4	FMC
Orion Little Giant II	20	70	3	C	3.5	8	47	Yes	BaK-4	FMC
Orion Mini Giant	15	63	3.7	C	4.2	19	40	Yes	BaK-4	FMC
Orion Vista	7	50	6	C	7.1	22	28	Yes	BaK-4	FMC
Orion Vista	8	42	6.5	C	5.3	18	22	Yes	BaK-4	FMC

Brand/Model	Magnifi-cation	Aper-ture (mm)	Field of View (degrees)	Type of Focusing	Exit Pupil (mm)	Eye Relief (mm)	Weight (ounces)	Tripod Adaptable?	Type of Prisms	Coat-ings[2]
Orion Vista	10	50	5.3	C	5	16	28	Yes	BaK-4	FMC
Parks Deluxe Giant[3]	10	70	5	C	7	17	62	Yes	BaK-4	FMC
Parks Deluxe Giant[3]	15	70	4.4	C	4.7	20	95	Yes	BaK-4	FMC
Parks ZWCF[3]	10	52	6.5	C	5.2	13	29	Yes	BaK-4	FMC
Pentax PCF WP II	20	60	2.2	C	3	21	45	Yes	BaK-4	FMC
Stellarvue B63	15	63	4.3	C	4.2	16	60	Yes	BaK-4	FMC
Swift Audubon 820[3]	8.5	44	8.2	C	5.2	15	25	No	BaK-4	FMC
Swift Sea Wolf 713[3]	7	50	7.1	C	7.1	19	33	Yes	BK-7	FC
Swift UltraLite 961[3]	8	42	6.5	C	5.3	18	25	Yes	BaK-4	FMC
Swift UltraLite 962[3]	10	42	6.5	C	4.2	18	25	Yes	BaK-4	FMC

Price Range: $301 to $500

Brand/Model	Magnifi-cation	Aper-ture (mm)	Field of View (degrees)	Type of Focusing	Exit Pupil (mm)	Eye Relief (mm)	Weight (ounces)	Tripod Adaptable?	Type of Prisms	Coat-ings[2]
Apogee RA-70-SA	16	70	n/s	I	4.4	n/s	n/s	Yes	n/s	FMC
Canon IS	10	30	6	C	3	15	22	No	BaK-4	FMC
Carton Adlerblick F592	10	50	5.3	C	5	16	27	Yes	BaK-4	FMC
Carton Adlerblick F596	12	50	5.3	C	4.2	11	28	Yes	BaK-4	FMC
Celestron SkyMaster	25	100	3	I	4	15	157	Yes	BaK-4	MC
Celestron Ultima DX	8	56	5.8	C	7	18	35	Yes	BaK-4	FMC
Celestron Ultima DX	9	63	5	C	7	17	40	Yes	BaK-4	FMC
Fujinon MT-SX	7	50	7.5	I	7.1	12	48	Yes	BaK-4	FMC
Fujinon MT-SX	10	70	5.3	I	7	12	76	Yes	BaK-4	FMC
Oberwerk 22 × 100	22	100	2.8	C	4.6	18	141	Yes	BaK-4	FMC
Oberwerk 25 × 100 IF	25	100	2.5	I	4	18	160	Yes	BaK-4	FMC
Orion GiantView	25	100	2.5	I	4	18	162	Yes	BaK-4	FMC
Orion MegaView	20	80	3.5	C	4	16	84	Yes	BaK-4	FMC
Orion MegaView	30	80	2.3	C	2.7	14	64	Yes	BaK-4	FMC
Parks Deluxe Giant[3]	11	80	4.5	C	7.3	20	82	Yes	BaK-4	FMC
Parks Deluxe Giant[3]	15	80	3.5	C	5.3	17	82	Yes	BaK-4	FMC
Parks Deluxe Giant[3]	20	80	3.5	C	4	16	82	Yes	BaK-4	FMC
Steiner Military/Marine[3]	7	50	7	I	7.1	22	37	Yes	BaK-4	FMC
Steiner Military/Marine[3]	10	50	6.2	I	5	17	36	Yes	BaK-4	FMC
Stellarvue B85	20	85	3.2	I	4.3	17	128	Yes	BaK-4	FMC
Swift Sea King 716R[3]	7	50	7	I	7.1	19	38	Yes	BaK-4	FC
Vixen Foresta	8	42	8.8	C	5.3	15	27	Yes	BaK-4	FMC
Vixen Foresta	10	42	6.5	C	4.2	17	26	Yes	BaK-4	FMC
Vixen Ultima 7 × 42ZCF	7	42	7	C	6	25	22	Yes	BaK-4	FMC
Vixen Ultima 8 × 42ZCF	8	42	6.6	C	5.3	20	22	Yes	BaK-4	FMC
Vixen Ultima 9 × 63ZCF	9	63	5.4	C	7	21	37	Yes	BaK-4	FMC
Vixen Ultima 10 × 42ZWCF	10	42	6.6	C	4.2	15	22	Yes	BaK-4	FMC

Price Range: $501 to $1,000

Brand/Model	Magnifi-cation	Aper-ture (mm)	Field of View (degrees)	Type of Focusing	Exit Pupil (mm)	Eye Relief (mm)	Weight (ounces)	Tripod Adaptable?	Type of Prisms	Coat-ings[2]
Apogee RA-88-SA	20 / 32	88	Varies	I	4.4/2.8	Varies	14 lb	Yes	n/s	FMC
Canon IS II	12	36	5	C	3	15	23	No	BaK-4	FMC
Fujinon FMT-SX	7	50	7.5	I	7.1	23	50	Yes	BaK-4	FMC
Fujinon FMT-SX	10	70	5.3	I	7	23	76	Yes	BaK-4	FMC
Fujinon FMT-SX	16	70	4	I	4.4	16	76	Yes	BaK-4	FMC
Fujinon Techno-Stabi	14	40	4	C	2.9	10	43	No	Roof	FMC

Brand/Model	Magnification	Aperture (mm)	Field of View (degrees)	Type of Focusing	Exit Pupil (mm)	Eye Relief (mm)	Weight (ounces)	Tripod Adaptable?	Type of Prisms	Coatings[2]
Miyauchi Binon	7	50	9.5	C	7.1	22	45	Yes	BaK-4	FMC
Miyauchi Pleiades Bs-iC	22	60	3	I	2.8	14	70	Yes	BaK-4	FMC
Nikon Prostar	7	50	7.3	I	7.1	16	52	[4]	BaK-4	FMC
Nikon Superior E	10	42	6	C	4.2	17	24	[4]	BaK-4	FMC
Nikon Superior E	12	50	5	C	4.2	17	32	[4]	BaK-4	FMC
Steiner Nighthunter XP[3]	7	50	7.1	I	7.1	21	33	No	BaK-4	FMC
Steiner Nighthunter XP[3]	8	56	6.3	I	7	21	41	No	BaK-4	FMC
Steiner Nighthunter XP[3]	10	50	6	I	5	21	35	No	BaK-4	FMC
Steiner Nighthunter XP[3]	12	56	4.9	I	4.7	22	41	No	BaK-4	FMC
Steiner Observer[3]	25	80	4.7	I	3.2	16	56	Yes	BaK-4	FMC
Vixen Giant ARK	16	80	4.3	C	5	17	84	Yes	BaK-4	MC
Vixen Giant ARK	20	80	3.5	C	6.7	16	84	Yes	BaK-4	MC
Vixen Giant ARK	12	80	4.1	C	6.7	18	84	Yes	BaK-4	MC
Vixen Giant ARK	30	80	2.3	C	6.7	18	84	Yes	BaK-4	MC
Vixen HFT-A	20	125	3	I	6.2	20	24 lb	Yes	BaK-4	FMC
Vixen HFT-A	30	125	1.6	I	4.2	20	24 lb	Yes	BaK-4	FMC
Vixen Ultima ED6.5×44ZCF	6.5	44	7	C	6.8	13	25	Yes	BaK-4	FMC
Vixen Ultima ED9.5×44ZCF	9.5	44	6	C	4.6	13	25	Yes	BaK-4	FMC
Zeiss B/GA T*I.F. ClassiC	7	50	7.4	I	7.1	21	42	Yes	BaK-4	FMC

Price Range: Over $1,000

Brand/Model	Magnification	Aperture (mm)	Field of View (degrees)	Type of Focusing	Exit Pupil (mm)	Eye Relief (mm)	Weight (ounces)	Tripod Adaptable?	Type of Prisms	Coatings[2]
Astromeccanica 120 Binoscope	Varies	120	Varies	I	Varies	Varies	30 lb	Yes	Mirrors	n/s
Astromeccanica 150 Binoscope	Varies	152	Varies	I	Varies	Varies	43 lb	Yes	Mirrors	n/s
Canon IS L IS WP	10	42	6.5	C	4.2	16	37	Yes	BaK-4	FMC
Canon IS All Weather	18	50	3.7	C	2.8	15	42	Yes	BaK-4	FMC
Fujinon MT-SX	25	150	2.7	I	6	19	41 lb	Yes	BaK-4	FMC
Fujinon EM-SX	25	150	2.7	I	6	19	43 lb	Yes	BaK-4	FMC
Fujinon EDMT-SX	25	150	2.7	I	6	19	41 lb	Yes	BaK-4	FMC
Fujinon EDMT-SX	40	150	1.7	I	3.8	15	41 lb	Yes	BaK-4	FMC
JMI RB-66	Varies	152	Varies	I	Varies	Varies	59 lb	Yes	Mirrors	n/s
JMI RB-10	Varies	254	Varies	I	Varies	Varies	91 lb	Yes	Mirrors	n/s
JMI RB-16	Varies	406	Varies	I	Varies	Varies	204 lb	Yes	Mirrors	n/s
Kowa Highlander	32	82	2.2	I	2.6	17	13.6 lb	Yes	BaK-4	FMC
Kowa Highlander Prominar	32	82	2.2	I	2.6	17	13.6 lb	Yes	BaK-4	FMC
Miyauchi Saturn II NBA-71	22	71	2.3	I	3.2	18	96	Yes	BaK-4	FMC
Miyauchi Saturn III	40	100	1.5	I	2.5	18	13 lb	Yes	BaK-4	FMC
Miyauchi Galaxy Bj-iCE	26	100	2.5	I	3.9	18	13 lb	Yes	BaK-4	FMC
Miyauchi BR-141	25	141	2.6	I	5.6	25	26 lb	Yes	BaK-4	FMC
Nikon Bino-Telescope	20	120	3	I	6	21	34 lb	[4]	BaK-4	FMC
Nikon Astroluxe	10	70	5.1	I	7	16	70	[4]	BaK-4	FMC
Nikon Astroluxe XL	18	70	3.1	I	3.9	15	68	[4]	BaK-4	FMC
Oberwerk Long-Range Observation	25 / 40	100	2.5/1.5	I	4 / 2.5	14 / 8	26.5 lb	Yes	BaK-4	FMC
Parks Deluxe Giant	25	100	2.6	C	4	9	123	Yes	BaK-4	FMC

Brand/Model	Magnifi-cation	Aper-ture (mm)	Field of View (degrees)	Type of Focusing	Exit Pupil (mm)	Eye Relief (mm)	Weight (ounces)	Tripod Adaptable?	Type of Prisms	Coat-ings[2]
Steiner Senator[3]	15	80	3.7	I	5.3	13	56	Yes	BaK-4	FMC
Steiner Senator[3]	20	80	3.1	I	4	13	56	Yes	BaK-4	FMC
Zeiss B/GA T*P* ClassiC	8	56	6.3	C	7	18	36	Yes	Roof	FMC
Zeiss Victory BT*P* AOS AK	8	56	7.5	C	7	18	41	Yes	Roof	FMC
Zeiss Victory BT*P* AOS AK	10	56	6.3	C	5.6	16	42	Yes	Roof	FMC
Zeiss Victory BT*P* AOS AK	12	56	5.1	C	4.7	16	46	Yes	Roof	FMC
Zeiss BS/GA-T* Stabilized	20	60	2.9	C	3	14	59	Yes	BaK-4	FMC

1. Type of focusing: C, center focus; I, individual focus.
2. Coatings: The optical coatings applied to the binoculars' elements. C, coated (e.g., single-layer magnesium fluoride, likely only on the outer surfaces of the objectives and eyepieces); FC, fully coated (e.g., single coating on all optical surfaces); MC, multicoated (e.g., multiple coatings on some optical surfaces); FC+MCobj, fully coated optics with multicoated objective lenses; FMC, fully multicoated (e.g., multiple coatings on all optical surfaces); n/s, not specified by the manufacturer.
3. Find a capsule review of this model in the Star Ware section of the author's Web site, www.philharrington.net.
4. A special adapter from the manufacturer is required to attach the binoculars.

Telescopes

Brand/Model	Aperture (inches)	Focal Ratio	Tube Material	Type of Mounting	Total Weight/ Heaviest Assembly (pounds)
### Achromatic Refractors					
Price Range: Under $300					
Celestron Firstscope 90 AZ	3.5	11	Metal	Alt-az	19
Celestron Firstscope 90 EQ	3.5	11	Metal	GEM	22
Celestron 102 Wide View	4	5	Metal	Sold separately	5
Konus Konusmotor 90[1]	3.5	11.1	Metal	GEM	29
Meade ETX-80AT-TC[1]	3.1	5	Plastic	GoTo alt-az	12
Orion Telescopes ShortTube 80	3.1	5	Metal	None or GEM	16 (12)
Orion Telescopes AstroView 80 EQ	3.1	11.4	Metal	GEM	24
Orion Telescopes Explorer 80	3.1	11.4	Metal	Alt-az	15
Sky-Watcher 804AZ3	3.1	5	Metal	Alt-az	20 (17)
Sky-Watcher 909	3.5	10	Metal	Alt-az or GEM	26
Price Range: $300 to $600					
Celestron StarSeeker 80 GoTo	3.1	5	Metal	GoTo alt-az	10
Celestron NexStar 80SLT	3.1	11	Metal	GoTo alt-az	14
Celestron NexStar 102SLT	3.9	6.5	Metal	GoTo alt-az	14
Celestron C4-R	4	10	Metal	GEM	36
Meade DS-2080AT-LNT	3.1	8.8	Metal	GoTo alt-az	17
Orion Telescopes Express 80	3.1	6	Metal	Sold separately	6
Orion Telescopes Explorer 90	3.5	10.1	Metal	Alt-az	16
Orion Telescopes AstroView 90 EQ	3.5	10.1	Metal	GEM	24
Orion Telescopes AstroView 100 EQ	3.9	6	Metal	GEM	29 (22)
Orion Telescopes AstroView 120ST	4.7	5	Metal	GEM	36 (28)
Orion Telescopes AstroView 120	4.7	8.3	Metal	GEM	39 (28)

Brand/Model	Aperture (inches)	Focal Ratio	Tube Material	Type of Mounting	Total Weight/ Heaviest Assembly (pounds)
Sky-Watcher 1025AZ3	4	5	Metal	Alt-az	22
Sky-Watcher 1021EQ3-2	4	9.8	Metal	GEM	37
Stellarvue Nighthawk 2	3.1	6	Metal	Sold separately	5
Stellarvue Nighthawk Next Generation	3.1	7	Metal	Sold separately	6
Stellarvue Nighthawk BV	3.1	9.4	Metal	Sold separately	9
Stellarvue SV80D	3.1	9.4	Metal	Sold separately	9
Vixen 80SS	3.1	5	Metal	Sold separately	5.1
Vixen 80M	3.1	11.4	Metal	Sold separately	5.5
William Optics ZenithStar 80	3.1	6	Metal	Sold separately	5

Price Range: $600 to $1,000

Brand/Model	Aperture (inches)	Focal Ratio	Tube Material	Type of Mounting	Total Weight/ Heaviest Assembly (pounds)
Apogee Widestar[1]	4	6.4	Metal	Sold separately	7
Borg 76SWII[1]	3	6.5	Metal	Alt-az	10
Celestron C6-R	6	8	Metal	GEM	57
Meade AR-5	5	9	Metal	GoTo GEM	49
Orion Telescopes Express 80 EQ	3.1	6	Metal	GEM	25 (19)
Orion Telescopes SkyView Pro 120 EQ	4.7	8.3	Metal	GEM	41 (30)
Sky-Watcher 1206EQ5	4.7	5	Metal	GEM	36 (28)
Sky-Watcher 1201EQ5	4.7	8.3	Metal	GEM	40

Price Range: $1,000 to $2,000

Brand/Model	Aperture (inches)	Focal Ratio	Tube Material	Type of Mounting	Total Weight/ Heaviest Assembly (pounds)
Borg 100SWII[1]	3.9	6.4	Metal	Alt-az	11
Celestron C6-RGT	6	8	Metal	GoTo GEM	57
D&G Optical[1]	5	12 or 15	Metal	Sold separately	19–20
D&G Optical[1]	6	12 or 15	Metal	Sold separately	26–28
Meade AR-6	8		Metal	GoTo GEM	72

Price Range: Over $2,000

Brand/Model	Aperture (inches)	Focal Ratio	Tube Material	Type of Mounting	Total Weight/ Heaviest Assembly (pounds)
D&G Optical[1]	8	12 or 15	Metal	Sold separately	37–39
D&G Optical[1]	10	12 or 15	Metal	Sold separately	75–80
D&G Optical[1]	11	12 or 15	Metal	Sold separately	82–87
Sky-Watcher 15075HEQ5	6	5	Metal	GEM	40 (24)
Sky-Watcher 15012-EQ6	6	8	Metal	GEM	46 (26)
Vixen 102M	4	9.8	Metal	GEM	48

Apochromatic Refractors

Price Range: Under $1,000

Brand/Model	Aperture (inches)	Focal Ratio	Tube Material	Type of Mounting	Total Weight/ Heaviest Assembly (pounds)
Celestron C80ED-R	3.1	7.5	Metal	GEM	42
Celestron Onyx 80EDF	3.1	6.25	Metal	Sold separately	6
Meade Series 5000 80 ED[1]	3.1	6		Sold separately	6
Orion Telescopes 80ED	3.1	7.5	Metal	Sold separately	6
Orion Telescopes 100ED	3.9	9	Metal	Sold separately	7
Orion Telescopes SkyView Pro 80ED EQ	3.1	7.5	Metal	GEM	39 (33)
Vixen ED80Sf	3.1	7.5	Metal	Alt-az	8 (OTA)
William Optics Megrez 80 II ED	3.1	7	Metal	Sold separately	5

Brand/Model	Aperture (inches)	Focal Ratio	Tube Material	Type of Mounting	Total Weight/ Heaviest Assembly (pounds)
Price Range: $1,000 to $2,000					
Borg 76EDSWII[1]	3	6.6	Metal	Alt-az	10
Borg 100ED[1]	4	6.4 (f/4 opt)	Metal	Sold separately	5 (OTA)
Celestron C80ED-RGT	3.1	7.5	Metal	GoTo GEM	42
Celestron C100ED-R	3.9	9	Metal	GEM	50
Celestron C100ED-RGT	3.9	9	Metal	GoTo GEM	50
Meade Series 5000 127 ED[1]	5	7.5		Sold separately	16
Orion Telescopes Sirius 80ED EQ-G	3.1	7.5	Metal	GoTo GEM	49 (32)
Orion Telescopes Sirius 100ED EQ-G	3.9	9	Metal	GoTo GEM	51 (32)
Orion Telescopes SkyView Pro 100ED EQ	3.9	9	Metal	GEM	41 (33)
Orion Telescopes 120ED	4.7	7.5	Metal	Sold separately	10
Pentax 75 SDHF[1]	3	6.7	Metal	Sold separately	5
Sky-Watcher ED80 Pro-Series	3.1	7.5	Metal	GoTo GEM	5 (OTA)
Stellarvue SV80S	3.1	6	Metal	Sold separately	7
Stellarvue Nighthawk T	3.1	7.5	Metal	Sold separately	7
Tele Vue TV-76	3	6.3	Metal	Sold separately	6
Tele Vue TV-85	3.3	7	Metal	Sold separately	8
Vixen ED81S	3.1	7.7	Metal	Alt-az	5 (OTA)
William Optics Megrez II 80 Super Apo	3.1	6	Metal	Sold separately	6
William Optics ZenithStar 105 ED	4.1	7	Metal	Sold separately	10
Price Range: $2,000 to $4,000					
Borg 100EDSWII[1]	4	6.4	Metal	Alt-az	12
Borg 125ED[1]	4.9	6.4 (f/4 opt)	Metal	Sold separately	n/s
Orion Telescopes SkyView Pro 120ED EQ	4.7	7.5	Metal	GEM	56 (46)
Orion Telescopes Sirius 120ED EQ-G	4.7	7.5	Metal	GoTo GEM	70 (60)
Sky-Watcher ED100 Pro-Series	3.9	9	Metal	GoTo GEM	7 (OTA)
Sky-Watcher ED120 Pro-Series	4.7	7.5	Metal	GoTo GEM	11 (OTA)
Stellarvue SV90T	3.5	7	Metal	Sold separately	9
Stellarvue SV4	4	6.4	Metal	Sold separately	13
Stellarvue SV102T	4	7.8	Metal	Sold separately	12
Takahashi TSA-102S	4	8	Metal	Sold separately	11
Tele Vue NP-101	4	5.4	Metal	Sold separately	10
Tele Vue TV-102	4	8.6	Metal	Sold separately	9
TMB 80	3.1	6 or 7.5	Metal	Sold separately	8
TMB 100	3.9	8	Metal	Sold separately	10
TMB 105	4.1	6.2	Metal	Sold separately	15
Vixen ED103S	4	7.7	Metal	GEM	48
William Optics Fluoro-Star 110	4.3	6.5	Metal	Sold separately	14
Price Range: Over $4,000					
Astro-Physics 160EDF	6.3	7.5	Metal	Sold separately	27
Pentax 100 SDUF II[1]	3.9	4	Metal	Sold separately	9
Pentax 105 SDP[1]	4.1	6.4	Metal	Sold separately	12
Stellarvue SV5S[1]	5	7	Metal	Sold separately	21
Stellarvue SV5L[1]	5	9.4	Metal	Sold separately	24

Brand/Model	Aperture (inches)	Focal Ratio	Tube Material	Type of Mounting	Total Weight/ Heaviest Assembly (pounds)
Takahashi FCL-90 (Sky-90 II)	3.5	5.6	Metal	OTA or GEM	7
Takahashi FS-102II	4	8	Metal	GEM	12 (OTA)
Tele Vue NP-127	5	5.2	Metal	Sold separately	14
TMB 115	4.5	7	Metal	Sold separately	23
TMB 130	5.1	6 or 9.25	Metal	Sold separately	24
TMB 152	6	7.9	Metal	Sold separately	39
Vixen ED115S	4.5	7.7	Metal	GEM	52
Price Range: Way over $4,000					
Pentax 125SDP[1]	4.9	6.4	Metal	Sold separately	20
Takahashi FSQ-106	4.2	5	Metal	GEM	13 (OTA)
Takahashi TOA-130	5.1	7.7	Metal	GEM	22 (OTA)
Takahashi TOA-150	5.9	7.3	Metal	GEM	~25 (OTA)
Takahashi FCT-200	7.9	10	Metal	GEM	137 (OTA)
Takahashi FCT-250	9.8	10	Metal	GEM	187 (OTA)
TMB 175	6.9	8	Metal	Sold separately	42
TMB 180	7.1	7	Metal	Sold separately	~40
TMB 203	8	7 or 9	Metal	Sold separately	62
TMB 229	9	9	Metal	Sold separately	68
TMB 254	10	9	Metal	Sold separately	135

Exotic Refractors

Brand/Model	Aperture (inches)	Focal Ratio	Tube Material	Type of Mounting	Total Weight/ Heaviest Assembly (pounds)
Price Range: Under $1,000					
Coronado Personal Solar Telescope (PST)	1.6	10	Metal	Sold separately	3
Coronado Personal Solar Telescope (PST) Ca-K	1.6	10	Metal	Sold separately	3
Price Range: $1,000 to $2,000					
Coronado SolarMax 40	1.6	10	Metal	Sold separately	3
Price Range: $3,000 to $4,000					
Coronado SolarMax 60	2.4	6.6	Metal	Sold separately	6
Coronado SolarMax 70 Hydrogen-Alpha	2.8	5.7	Metal	Sold separately	7
Coronado SolarMax 70 Calcium-K	2.8	5.7	Metal	Sold separately	7
Price Range: Over $9,000					
Coronado SolarMax 90	3.5	8.8	Metal	Sold separately	23

Newtonian Reflectors

Brand/Model	Aperture (inches)	Focal Ratio	Tube Material	Type of Mounting	Total Weight/ Heaviest Assembly (pounds)
Price Range: Under $300					
Celestron Firstscope 114 EQ	4.5	8	Metal	GEM	19
Celestron Star Hopper 6	6	8	Metal	Dob	45 (29)
Edmund Astroscan	4.13	4.2	Plastic	Ball	13

Brand/Model	Aperture (inches)	Focal Ratio	Tube Material	Type of Mounting	Total Weight/ Heaviest Assembly (pounds)
Guan Sheng Optical GSO-580	6	8	Metal	Dob	~30
Konus Konusmotor 114[1]	4.5	8	Metal	GEM	~20
Konus Konusmotor 500[1]	4.5	4.4	Metal	GEM	~25
Orion StarBlast	4.5	4	Metal	Alt-az	13
Orion StarBlast EQ	4.5	4	Metal	GEM	18
Orion SkyQuest XT4.5	4.5	8	Metal	Dob	18
Orion SpaceProbe 130ST	5.1	5	Metal	GEM	27
Orion SpaceProbe 130	5.1	6.9	Metal	GEM	30
Orion SkyQuest XT6 Classic	6	8	Metal	Dob	34 (24)
Sky-Watcher 1145EQ1	4.5	4.4	Metal	GEM	18
Sky-Watcher 1149EQ2	4.5	8	Metal	GEM	37
Sky-Watcher 153 Dob	6	7.8	Metal	Dob	31 (18)
Stargazer Steve 4.25 Deluxe Planetary Kit	4.25	10	Cardboard	Dob	16

Price Range: $300 to $500

Brand/Model	Aperture (inches)	Focal Ratio	Tube Material	Type of Mounting	Total Weight/ Heaviest Assembly (pounds)
Celestron StarSeeker 114 GoTo	4.5	4	Metal	GoTo alt-az	14
Celestron StarSeeker 130 GoTo	5.1	5	Metal	GoTo alt-az	17
Celestron NexStar 130 SLT	5.1	5	Metal	GoTo alt-az	18
Celestron C6-N	6	5	Metal	GEM	35
Celestron Star Hopper 8	8	6	Metal	Dob	57 (33)
Guan Sheng Optical GSO-680	8	6	Metal	Dob	43 (23)
Meade N-6	6	5	Metal	GEM	47
Meade LightBridge 8 Standard	8	6	Truss	Dob	44 (19)
Novosibirsk TAL-1[1]	4.3	7.3	Metal	GEM	42
Novosibirsk TAL-120[1]	4.7	6.7	Metal	GEM	44
Orion AstroView 6 EQ	6	5	Metal	GEM	39
Orion SkyQuest XT6 IntelliScope	6	8	Metal	Dob (opt. push-to controller)	34
Orion SkyQuest XT8 Classic	8	5.9	Metal	Dob	41 (23)
Orion SkyQuest XT8 IntelliScope	8	5.9	Metal	Dob (opt. push-to controller)	42
Sky-Watcher 13065PEQ2	5.1	5	Metal	GEM	48
Sky-Watcher 1309EQ2	5.1	7	Metal	GEM	29
Sky-Watcher 203 Dob	8	5.9	Metal	Dob	35 (18)
Stargazer Steve 6 Kit	6	8	Cardboard	Dob	24
Stargazer Steve SGR-4	4.25	7.1	Cardboard	Alt-az	19

Price Range: $500 to $1,000

Brand/Model	Aperture (inches)	Focal Ratio	Tube Material	Type of Mounting	Total Weight/ Heaviest Assembly (pounds)
Celestron C6-NGT	6	5	Metal	GoTo GEM	54
Celestron C6-NHD	6	5	Metal	GEM	54
Celestron C-8N	8	5	Metal	GEM	67
Celestron C10-N	10	4.7	Metal	GEM	83
Celestron Star Hopper 10	10	5	Metal	Dob	59 (36)
Celestron Star Hopper 12	12	5	Metal	Dob	~92 (52)
Guan Sheng Optical GSO-880	10	5	Metal	Dob	59 (36)
Guan Sheng Optical GSO-980	12	5	Metal	Dob	~80
Konus Konusky-200 Motor	8	5	Metal	GEM	~70

Brand/Model	Aperture (inches)	Focal Ratio	Tube Material	Type of Mounting	Total Weight/ Heaviest Assembly (pounds)
Meade LightBridge 8 Deluxe	8	6	Truss	Dob	44 (19)
Meade LightBridge 10 Standard	10	5	Truss	Dob	65 (30)
Meade LightBridge 10 Deluxe	10	5	Truss	Dob	65 (30)
Meade LightBridge 12 Standard	12	5	Truss	Dob	80 (36)
Novosibirsk TAL-150P[1]	6	5	Metal	GEM	55
Novosibirsk TAL-150P8[1]	6	8	Metal	GEM	88
Novosibirsk TAL-2	6	8	Metal	GEM	88
Orion Optics (UK) Europa 150[1]	5.9	5, 8, or 11	Metal	GEM	33 (24)
Orion Optics (UK) OD150[1]	5.9	8 or 11	Metal	Dob	n/s
Orion SkyQuest XT10 Classic	10	4.7	Metal	Dob	55 (35)
Orion SkyQuest XT10 IntelliScope	10	4.7	Metal	Dob (opt. push-to controller)	55 (35)
Orion SkyQuest XT12 Classic	12	4.9	Metal	Dob	81 (50)
Orion SkyQuest XT12 IntelliScope	12	4.9	Metal	Dob (opt. push-to controller)	83 (50)
Orion SkyView Pro 8 EQ	8	4.9	Metal	GEM	64
Parks Astrolight[1]	6	6 or 8	Fiberglass	GEM	42
Parks Companion[1]	4.5	5	Fiberglass	Sold separately	n/s
Parks Precision[1]	6	6 or 8	Fiberglass	GEM	65
Sky-Watcher 15075PEQ3-2	6	5	Metal	GEM	40
Sky-Watcher 2001PEQ5	8	5	Metal	GEM	64
Sky-Watcher 254 Dob	10	4.7	Metal	Dob	45 (28)
Stargazer Steve 6 Truss Tube Kit	6	5	Truss	Dob	24 (15)
Vixen R135S	5.3	5.3	Metal	Sold separately	8 (OTA)

Price Range: $1,000 to $2,000

Brand/Model	Aperture (inches)	Focal Ratio	Tube Material	Type of Mounting	Total Weight/ Heaviest Assembly (pounds)
Celestron C8-NGT	8	5	Metal	GoTo GEM	67
Celestron C10-NGT	10	4.7	Metal	GoTo GEM	83
Discovery PDHQ	8	7	Cardboard	Dob	41 (24)
Discovery PDHQ	10	6	Cardboard	Dob	76 (38)
Discovery PDHQ	12.5	5	Cardboard	Dob	99 (54)
LiteBox 10 (optics not included)[1]	10	4.5 to 6	Truss	Dob	41 plus optics
LiteBox 12.5 (optics not included)[1]	12.5	5 to 6	Truss	Dob	55 plus optics
MC Telescopes Truss Dob 10	10	5 to 6	Truss	Dob	60
Meade LightBridge 12 Deluxe	12	5	Truss	Dob	80 (36)
Orion Telescopes Atlas 8EQ	8	4.9	Metal	GEM	95 (54)
Orion Telescopes Atlas 8EQ-G	8	4.9	Metal	GoTo GEM	95 (54)
Orion Telescopes Sirius 8 EQ	8	4.9	Metal	GEM	84 (32)
Orion Telescopes Sirius 8 EQ-G	8	4.9	Metal	GoTo GEM	84 (32)
Orion Telescopes Atlas 10 EQ	10	4.7	Metal	GEM	117 (54)
Orion Telescopes Atlas 10 EQ-G	10	4.7	Metal	GoTo GEM	117 (54)
Orion Optics (UK) DX150[1]	5.9	5 or 8	Metal	GEM	n/s
Orion Optics (UK) DX200[1]	7.9	4.5 or 6	Metal	GEM	n/s
Orion Optics (UK) Europa 200[1]	7.9	4.5, 6, or 8	Metal	GEM	47 (33)
Orion Optics (UK) OD200[1]	7.9	6 or 8	Metal	Dob	n/s
Orion Optics (UK) DX250[1]	9.8	4.8	Metal	GEM	n/s
Orion Optics (UK) Europa 250[1]	9.8	4.8	Metal	GEM	57 (39)

Brand/Model	Aperture (inches)	Focal Ratio	Tube Material	Type of Mounting	Total Weight/ Heaviest Assembly (pounds)
Orion Optics (UK) OD250[1]	9.8	4.8 or 6.3	Metal	Dob	n/s
Orion Optics (UK) OD300[1]	11.8	4 or 5.3	Metal	Dob	n/s
Orion Optics (UK) OD350[1]	13.8	4.7	Metal	Dob	n/s
Parks Precision[1]	8	6	Fiberglass	GEM	85
Parks Superior[1]	8	6	Fiberglass	GEM	185

Price Range: $2,000 to $3,000

Brand/Model	Aperture (inches)	Focal Ratio	Tube Material	Type of Mounting	Total Weight/ Heaviest Assembly (pounds)
Discovery Truss Design	12.5	5	Truss	Dob	80 (49)
Discovery PDHQ	15	4.2 or 5	Cardboard	Dob	126 (71) f/5 model
LiteBox 14 (optics not included)[1]	14	4.5 to 6	Truss	Dob	62 plus optics
LiteBox 15 (optics not included)[1]	15	4.5 to 6	Truss	Dob	65 plus optics
LiteBox 18 (optics not included)[1]	18	4.5 to 6	Truss	Dob	75 plus optics
Mag One PortaBall 8	8	5 or 6	Truss	Ball	31 (24)
MC Telescopes Truss Dob 12.5	12	5 to 6	Truss	Dob	71
Meade LightBridge 16 Deluxe	16	4.5	Truss	Dob	128
Sky-Watcher 25012EQ6	10	4.7	Metal	GEM	117 (54)
Stargazer Steve 10 Truss Tube Kit	10	5	Truss	Dob	43 (8)
Starmaster Versa 8	8	5.6	Multi-pole	Sold separately	30
Starsplitter Compact IV 8	8	6	Multi-pole	Dob	35 (30)
Starsplitter Compact II 10	10	6	Truss	Dob	50 (40)
Starsplitter Compact IV 10	10	6	Multi-pole	Dob	41 (35)
Starsplitter Compact II 12.5	12.5	4.8 or 6	Truss	Dob	62 (50)
Starsplitter Compact IV 12.5	12.5	4.8	Multi-pole	Dob	58 (50)
TScope 13	13	4.6	Truss	Dob	65 (30)
TScope 14	14	4.7	Truss	Dob	65 (30)
Vixen R150S	5.9	5	Metal	GEM	52 (11 OTA)
Vixen R200SS	7.9	4	Metal	GEM	65 (12 OTA)

Price Range: $3,000 to $5,000

Brand/Model	Aperture (inches)	Focal Ratio	Tube Material	Type of Mounting	Total Weight/ Heaviest Assembly (pounds)
AstroSystems TeleKit 10	10	6	Truss	Dob	75
AstroSystems TeleKit 12.5	12.5	5	Truss	Dob	95
AstroSystems TeleKit 15	15	5	Truss	Dob	145
AstroSystems TeleKit 16	16	4.5	Truss	Dob	145
Discovery Truss Design	15	4.2 or 5	Truss	Dob	98 (57) f/5 model
Discovery PDHQ	17.5	4.1 or 5	Cardboard	Dob	194 (96) f/5 model
Discovery Truss Design	17.5	4.1 or 5	Truss	Dob	125 (75)
Discovery PDHQ	20	5	Cardboard	Dob	n/s
JMI NGT-12.5	12.5	4.5	Truss	Split-ring EQ	120 (40)
JMI NTT-12	12	5	Truss	GoTo alt-az	~125
Mag One PortaBall 10	10	5	Truss	Ball	45 (35)
Mag One PortaBall 12.5	12.5	5	Truss	Ball	62 (50)
MC Telescopes Double Truss 12.5	12.5	5	Truss (double)	Dob	58
MC Telescopes Truss Dob 15	15	4.5 to 6	Truss	Dob	88
MC Telescopes Truss Dob 16	16	4.5 to 6	Truss	Dob	99

Brand/Model	Aperture (inches)	Focal Ratio	Tube Material	Type of Mounting	Total Weight/ Heaviest Assembly (pounds)
MC Telescopes Truss Dob 18	18	4.5 to 6	Truss	Dob	134
NightSky Scopes 12-5	12.5	5	Truss	Dob	80 (45)
NightSky Scopes 14-5	14.5	4.3	Truss	Dob	105 (60)
NightSky Scopes 16	16	4.5	Truss	Dob	117 (37)
Obsession 12.5	12.5	5	Truss	Dob	46 (39)
Obsession 15	15	4.5	Truss	Dob	88 (60)
Orion Optics (UK) SPX200[1]	7.9	4.5, 6, or 8	Metal	GEM	60 (46)
Orion Optics (UK) SPX250[1]	9.8	4.8 or 6.3	Metal	GEM	70 (50)
Orion Optics (UK) DX300[1]	11.8	4	Metal	GEM	n/s
Parallax Instruments PI200[1]	8	7.5	Metal	Sold separately	42
Parallax Instruments PI250[1]	10	6.5	Metal	Sold separately	58
Parks Superior[1]	10	5	Fiberglass	GEM	205
Parks Superior[1]	12.5	5	Fiberglass	GEM	265
Parks Superior Nitelight[1]	10	4	Fiberglass	GEM	191
Plettstone 12.5[1]	12.5	5	Multi-pole	Dob	50
Plettstone 15[1]	15	4.5 or 5	Multi-pole	Dob	75
Starmaster Versa 11	11	4.5	Multi-pole	Sold separately	35
Starmaster Truss 14.5	14.5	4.3	Truss	Dob	91 (33)
Starsplitter Compact II 15	15	4.5	Truss	Dob	72 (60)
Starsplitter II 15	15	4.5	Truss	Dob	80 (70)
Starsplitter II 16	16	4.7	Truss	Dob	90 (80)
TScope Farstar 13	13	4.6	Truss	Dob	64 (33)
TScope Farstar 14	14	4.7	Truss	Dob	65 (34)
TScope Type 3 Deluxe 14	14	4.7	Truss	Dob	70 (44)
TScope Farstar 15	15	4.5 or 5	Truss	Dob	70 (35)

Price Range: Over $5,000

Brand/Model	Aperture (inches)	Focal Ratio	Tube Material	Type of Mounting	Total Weight/ Heaviest Assembly (pounds)
AstroSystems TeleKit 17.5	17.5	5	Truss	Dob	160
AstroSystems TeleKit 18	18	5	Truss	Dob	160
AstroSystems TeleKit 20	20	4.5 or 5	Truss	Dob	180
AstroSystems TeleKit 22	22	4.5 or 5	Truss	Dob	210
AstroSystems TeleKit 24	24	5	Truss	Dob	260
AstroSystems TeleKit 25	25	4.5	Truss	Dob	285
Discovery Truss Design	20	4.5	Truss	Dob	150
Discovery Truss Design	24	4.5	Truss	Dob	185
JMI NGT-18	18	4.5	Truss	Split-ring EQ	245 (75)
JMI NTT-25	25	5	Truss	GoTo Alt-az	~300
JMI NTT-30	30	4.2	Truss	GoTo Alt-az	~350
Mag One PortaBall 14.5	14.5	4.3	Truss	Ball	72 (60)
MC Telescopes Double Truss 16	16	4.5	Truss (double)	Dob	76
MC Telescopes Truss Dob 20	20	4.5 to 6	Truss	Dob	170
NightSky Scopes 18	18	4.2	Truss	Dob	138 (50)
NightSky Scopes 20	20	4.5	Truss	Dob	~170
NightSky Scopes 22	22	4.5	Truss	Dob	n/s
Obsession 18	18	4.5	Truss	Dob	109 (75)
Obsession 20	20	5	Truss	Dob	141 (90)
Obsession 25	25	4 or 5	Truss	Dob	240 (110)

Brand/Model	Aperture (inches)	Focal Ratio	Tube Material	Type of Mounting	Total Weight/ Heaviest Assembly (pounds)
Obsession 30	30	4.5	Truss	Dob	395 (175)
Orion Optics (UK) SPX300[1]	11.8	4 or 5.3	Metal	GEM	88 (62)
Parallax Instruments PI320[1]	12.5	6	Metal	Sold separately	70
Parallax Instruments PI400[1]	16	6	Metal	Sold separately	110
Parks Superior Nitelight[1]	12.5	4	Fiberglass	GEM	231
Parks Observatory[1]	12.5	5	Fiberglass	GEM	515 (375)
Parks Superior Nitelight[1]	16	4	Fiberglass	GEM	615
Parks Observatory[1]	16	5	Fiberglass	GEM	755 (375)
Plettstone 18[1]	18	4.2	Multi-pole	Dob	95 (45)
Starmaster Truss 16	16	4.3	Truss	Dob	116 (42)
Starmaster Truss 18	18	4.3	Truss	Dob	132 (50)
Starmaster Truss 20	20	4.3	Truss	Dob	149 (59)
Starmaster Truss 24	24	4.2	Truss	Dob	186 (79)
Starmaster Truss 28	28	3.7 or 4.1	Truss	Dob	n/s
Starsplitter II 18	18	4.5	Truss	Dob	102 (90)
Starsplitter II Light 20	20	4	Truss	Dob	134 (110)
Starsplitter II 20	20	5	Truss	Dob	134 (120)
Starsplitter II Light 22	22	4.5	Truss	Dob	145 (130)
Starsplitter II Light 24	24	4.5	Truss	Dob	160 (140)
Starsplitter II 25	25	5	Truss	Dob	240 (220)
Starsplitter II Light 28	28	4.5	Truss	Dob	244 (220)
Starsplitter II 30	30	4.5	Truss	Dob	375 (350)
TScope Type 3 Deluxe 16	16	4.5	Truss	Dob	87 (56)

Cassegrain Reflectors

Price Range: $5,000 to $10,000

Brand/Model	Aperture (inches)	Focal Ratio	Tube Material	Type of Mounting	Total Weight/ Heaviest Assembly (pounds)
Parallax Instruments PI250C[1]	10	15	Metal or carbon fiber	Sold separately	40
Parks H.I.T. Penta Series[1]	8	4 and 16	Fiberglass	GEM	n/s
Parks H.I.T. Superior[1]	10	4 and 16	Fiberglass	GEM	203 (70)
Parks H.I.T. Superior[1]	12.5	4 and 16	Fiberglass	GEM	249 (70)
Takahashi Mewlon M-180PZ	7	12	Metal	GEM	40 (25)
Takahashi Mewlon M-210UH	8.3	11.5	Metal	GEM	74 (34)

Price Range: $10,000 to $20,000

Brand/Model	Aperture (inches)	Focal Ratio	Tube Material	Type of Mounting	Total Weight/ Heaviest Assembly (pounds)
OGS RC10	10	9	Metal or carbon fiber	GEM	
OGS RC12.5	12.5	9	Metal or carbon fiber	GEM	
Parallax Instruments PI250R[1]	10	9	Metal or carbon fiber	Sold separately	40
Parallax Instruments PI320C[1]	12.5	16	Metal or carbon fiber	Sold separately	60
Parallax Instruments PI320R[1]	12.5	9	Metal or carbon fiber	Sold separately	60

Brand/Model	Aperture (inches)	Focal Ratio	Tube Material	Type of Mounting	Total Weight/ Heaviest Assembly (pounds)
RC Optical Systems	10	9	Carbon fiber or truss	Sold separately	34
RC Optical Systems	12.5	9	Carbon fiber or truss	Sold separately	50
Takahashi CN-212	8.3	3.9 and 12.4	Metal	GEM	55 (34)
Takahashi Mewlon M-250US	9.8	12	Metal	GEM	62 (34)

Price Range: Over $20,000

Brand/Model	Aperture (inches)	Focal Ratio	Tube Material	Type of Mounting	Total Weight/ Heaviest Assembly (pounds)
OGS RC14.5	14.25	7.9	Metal or carbon fiber	GEM	70
OGS RC16	16	8.4	Metal or carbon fiber	GEM	85
OGS RC20	20	8.1	Metal or carbon fiber	GEM	150
OGS RC24	24	8	Metal or carbon fiber	GEM	210
OGS RC32	32	7.6	Metal or carbon fiber	GEM	340
Parallax Instruments PI400C[1]	16	14.25	Metal or carbon fiber	Sold separately	85
Parallax Instruments PI400R[1]	16	8.4	Metal or carbon fiber	Sold separately	85
Parallax Instruments PI500C[1]	20	16	Metal or carbon fiber	Sold separately	130
Parallax Instruments PI500R[1]	20	8.1	Metal or carbon fiber	Sold separately	130
Parks H.I.T. Superior 101-10460[1]	16	4 and 15	Fiberglass	GEM	655 (306)
RC Optical Systems	14	7.9	Carbon fiber or truss	Sold separately	70
RC Optical Systems	16	8.4	Carbon fiber or truss	Sold separately	89
RC Optical Systems	20	8.1	Carbon fiber	Sold separately	150
Takahashi BRC-250[1]	10	5	Metal	GEM	34 (OTA)
Takahashi Mewlon M-300SQI	11.8	11.9	Metal	GEM	89 (55)

Exotic Reflectors

Price Range: Under $1,000

Brand/Model	Aperture (inches)	Focal Ratio	Tube Material	Type of Mounting	Total Weight/ Heaviest Assembly (pounds)
DGM Optics OA 3.6 ATS	3.6	11.1	PVC or metal	Dob	29 (18)
DGM Optics OA 4.0	4	10.4	PVC or metal	Dob	29 (18)

Price Range: $1,000 to $3,000

Brand/Model	Aperture (inches)	Focal Ratio	Tube Material	Type of Mounting	Total Weight/ Heaviest Assembly (pounds)
DGM Optics OA 5.1 ATS	5.1	10.8	Metal	Dob	45 (28)
DGM Optics OA 6.5 ATS	6.5	10.4	Metal	Dob	63 (40)

Brand/Model	Aperture (inches)	Focal Ratio	Tube Material	Type of Mounting	Total Weight/ Heaviest Assembly (pounds)
Schmidt-Cassegrain Catadioptrics					
Price Range: Under $1,000					
Celestron NexStar 5SE	5	10	Metal	GoTo Single-arm Fork	28 (18)
Celestron C6-S	6	10	Metal	GEM	52
Price Range: $1,000 to $2,000					
Celestron C6-SGT	6	10	Metal	GoTo GEM	52
Celestron NexStar 6SE	6	10	Metal	GoTo Single-arm Fork	30 (21)
Celestron C8-S	8	10	Metal	GEM	54
Celestron C8-SGT	8	10	Metal	GoTo GEM	54
Celestron CPC 800	8	10	Metal	GoTo Fork	61 (42)
Celestron NexStar 8SE	8	10	Metal	GoTo Single-arm Fork	33 (24)
Celestron C9.25-S	9.25	10	Metal	GEM	62
Celestron C9.25-SGT	9.25	10	Metal	GoTo GEM	62
Celestron C11-S	11	10	Metal	GEM	91
Celestron C11-SGT	11	10	Metal	GoTo GEM	91
Meade LX90 GPS	8	10	Metal	GoTo Fork	53 (33)
Meade SC-8	8	10	Metal	GoTo GEM	65
Orion Telescopes Sirius 8 XLT EQ-G	8	10	Metal	GoTo GEM	73 (32)
Orion Telescopes SkyView Pro 8 XLT	8	10	Metal	GEM	53
Orion Telescopes SkyView Pro 9.25 XLT	9.25	10	Metal	GEM	61
Price Range: $2,000 to $3,000					
Celestron CPC 925	9.25	10	Metal	GoTo Fork	77 (58)
Celestron CPC 1100	11	10	Metal	GoTo Fork	84 (65)
Meade LX90 GPS	10	10	Metal	GoTo Fork	61 (41)
Meade LX90 GPS	12	10	Metal	GoTo Fork	80 (58)
Orion Telescopes Sirius 9.25 XLT EQ-G	9.25	10	Metal	GoTo GEM	83 (32)
Orion Telescopes Atlas 11 XLT	11	10	Metal	GEM	116
Orion Telescopes Atlas 11 XLT EQ-G	11	10	Metal	GoTo GEM	116
Price Range: Over $3,000					
Celestron CGE 800	8	10	Carbon fiber	GoTo GEM	113
Celestron CGE 925	9.25	10	Carbon fiber	GoTo GEM	134
Celestron CGE 1100	11	10	Carbon fiber	GoTo GEM	142
Celestron CGE 1400	14	11	Carbon fiber	GoTo GEM	184
Schmidt-Newtonian Catadioptrics					
Price Range: Under $1,000					
Meade SN-6	6	5	Metal	GoTo GEM	48
Meade SN-8	8	4	Metal	GoTo GEM	69

Brand/Model	Aperture (inches)	Focal Ratio	Tube Material	Type of Mounting	Total Weight/ Heaviest Assembly (pounds)
Price Range: $1,000 to $2,000					
Meade SN-10	10	4	Metal	GoTo GEM	85

Maksutov Catadioptrics

(Note: MC: Maksutov-Cassegrain; MN: Maksutov-Newtonian)

Brand/Model	Aperture (inches)	Focal Ratio	Tube Material	Type of Mounting	Total Weight/ Heaviest Assembly (pounds)
Price Range: Under $1,000					
Celestron NexStar 4SE	4	13	Metal	GoTo Single-arm Fork	21 (11)
Celestron C130-M (MC)	5	13	Metal	GEM	51
Intes Micro Alter M503 (MC)	5	10	Metal	Sold separately	7
Intes Micro MN56	5	6	Metal	Sold separately	11
Intes Micro MN55	5.5	5.5	Metal	Sold separately	11
Konus MotorMax 90 MC[1]	3.5	13.3	Metal	GEM	20
Konus MotorMax 130 MC[1]	5	15.4	Metal	GEM	47
Konus MotorMax 150 MC[1]	6	12	Metal	GEM	68 (46)
Meade ETX-90PE (MC)	3.5	13.8	Metal	GoTo Fork	21 (12)
Meade ETX-105PE (MC)	4.1	14	Metal	GoTo Fork	27 (14)
Orion SkyView Pro 127 (MC)	5	12.1	Metal	GEM	43 (34)
Orion SkyView Pro 150 (MC)	5.9	12	Metal	Sold separately	12
Orion StarMax 90 EQ (MC)	3.5	13.9	Metal	GEM	15 (12)
Orion StarMax 102 EQ (MC)	4	12.7	Metal	GEM	22 (17)
Orion StarMax 127 EQ (MC)	5	12.1	Metal	GEM	36 (25)
Price Range: $1,000 to $2,000					
Celestron C130-MGT (MC)	5	13	Metal	GoTo GEM	51
Intes Micro Alter M603 (MC)	6	10	Metal	Sold separately	13
Intes Micro Alter M6511 (MC)	6	11	Metal	Sold separately	15
Intes Micro MN65	6	5.5	Metal	Sold separately	16
Intes Micro MN66	6	6	Metal	Sold separately	16
Meade ETX-125PE	5	15	Metal	GoTo Fork	28 (15)
Orion (UK) OMC140 GP (MC)[1]	5.5	14.3	Metal	GEM	40 (32)
Price Range: $2,000 to $4,000					
Intes Micro Alter M703 (MC)	7	10	Metal	Sold separately	14
Intes Micro Alter M715 (MC)	7	15	Metal	Sold separately	15
Intes Micro MN76	7	6	Metal or carbon fiber	Sold separately	30
Questar Field 3.5 (MC)[1]	3.5	14.4	Metal	Sold separately	4
Price Range: $4,000–$10,000					
Intes Micro Alter M1008 (MC)	10	12.5	Metal	Sold separately	38
Intes Micro Alter M809 (MC)	8	10	Metal	Sold separately	22
Intes Micro Alter M815 (MC)	8	15	Metal	Sold separately	22
Intes Micro MN86	8	6	Metal	Sold separately	40

Brand/Model	Aperture (inches)	Focal Ratio	Tube Material	Type of Mounting	Total Weight/ Heaviest Assembly (pounds)
Orion (UK) OMC140 SP (MC)[1]	5.5	14.3	Metal	GoTo GEM	44 (36)
Orion (UK) OMC200 GP (MC)[1]	7.9	20	Metal	GEM	66 (47)
Orion (UK) OMC200 SP (MC)[1]	7.9	20	Metal	GoTo GEM	66 (47)
Questar Duplex 3.5 (MC)[1]	3.5	14.4	Metal	Fork	7
Questar Standard 3.5 (MC)[1]	3.5	14.4	Metal	Fork	7
TEC MC200/15[1]	8	15.5	Metal	Sold separately	21
TEC MC250/12[1]	10	12	Metal	Sold separately	37
TEC MC250/20[1]	10	20	Metal	Sold separately	37
Price Range: Over $10,000					
Intes Micro Alter M1408 (MC)	14	12	Metal	Sold separately	n/s
Intes Micro Alter M1608 (MC)	16	12	Metal	Sold separately	n/s
Questar Astro 7 (MC)[1]	7	13.4	Metal	Fork	19
Questar Classic 7 (MC)[1]	7	14.3	Metal	Fork	26

Exotic Catadioptrics

Brand/Model	Aperture (inches)	Focal Ratio	Tube Material	Type of Mounting	Total Weight/ Heaviest Assembly (pounds)
Price Range: $1,000 to $2,000					
Novosibirsk TAL-150K[1]	5.9	10.3	Metal	GEM	51 (35)
Novosibirsk TAL-200K[1]	7.9	10	Metal	GEM	66 (40)
Price Range: $2,000 to $4,000					
Meade LX200R	8	10	Metal	GoTo Fork	73
Meade LX200R	10	10	Metal	GoTo Fork	90
Novosibirsk TAL-250K[1]	9.8	8.5	Metal	GEM	106 (71)
Vixen VC200L	7.9	9	Metal	GoTo GEM	45 (30)
Vixen VMC200L	7.9	9.8	Metal	GoTo GEM	45 (30)
Price Range: Over $4,000					
Meade RCX400	10	8	Carbon fiber	GoTo Fork	140 (84)
Meade LX200R	12	10	Metal	GoTo Fork	125
Meade RCX400	12	8	Carbon fiber	GoTo Fork	150 (94)
Meade LX200R	14	10	Metal	GoTo Fork	166
Meade RCX400	14	8	Carbon fiber	GoTo Fork	191 (135)
Meade LX200R	16	10	Metal	GoTo Fork	318
Meade RCX400	16	8	Carbon fiber	GoTo Fork	>300
Meade RCX400 Max[1]	16	8	Carbon fiber	GoTo Fork	648 (140)
Meade RCX400 Max[1]	20	8	Carbon fiber	GoTo Fork	733 (190)
Vixen VMC260L	10.2	11.6	Metal	Sold separately	27 (OTA)
Vixen VMC330L	13	13.1	Metal	Sold separately	44 (OTA)

1. *Find a capsule review of this model in the* Star Ware *section of the author's Web site, www.philharrington.net.*
2. *Tube material:*
    Carbon fiber: Solid-wall tube; material is very lightweight, yet strong
    Cardboard: Solid-wall cardboard tube
    Fiberglass: Solid-wall fiberglass tube

Metal: Solid-wall metal tube

Multi-pole: Open design built from 3 to 8 parallel poles

Plastic: Solid-wall plastic tube

Truss: Open triangular truss built from 6 or 8 aluminum poles

Truss (double): Similar to an open truss tube, except that mirror box is also made from aluminum truss poles

3. Mount:

Alt-az: Altitude-azimuth mount

Ball: Spherical end of telescope tube rides in a concave base

GoTo: Computer-driven mount

Dob: Dobsonian-style altitude-azimuth mount

Fork: Fork equatorial mount

GEM: German equatorial mount

Single-arm Fork: Fork equatorial mount, but telescope is supported by only one arm

Split-ring EQ: Spli-ring equatorial mount

n/s: Not supplied; mounting must be purchased separately

4. "Total weight" includes both telescope and mounting, while "Heaviest Assembly" lists the weight of the heaviest single part that a user is likely to carry out into the field. If the weights are the same in both columns, then the telescope and mounting are usually transported together. In addition, the following abbreviations apply:

n/s: Not supplied by manufacturer

n/a: Not applicable, usually denoting a telescope that is sold without a mounting; instead, a mounting may be purchased separately, which, depending on the mount used, may cause the total instrument weight to vary. In these cases, the weight quoted in the "Heaviest Assembly" is that of the optical tube assembly (OTA). Bear in mind that the mounting will probably be heavier.

All technical specifications were obtained from manufacturer data and have not been independently verified.

Appendix B
Eyepiece Marketplace

Company	Focal Length	Design	Apparent Field of View	Eye Relief	Eyecup	Parfocal
1.25-Inch Barrel: Under $50						
Apogee Super Abbe	4.8	Orthoscopic	45	n/s	Yes	n/s
Apogee Super Abbe	7.7	Orthoscopic	45	n/s	Yes	n/s
Apogee Super Abbe	10.5	Orthoscopic	45	n/s	Yes	n/s
Apogee Super Abbe	16.8	Orthoscopic	45	n/s	Yes	n/s
Apogee Super Abbe	24	Orthoscopic	45	n/s	Yes	n/s
Celestron E-Lux	6	Plössl	50	5	Yes	No
Celestron E-Lux	10	Plössl	50	7	Yes	No
Celestron E-Lux	25	Plössl	50	22	Yes	No
Celestron E-Lux	40	Plössl	43	31	Yes	No
Celestron Omni	4	Plössl	52	6	Yes	No
Celestron Omni	6	Plössl	52	5	Yes	No
Celestron Omni	9	Plössl	52	6	Yes	No
Celestron Omni	12	Plössl	52	8	Yes	No
Celestron Omni	15	Plössl	52	13	Yes	No
Celestron Omni	20	Plössl	52	20	Yes	No
Celestron Omni	25	Plössl	52	22	Yes	No
Discovery	4	Plössl	n/s	n/s	Yes	n/s
Discovery	6.5	Plössl	n/s	n/s	Yes	n/s
Discovery	10	Plössl	n/s	n/s	Yes	n/s
Discovery	12.5	Plössl	n/s	n/s	Yes	n/s
Discovery	15	Plössl	n/s	n/s	Yes	n/s
Discovery	20	Plössl	n/s	n/s	Yes	n/s
Discovery	25	Plössl	n/s	n/s	Yes	n/s
Discovery	30	Plössl	n/s	n/s	Yes	n/s
Discovery	40	Plössl	n/s	n/s	Yes	n/s
Edmund Scientific Plössl	12	Plössl	50	n/s	No	No

Company	Focal Length	Design	Apparent Field of View	Eye Relief	Eyecup	Parfocal
Edmund Scientific Plössl	15	Plössl	50	n/s	No	No
Edmund Scientific Plössl	21	Plössl	50	n/s	No	No
Edmund Scientific Plössl	28	Plössl	50	n/s	No	No
Guan Sheng Optical SuperView	5	5-element hybrid	65	10	Yes	No
Guan Sheng Optical SuperView	10	5-element hybrid	65	11	Yes	No
Guan Sheng Optical SuperView	15	5-element hybrid	65	13	Yes	No
Guan Sheng Optical SuperView	20	5-element hybrid	67	18	Yes	No
Hands-on Optics GTO Plössl	4	Plössl	52	4	Yes	No
Hands-on Optics GTO Plössl	6	Plössl	52	5	Yes	No
Hands-on Optics GTO Plössl	9	Plössl	52	7	Yes	No
Hands-on Optics GTO Plössl	15	Plössl	52	12	Yes	No
Hands-on Optics GTO Plössl	25	Plössl	52	20	Yes	No
Hands-on Optics GTO Plössl	32	Plössl	51	24	Yes	No
Hands-on Optics GTO Plössl	40	Plössl	44	32	Yes	No
Meade 4000 Super Plössl	6.4	4-element hybrid	52	3	Yes	Yes
Meade 4000 Super Plössl	9.7	4-element hybrid	52	5	Yes	Yes
Meade 4000 Super Plössl	12.4	4-element hybrid	52	7	Yes	Yes
Meade 4000 Super Plössl	15	4-element hybrid	52	9	Yes	Yes
Meade 4000 Super Plössl	20	4-element hybrid	52	13	Yes	Yes
Meade 4000 Super Plössl	26	4-element hybrid	52	18	Yes	Yes
Orion Explorer II	6	Kellner	50	4	No	Yes
Orion Explorer II	10	Kellner	50	5	No	Yes
Orion Explorer II	13	Kellner	50	7	No	Yes
Orion Explorer II	17	Kellner	50	12	No	Yes
Orion Explorer II	20	Kellner	50	15	No	Yes
Orion Explorer II	25	Kellner	50	15	No	Yes
Orion Sirius Plössl	6.3	Plössl	50	5	Yes	Yes
Orion Sirius Plössl	7.5	Plössl	50	5	Yes	Yes
Orion Sirius Plössl	10	Plössl	50	7	Yes	Yes
Orion Sirius Plössl	12.5	Plössl	50	8	Yes	Yes
Orion Sirius Plössl	17	Plössl	50	13	Yes	Yes
Orion Sirius Plössl	20	Plössl	50	14	Yes	Yes
Orion Sirius Plössl	26	Plössl	50	18	Yes	Yes
Orion Sirius Plössl	32	Plössl	50	25	Yes	Yes
Orion Sirius Plössl	40	Plössl	43	22	Yes	No

1.25-Inch Barrel: $50 to $100

Company	Focal Length	Design	Apparent Field of View	Eye Relief	Eyecup	Parfocal
Celestron Omni	32	Plössl	52	22	Yes	No
Celestron Omni	40	Plössl	43	31	Yes	No
Celestron X-Cel	2.3	6-element hybrid	55	20	Yes	Yes
Celestron X-Cel	5	6-element hybrid	55	20	Yes	Yes
Celestron X-Cel	8	6-element hybrid	55	20	Yes	Yes
Celestron X-Cel	10	6-element hybrid	55	20	Yes	Yes
Celestron X-Cel	12.5	6-element hybrid	55	20	Yes	Yes
Celestron X-Cel	18	6-element hybrid	55	20	Yes	Yes

Company	Focal Length	Design	Apparent Field of View	Eye Relief	Eyecup	Parfocal
Celestron X-Cel	21	6-element hybrid	55	20	Yes	Yes
Celestron X-Cel	25	6-element hybrid	55	20	Yes	Yes
Celestron Zoom	8–24	7-element zoom	60–40	15–18	Yes	No
Discovery Series 70	10	Plössl	n/s	n/s	Yes	n/s
Discovery Series 70	15	Plössl	n/s	n/s	Yes	n/s
Discovery Series 70	20	Plössl	n/s	n/s	Yes	n/s
Discovery Zoom	7–23	7-element zoom	65–36	17	Yes	No
Discovery Zoom	8–24	6-element zoom	54–33	10–11	Yes	No
Edmund Optics RKE	8	RKE	45	8	No	No
Edmund Optics RKE	12	RKE	45	11	No	No
Edmund Optics RKE	15	RKE	45	13	No	No
Edmund Optics RKE	21	RKE	45	19	No	No
Edmund Optics RKE	28	RKE	45	25	No	No
Edmund Scientific Orthoscopic	4	Orthoscopic	41	3	No	No
Edmund Scientific Orthoscopic	25	Orthoscopic	47	22	No	No
Edmund Scientific Plössl	8	Plössl	50	10	No	No
Meade 4000 QX	15	5-element hybrid	70	5	Yes	No
Meade 4000 QX	20	5-element hybrid	70	10	Yes	No
Meade 4000 Super Plössl	32	4-element hybrid	52	20	Yes	Yes
Meade 4000 Super Plössl	40	4-element hybrid	44	30	Yes	No
Meade 5000 Plössl	5.5	6-element hybrid	60	7	Yes	Yes
Meade 5000 Plössl	9	5-element hybrid	60	7	Yes	Yes
Meade 5000 Plössl	14	5-element hybrid	60	10	Yes	Yes
Meade 5000 Plössl	20	5-element hybrid	60	15	Yes	Yes
Meade 5000 Plössl	26	5-element hybrid	60	19	Yes	Yes
Orion Epic ED-2	3.7	6-element hybrid	55	20	Yes	Yes
Orion Epic ED-2	5.1	6-element hybrid	55	20	Yes	Yes
Orion Epic ED-2	7.5	6-element hybrid	55	20	Yes	Yes
Orion Epic ED-2	9.5	6-element hybrid	55	20	Yes	Yes
Orion Epic ED-2	12.3	6-element hybrid	55	20	Yes	Yes
Orion Epic ED-2	14	6-element hybrid	55	20	Yes	Yes
Orion Epic ED-2	18	6-element hybrid	55	20	Yes	Yes
Orion Epic ED-2	22	6-element hybrid	55	20	Yes	Yes
Orion Epic ED-2	25	6-element hybrid	55	20	Yes	Yes
Orion Expanse	6	5-element hybrid	66	15	Yes	Yes
Orion Expanse	9	6-element hybrid	66	15	Yes	Yes
Orion Expanse	15	4-element hybrid	66	15	Yes	Yes
Orion Expanse	20	4-element hybrid	66	18	Yes	Yes
Orion HighLight Plössl	6.3	Plössl	52	4	Yes	Yes
Orion HighLight Plössl	7.5	Plössl	52	5	Yes	Yes
Orion HighLight Plössl	10	Plössl	52	7	Yes	Yes
Orion HighLight Plössl	12.5	Plössl	52	8	Yes	Yes
Orion HighLight Plössl	17	Plössl	52	13	Yes	Yes
Orion HighLight Plössl	20	Plössl	52	16	Yes	Yes
Orion HighLight Plössl	26	Plössl	52	18	Yes	Yes
Orion HighLight Plössl	32	Plössl	52	20	Yes	Yes
Orion HighLight Plössl	40	Plössl	43	22	Yes	Yes

Company	Focal Length	Design	Apparent Field of View	Eye Relief	Eyecup	Parfocal
Orion Ultrascopic	7.5	5-element hybrid	52	5	Yes	Yes
Orion Ultrascopic	10	5-element hybrid	52	6	Yes	Yes
Orion Ultrascopic	15	5-element hybrid	52	10	Yes	Yes
Orion Ultrascopic	20	5-element hybrid	52	13	Yes	Yes
Parks Kellner	6	Kellner	40	n/s	No	No
Parks Kellner	12	Kellner	40	n/s	No	No
Parks Kellner	25	Kellner	40	20	No	No
ScopeStuff UltraWide	6	5-element hybrid	66	15	Yes	Yes
ScopeStuff UltraWide	9	6-element hybrid	66	15	Yes	Yes
ScopeStuff UltraWide	15	4-element hybrid	66	15	Yes	Yes
ScopeStuff UltraWide	20	4-element hybrid	66	18	Yes	Yes
Sky Instruments Antares Elite Plössl	5	7-element hybrid	52	4	Yes	Yes
Sky Instruments Antares Elite Plössl	7.5	5-element hybrid	52	5	Yes	Yes
Sky Instruments Antares Elite Plössl	10	5-element hybrid	52	7	Yes	Yes
Sky Instruments Antares Elite Plössl	15	5-element hybrid	52	10	Yes	Yes
Sky Instruments Antares Elite Plössl	20	5-element hybrid	52	13	Yes	Yes
Sky Instruments Antares Elite Plössl	25	5-element hybrid	52	17	Yes	Yes
Sky Instruments Antares Orthoscopic	6	Orthoscopic	40	5	Yes	n/s
Sky Instruments Antares Orthoscopic	7	Orthoscopic	40	6	Yes	n/s
Sky Instruments Antares Orthoscopic	9	Orthoscopic	40	7	Yes	n/s
Sky Instruments Antares Orthoscopic	12	Orthoscopic	40	10	Yes	n/s
Sky Instruments W70	5.8	6-element hybrid	70	15	Yes	No
Sky Instruments W70	8.6	5-element hybrid	70	15	Yes	No
Sky Instruments W70	14	4-element hybrid	70	15	Yes	No
Sky Instruments W70	19	5-element hybrid	70	15	Yes	No
Sky Instruments W70	25	8-element hybrid	70	15	Yes	No
Tele Vue Plössl	8	Plössl	50	6	Yes	Yes
Tele Vue Plössl	11	Plössl	50	8	Yes	Yes
Tele Vue Plössl	15	Plössl	50	10	Yes	Yes
Tele Vue Plössl	20	Plössl	50	14	Yes	Yes
Tele Vue Plössl	25	Plössl	50	17	Yes	Yes
William Optics Swan	9	5-element hybrid	72	12	Yes	Yes
William Optics Swan	15	5-element hybrid	72	14	Yes	Yes
William Optics Swan	20	5-element hybrid	72	17	Yes	Yes

1.25-Inch Barrel: $100 to $200

Company	Focal Length	Design	Apparent Field of View	Eye Relief	Eyecup	Parfocal
Baader Genuine Ortho	5	Orthoscopic	40	4	Yes	Yes
Baader Genuine Ortho	6	Orthoscopic	40	5	Yes	Yes
Baader Genuine Ortho	7	Orthoscopic	40	6	Yes	Yes

Company	Focal Length	Design	Apparent Field of View	Eye Relief	Eyecup	Parfocal
Baader Genuine Ortho	9	Orthoscopic	40	7	Yes	Yes
Baader Genuine Ortho	12.5	Orthoscopic	40	10	Yes	Yes
Baader Genuine Ortho	18	Orthoscopic	40	15	Yes	Yes
Celestron Axiom	15	7-element hybrid	70	7	Yes	Yes
Celestron Axiom	19	7-element hybrid	70	10	Yes	Yes
Celestron Axiom	23	7-element hybrid	70	10	Yes	Yes
Kokusai Kohki Wide Scan III	13	5-element hybrid	84	8	Yes	No
Kokusai Kohki Wide Scan III	16	5-element hybrid	84	9	Yes	No
Kokusai Kohki Wide Scan III	20	5-element hybrid	84	12	Yes	No
Meade 5000 Super Wide Angle	16	6-element hybrid	68	12	Yes	Yes
Orion Ultrascopic	3.8	7-element hybrid	52	5	Yes	Yes
Orion Ultrascopic	5	7-element hybrid	52	6	Yes	Yes
Orion Ultrascopic	25	5-element hybrid	52	17	Yes	Yes
Orion Ultrascopic	30	5-element hybrid	52	21	Yes	Yes
Orion Ultrascopic	35	5-element hybrid	49	25	Yes	No
Parks Gold Series	3.8	5-element hybrid	52	6	Yes	Yes
Parks Gold Series	5	5-element hybrid	52	8	Yes	Yes
Parks Gold Series	7.5	5-element hybrid	52	6	Yes	Yes
Parks Gold Series	10	5-element hybrid	52	8	Yes	Yes
Parks Gold Series	15	5-element hybrid	52	12	Yes	Yes
Parks Gold Series	20	5-element hybrid	52	14	Yes	Yes
Parks Gold Series	25	5-element hybrid	52	16	Yes	Yes
Parks Gold Series	30	5-element hybrid	52	24	Yes	Yes
Parks Gold Series	35	5-element hybrid	49	25	Yes	No
Sky Instruments Speers-Waler	7.5	8-element hybrid	82	12	Yes	Yes
Sky Instruments Speers-Waler	10	8-element hybrid	72	12	Yes	Yes
Sky Instruments Speers-Waler	10	8-element hybrid	82	12	Yes	Yes
Sky Instruments Speers-Waler	14	8-element hybrid	72	12	Yes	Yes
Sky Instruments Speers-Waler	14	8-element hybrid	82	12	Yes	Yes
Sky Instruments Speers-Waler-L	14	8-element hybrid	72	12	Yes	Yes
Sky Instruments Speers-Waler	18	8-element hybrid	82	12	Yes	Yes
Takahashi LE-ED	5	5-element hybrid	52	10	Yes	Yes
Takahashi LE-ED	7.5	5-element hybrid	52	10	Yes	Yes
Takahashi LE	12.5	5-element hybrid	52	9	Yes	Yes
Takahashi LE	18	5-element hybrid	52	13	Yes	Yes
Takahashi LE	24	5-element hybrid	52	17	Yes	Yes
Takahashi LE	30	5-element hybrid	52	20	Yes	Yes
Tele Vue Plössl	32	Plössl	50	22	Yes	Yes
Tele Vue Plössl	40	Plössl	43	28	Yes	No
Vixen Lanthanum LV	2.5	8-element hybrid	45	20	Yes	Yes
Vixen Lanthanum LV	4	8-element hybrid	45	20	Yes	Yes

Company	Focal Length	Design	Apparent Field of View	Eye Relief	Eyecup	Parfocal
Vixen Lanthanum LV	5	6-element hybrid	45	20	Yes	Yes
Vixen Lanthanum LV	6	6-element hybrid	45	20	Yes	Yes
Vixen Lanthanum LV	7	6-element hybrid	45	20	Yes	Yes
Vixen Lanthanum LV	9	6-element hybrid	50	20	Yes	Yes
Vixen Lanthanum LV	10	6-element hybrid	50	20	Yes	Yes
Vixen Lanthanum LV	12	6-element hybrid	50	20	Yes	Yes
Vixen Lanthanum LV	15	6-element hybrid	50	20	Yes	Yes
Vixen Lanthanum LV	18	6-element hybrid	50	20	Yes	Yes
Vixen Lanthanum LV	20	6-element hybrid	50	20	Yes	Yes
Vixen Lanthanum LV	25	6-element hybrid	50	20	Yes	Yes
Vixen Lanthanum LV	40	7-element hybrid	42	20	Yes	Yes
Vixen Lanthanum Zoom	8–24	7-element zoom	60–40	19	Yes	No

1.25-Inch Barrel: $200 to $300

Company	Focal Length	Design	Apparent Field of View	Eye Relief	Eyecup	Parfocal
Meade 4000 Zoom	8–24	7-element zoom	55–40	16	Yes	No
Meade 5000 Super Wide Angle	20	6-element hybrid	68	15	Yes	Yes
Meade 5000 Super Wide Angle	24	6-element hybrid	68	18	Yes	Yes
Meade 5000 Ultra Wide Angle	4.7	7-element hybrid	82	13	Yes	Yes
Meade 5000 Ultra Wide Angle	6.7	7-element hybrid	82	15	Yes	Yes
Meade 5000 Ultra Wide Angle	8.8	7-element hybrid	82	15	Yes	Yes
Meade 5000 Ultra Wide Angle	14	7-element hybrid	82	15	Yes	Yes
Meade 5000 Ultra Wide Angle	18	6-element hybrid	82	13	Yes	Yes
Orion Premium Zoom	8–24	7-element zoom	60–40	17–19	Yes	No
Sky Instruments Speers-Waler	5–8	9-element zoom	89–81	12	Yes	No
Takahashi Hi-LE-ED	2.8	6-element hybrid	42	6	Yes	Yes
Takahashi Hi-LE-ED	3.6	6-element hybrid	40	8	Yes	Yes
Tele Vue Nagler Type 6	2.5	7-element hybrid	82	12	Yes	Yes
Tele Vue Nagler Type 6	3.5	7-element hybrid	82	12	Yes	Yes
Tele Vue Nagler Type 6	5	7-element hybrid	82	12	Yes	Yes
Tele Vue Nagler Type 6	7	7-element hybrid	82	12	Yes	Yes
Tele Vue Nagler Type 6	9	7-element hybrid	82	12	Yes	Yes
Tele Vue Nagler Type 6	11	7-element hybrid	82	12	Yes	Yes
Tele Vue Nagler Type 6	13	7-element hybrid	82	12	Yes	Yes
Tele Vue Panoptic	15	6-element hybrid	68	10	Yes	Yes
Tele Vue Panoptic	19	6-element hybrid	68	13	Yes	Yes
Tele Vue Panoptic	24	6-element hybrid	68	15	Yes	Yes
Tele Vue Radian	3	7-element hybrid	60	20	Yes	Yes
Tele Vue Radian	4	7-element hybrid	60	20	Yes	Yes
Tele Vue Radian	5	7-element hybrid	60	20	Yes	Yes
Tele Vue Radian	6	7-element hybrid	60	20	Yes	Yes
Tele Vue Radian	8	6-element hybrid	60	20	Yes	Yes

Company	Focal Length	Design	Apparent Field of View	Eye Relief	Eyecup	Parfocal
Tele Vue Radian	10	6-element hybrid	60	20	Yes	Yes
Tele Vue Radian	12	6-element hybrid	60	20	Yes	Yes
Tele Vue Radian	14	6-element hybrid	60	20	Yes	Yes
Tele Vue Radian	18	6-element hybrid	60	20	Yes	Yes
Tele Vue Zoom	8–24	7-element zoom	55–40	15–20	Yes	No
TMB Super Monocentric	4	Monocentric	32	3	Yes	Yes
TMB Super Monocentric	5	Monocentric	32	4	Yes	Yes
TMB Super Monocentric	6	Monocentric	32	5	Yes	Yes
TMB Super Monocentric	7	Monocentric	32	6	Yes	Yes
TMB Super Monocentric	8	Monocentric	32	7	Yes	Yes
TMB Super Monocentric	9	Monocentric	32	8	Yes	Yes
TMB Super Monocentric	10	Monocentric	32	9	Yes	Yes
TMB Super Monocentric	12	Monocentric	32	10	Yes	Yes
TMB Super Monocentric	14	Monocentric	32	12	Yes	Yes
TMB Super Monocentric	16	Monocentric	32	14	Yes	Yes
Vernonscope Brandon	8	Brandon	50	6	Yes	Yes
Vernonscope Brandon	12	Brandon	50	10	Yes	Yes
Vernonscope Brandon	16	Brandon	50	13	Yes	Yes
Vernonscope Brandon	24	Brandon	50	19	Yes	Yes
Vernonscope Brandon	32	Brandon	50	26	Yes	Yes
William Optics UWAN	4	7-element hybrid	82	12	Yes	No
William Optics UWAN	7	7-element hybrid	82	12	Yes	No
William Optics UWAN	16	7-element hybrid	82	12	Yes	No

1.25-Inch Barrel: More than $300

Company	Focal Length	Design	Apparent Field of View	Eye Relief	Eyecup	Parfocal
Pentax XO	2.5	Orthoscopic	44	4	No	No
Pentax XO	5	Orthoscopic	44	4	No	No
Pentax XW	3.5	8-element hybrid	70	20	Yes	Yes
Pentax XW	5	8-element hybrid	70	20	Yes	Yes
Pentax XW	7	8-element hybrid	70	20	Yes	Yes
Pentax XW	10	7-element hybrid	70	20	Yes	Yes
Pentax XW	14	7-element hybrid	70	20	Yes	Yes
Pentax XW	20	6-element hybrid	70	20	Yes	Yes
Tele Vue Nagler Type 5	16	6-element hybrid	82	10	Yes	Yes
Tele Vue Nagler Zoom	2–4	5-element zoom	50	10	Yes	Yes
Tele Vue Nagler Zoom	3–6	5-element zoom	50	10	Yes	Yes

1.25-Inch/2-Inch Barrel: $100 to $200

Company	Focal Length	Design	Apparent Field of View	Eye Relief	Eyecup	Parfocal
Orion Stratus	3.5	8-element hybrid	68	20	Yes	Yes
Orion Stratus	5	8-element hybrid	68	20	Yes	Yes
Orion Stratus	8	8-element hybrid	68	20	Yes	Yes
Orion Stratus	13	8-element hybrid	68	20	Yes	Yes
Orion Stratus	17	8-element hybrid	68	20	Yes	Yes
Orion Stratus	21	8-element hybrid	68	20	Yes	Yes

1.25-Inch/2-Inch Barrel: More than $200

Company	Focal Length	Design	Apparent Field of View	Eye Relief	Eyecup	Parfocal
Collins I3 Piece	25	Night-vision Plössl	35	38	Yes	No
Tele Vue Nagler Type 4	12	6-element hybrid	82	17	Yes	No
Tele Vue Panoptic	22	6-element hybrid	68	15	Yes	Yes

Company	Focal Length	Design	Apparent Field of View	Eye Relief	Eyecup	Parfocal
Vixen Lanthanum LVW	3.5	8-element hybrid	65	20	Yes	Yes
Vixen Lanthanum LVW	5	8-element hybrid	65	20	Yes	Yes
Vixen Lanthanum LVW	8	8-element hybrid	65	20	Yes	Yes
Vixen Lanthanum LVW	13	8-element hybrid	65	20	Yes	Yes
Vixen Lanthanum LVW	17	8-element hybrid	65	20	Yes	Yes
Vixen Lanthanum LVW	22	8-element hybrid	65	20	Yes	Yes
2-Inch Barrel: Under $100						
Astrobuffet 1rpd	30	5-element hybrid	80	17	No	No
Celestron E-Lux	26	Kellner	56	20	Yes	No
Celestron E-Lux	32	Kellner	56	20	Yes	No
Celestron E-Lux	40	Kellner	50	20	Yes	No
Guan Sheng Optical SuperView	30	5-element hybrid	68	22	Yes	No
Guan Sheng Optical SuperView	42	5-element hybrid	65	30	Yes	No
Guan Sheng Optical SuperView	50	5-element hybrid	54	35	Yes	No
Meade 4000 QX	26	5-element hybrid	70	10	Yes	No
Meade 4000 QX	30	5-element hybrid	70	15	Yes	No
Meade 4000 QX	36	5-element hybrid	70	12	Yes	No
Meade 4000 Super Plössl	56	4-element hybrid	52	47	Yes	No
Orion DeepView	35	RKE	56	20	Yes	No
Orion DeepView	42	RKE	52	20	Yes	No
Sky Instruments Classic Erfle	30	Erfle	65	~13	Yes	No
Sky Instruments Classic Erfle	42	Erfle	65	~20	Yes	No
Sky Instruments Classic Erfle	52	Erfle	55	~25	Yes	No
2-Inch Barrel: $100 to $200						
Celestron Axiom	19	7-element hybrid	70	10	Yes	Yes
Discovery Series 70	32	Plössl	n/s	n/s	Yes	n/s
Hands-on Optics Proxima	31	Erfle	71	20	Yes	No
Meade 5000 Plössl	32	5-element hybrid	60	22	Yes	Yes
Meade 5000 Plössl	40	5-element hybrid	60	28	Yes	Yes
Orion Optiluxe	32	4-element hybrid	58	22	No	No
Orion Optiluxe	40	4-element hybrid	62	27	No	No
Orion Optiluxe	50	5-element hybrid	45	40	No	No
Orion Stratus	30	8-element hybrid	68	20	Yes	No
Parks Gold Series	50	5-element hybrid	50	30	No	No
Sky Instruments Classic Erfle	30	Erfle	75	15	Yes	No
Vixen Lanthanum LV	50	7-element hybrid	45	20	Yes	Yes
William Optics SWAN	25	5-element hybrid	72	21	Yes	Yes
William Optics SWAN	33	5-element hybrid	72	25	Yes	Yes
William Optics SWAN	40	5-element hybrid	70	28	Yes	Yes
2-Inch Barrel: $200 to $300						
Celestron Axiom	34	7-element hybrid	70	16	Yes	Yes
Celestron Axiom	40	7-element hybrid	70	21	Yes	Yes
Edmund Optics Erfle	32	Erfle	68	18	No	No

Company	Focal Length	Design	Apparent Field of View	Eye Relief	Eyecup	Parfocal
Kokusai Kohki Wide Scan III	30	5-element hybrid	84	18	Yes	No
Meade 5000 Super Wide Angle	28	6-element hybrid	68	22	Yes	Yes
Orion Stratus	35	8-element hybrid	68	20	Yes	No
Tele Vue Plössl	55	Plössl	50	38	Yes	No
Vixen Lanthanum LV	30	7-element hybrid	50	20	Yes	Yes
2-Inch Barrel: More than $300						
Celestron Axiom	50	7-element hybrid	50	38	Yes	Yes
Meade 5000 Super Wide Angle	34	6-element hybrid	68	26	Yes	Yes
Meade 5000 Super Wide Angle	40	6-element hybrid	68	31	Yes	Yes
Meade 5000 Ultra Wide Angle	24	6-element hybrid	82	18	Yes	Yes
Meade 5000 Ultra Wide Angle	30	6-element hybrid	82	22	Yes	Yes
Pentax XW	30	7-element hybrid	70	20	Yes	Yes
Pentax XW	40	6-element hybrid	70	20	Yes	Yes
Takahashi LE	50	5-element hybrid	50	40	Yes	Yes
Tele Vue Nagler Type 4	17	7-element hybrid	82	17	Yes	No
Tele Vue Nagler Type 4	22	7-element hybrid	82	19	Yes	No
Tele Vue Nagler Type 5	20	6-element hybrid	82	12	Yes	Yes
Tele Vue Nagler Type 5	26	6-element hybrid	82	16	Yes	No
Tele Vue Nagler Type 5	31	6-element hybrid	82	19	Yes	No
Tele Vue Panoptic	27	6-element hybrid	68	19	Yes	Yes
Tele Vue Panoptic	35	6-element hybrid	68	24	Yes	Yes
Tele Vue Panoptic	41	6-element hybrid	68	27	Yes	Yes
Vixen Lanthanum LVW	42	8-element hybrid	65	20	Yes	No
William Optics UWAN	28	6-element hybrid	82	18	Yes	No

All technical specifications were supplied by the manufacturers and have not been independently verified. Any value noted as n/s is unavailable from the manufacturer.

Apparent Field of View: The eyepiece's apparent field of view, expressed in degrees.

Eye Relief: The distance, in millimeters, that an observer's eye must be away from the eye lens in order to see the entire field.

Eyecups: Useful for preventing stray light from entering through the corner of the observer's eye.

Focal Length: The eyepiece's focal length, expressed in millimeters.

Parfocal: The eyepiece focuses at nearly the same distance as other eyepieces in the manufacturer's series. Occasionally, some eyepieces in a particular line may not be parfocal with all others in the same line, but instead only with one or two others.

Series: The optical design or trade name of the particular eyepiece.

Appendix C
The Astronomical
Yellow Pages

Manufacturers:

Here is a listing of all of the companies whose products are discussed throughout this book, along with a few other enterprises that, while not mentioned earlier, offer worthwhile services to the amateur astronomical community. You will also find this list as well as a list of astronomical equipment retailers on the *Star Ware* Web site, www.philharrington.net.

Adirondack Video Astronomy
    72 Harrison Avenue, Hudson Falls, NY 12839
    Phone: (877) 348-8433, (518) 747-4141
    Web site: www.astrovid.com
    Product(s): Video cameras
Anttler Optics
    103 Third Street, Suite B, Kalkaska, MI 49646
    Phone: (866) 409-6965, (231) 409-6965
    Web site: www.geckooptical.com/anttler_optics
    Product(s): Newtonian reflectors, eyepieces, accessories
Apogee, Inc
    P.O. Box 136, Union, IL 60180
    Phone: (815) 923-4188
    Web site: www.apogeeinc.com
    Product(s): Refractors, eyepieces, accessories
Astrobuffet
    P.O. Box 510, Pepperell, MA 01463
    Phone: (978) 835-0376

    Web site: www.astrobuffet.com
    Product(s): Eyepieces
Astro Haven
    5650 Production Way, Langley, BC, V3A 4N4
        Canada
    Phone: (604) 539-2208
    Web site: www.astrohaven.com
    Product(s): Observatory domes
Astro-Lite
    627 Ridge Road, Lewisberry, PA 17339
    Phone: (717) 938-9890
    Web site: www.astrolite-led.com
    Product(s): LED flashlights
Astromeccanica
    via Pernice 71 - 54100; Romagnano
        (MS) Italy
    Phone: 39-05-858-31130
    Web site: www.astromeccanica.it
    Product(s): Binoscopes

Astronomical Society of the Pacific
390 Ashton Avenue, San Francisco, CA 94112
Phone: (415) 337-1100
Web site: www.astrosociety.org
Product(s): *RealSky* software, *Mercury* magazine, posters, books

Astronomik
c/o Astro-shop; Eiffestraße 426; D-20537 Hamburg, Germany
Phone: 49-40-511-4348
Web site: www.astronomik.com
Product(s): Filters

Astro-Physics, Inc.
11250 Forest Hills Road, Rockford, IL 61115
Phone: (815) 282-1513
Web site: www.astro-physics.com
Product(s): Apochromatic refractors, mountings

AstroSystems, Inc.
124 N. Second Street, LaSalle, CO 80645
Phone: (970) 284-9471
Web site: www.astrosystems.biz
Product(s): TeleKits telescopes, collimation tools, accessories

Astrozap
P.O. Box 502, Lakewood, OH 44107
Phone: (330) 278-2507
Web site: www.astrozap.com
Product(s): Dew caps

Baader Planetarium
Zur Sternwarte, D-82291 Mammendorf, Germany
Phone: 49-81-45-8802
Web site: www.baader-planetarium.com
Product(s): Solar filter material, eyepieces

Barska
2215 E. Huntington Drive, Duarte, CA 91010
Phone: (888) 666-6769, (626) 357-2658
Web site: www.barska.com
Product(s): Binoculars

Bike Nashbar
6103 State Route 446, Canfield, OH 44406
Phone: (800) 888-2710
Web site: www.nashbar.com
Product(s): Cold-weather clothing and accessories (bike stuff, too!)

Blaho Company
P.O. Box 419, Pleasant Hill, OH 45359
Phone: (937) 676-5891
Web site: www.blaho.com
Product(s): Binocular mounts

Bogen Photo Corporation (see also Manfrotto, Gitzo)
565 East Crescent Avenue, P.O. Box 506, Ramsey, NJ 07446
Phone: (201) 818-9500, (212) 695-8166
Web site: www.bogenphoto.com
Product(s): Photographic tripods

Borg (Oasis Borg)
c/o Hutech Corporation, 23505 Crenshaw Boulevard #225, Torrance, CA, USA 90505
Phone: (877) 289-2674, (310) 325-5511
Web site: www.sciencecenter.net/hutech
Product(s): Refractors, light-pollution reduction filters

Burgess Optical Company
7756 Oak Ridge Highway, Knoxville, TN 37931
Phone: (865) 769-8777
Web site: www.burgessoptical.com
Product(s): Binocular viewer, red-dot finder, accessories

Bushnell Sports Optics
9200 Cody Street, Overland Park, KS 66214
Phone: (800) 423-3537, (913) 752-3400, (800) 361-5702 (Canada only)
Web site: www.bushnell.com
Product(s): Binoculars

Campmor
28 Parkway, Box 700, Upper Saddle River, NJ 07458
Phone: (800) 525-4784, (201) 825-8300
Web site: www.campmor.com
Product(s): Cold-weather clothing and accessories

Canon, Inc.
30-2, Shimomaruko 3-chome, Ohta-ku, Tokyo 146-8501, Japan
Phone: 81-33-758-2111
Canon USA
1 Canon Plaza, Lake Success, NY 1042
Phone: (800) 652-2666, (516) 328-5000

Web site: www.canon.com, www.canon.ca, www.canon.co.uk

Product(s): Binoculars, cameras

Carina Software

602 Morninghome Road, Danville, CA 94526

Phone: (925) 838-0695

Web site: www.carinasoft.com

Product(s): Voyager IV software

Carton Optical Canada, Inc.

6568 Des Merles Lane, Ottawa, Ontario K1C 7G9 Canada

Phone: (613) 830-7750

Web site: www.carton.ca

Product(s): Binoculars

Celestron International

2835 Columbia Street, Torrance, CA 90503

Phone: (310) 803-5955

Web site: www.celestron.com

Product(s): Binoculars, telescopes, eyepieces, accessories

Collins Electro Optics

9025 E. Kenyon Avenue, Denver, CO 80237

Phone: (303) 889-5910

Web site: www.ceoptics.com

Product(s): Electronic eyepiece

Coronado Technology Group

6001 Oak Canyon, Irvine, CA 92618

Phone: (919) 451-1450, (866) 786-9282

Web site: www.coronadofilters.com

Product(s): Hydrogen-alpha solar filters and telescopes

D & G Optical

2075 Creek Road, Manheim, PA 17545

Phone: (717) 665-2076

Web site: www.dgoptical.com

Product(s): Refractors, Cassegrain reflectors

Dark Sky, Inc.

9407 Aboite Road, Roanoke, IN 46783

Phone: (877) 327-5759

Web site: www.darkskypanels.com

Product(s): Panel observatory

DayStar

3857 Schaefer Avenue, Suite D, Chino, CA 91710

Phone: (909) 591-4673

Web site: www.daystarfilters.com

Product(s): Hydrogen-alpha solar filters

Denkmeier Optical, Inc.

12636 Sunset Avenue J2, Ocean City, MD 21842

Phone: (866) 340-4578, (410) 208-6014

Web site: http://deepskybinoviewer.com

Product(s): Binocular viewers, focal reducer

DewBuster

269 St. Andrews Boulevard, Laplace, LA 70068

Phone: (985) 652-1985

Web site: www.dewbuster.com

Product(s): Dew heater controller and accessories

Dew-Not

87 Cardinal Lane, Marstons Mills, MA 02648

Phone: (774) 238-0743

Web site: www.dew-not.com

Product(s): Dew shields

Dewzee Astronomy Products

P.O. Box 90340, San Jose, CA 95109

Phone: (408) 656-9815

Web site: www.dewzee.com

Product(s): Dew shields

DGM Optics

P.O. Box 750, Charlestown, NH 03603

Phone: (978) 874-2985

Web site: www.dgmoptics.com

Product(s): Off-axis reflectors, filters

Discovery Telescopes

615 South Tremont Street, Oceanside, CA 92054

Phone: (760) 967-6598, (877) 523-4400

Web site: www.discovery-telescopes.com

Product(s): Newtonian reflectors, short-tube refractor, eyepieces, accessories

Eastern Mountain Sports

1 Vose Farm Road, Peterborough, NH 03458

Phone: (888) 463-6367

Web site: www.ems.com

Product(s): Cold-weather clothing and accessories

Edmund Optics

101 East Gloucester Pike, Barrington, NJ 08007

Phone: (800) 363-1992, (856) 573-6250

Web site: www.edmundoptics.com

Product(s): Eyepieces, telescope mirrors

Edmund Scientific Co.

60 Pearce Avenue, Tonawanda, NY 14150-6711

Phone: (800) 728-6999

Web site: www.scientificsonline.com

Product(s): Binoculars, Newtonian reflector, eyepieces

Equatorial Platforms

15736 McQuiston Lane, Grass Valley, CA 95945

Phone: (530) 274-9113

Web site: www.equatorialplatforms.com

Product(s): Equatorial tracking tables

FAR Laboratories

P.O. Box 25, South Hadley, MA 01075

Phone: (800) 336-9054, (413) 467-9262

Web site: www.dynapod.com

Product(s): Handicapped-accessible telescope mounts and piers, eyepiece holders and accessories

Finger Lakes Instrumentation

7298 West Main Street, P.O. Box 19A, Lima, NY 14485

Phone: (585) 624-3760

Web site: www.fli-cam.com

Product(s): CCD cameras

Fuji Photo Optical Co. Ltd. (Fujinon, Inc.)

1-324 Uetake, Omiya City, Saitama 330-8624 Japan

Phone: 048-668-2152

Fujinon, Inc.

10 High Point Drive, Wayne, NJ 07470

Phone: (973) 633-5600

Web site: www.fujinon.co.jp, www.fujinonbinocular.com

Product(s): Binoculars, film

Galaxy Optics

P.O. Box 2045, Buena Vista, CO 81211

Phone: (719) 395-8242

Web site: www.galaxyoptics.com

Product(s): Telescope mirrors

Gitzo S.A.

ZA de Mondetour RN 10/Le Bois Paris, 28630 Nogent Le Phaye, France

Web site: www.gitzo.com

Product(s): Photographic tripods (see also Bogen)

Glatter Laser Collimators

3850 Sedgwick Avenue, 14F, Bronx, NY 10463

Phone: (718) 796-3203

Web site: www.collimator.com

Product(s): Laser collimators

Grabber Performance Group

4600 Danvers Drive Southeast, Grand Rapids, MI 49512

Phone: (800) 518-0938

Web site: www.grabberwarmers.com

Product(s): Hand and foot warmers

Guan Sheng Optical Co., Ltd.

No. 152, Huei An St, Chu Tung Town, Hsinchu Hsien, Taiwan, R.O.C.

Phone: 886-3-595-1510

Web site: www.gs-telescope.com

Product(s): Refractors, Newtonian reflectors, eyepieces

Hands-On Optics

26437 Ridge Road, Damascus, MD 20872

Phone: (866) 726-7371, (301) 482-0000

Web site: www.handsonoptics.com

Product(s): Eyepieces, accessories

Heat Factory

2390 Oak Ridge Way, Vista, CA 92083

Phone: (800) 993-4328, (760) 734-5300

Web site: www.heatfactory.com

Product(s): Hand and foot warmers

Helix Observing Accessories

P.O. Box 490, Gibsonia, PA 15044

Phone: (724) 316-0306

Web site: http://helix-mfg.com

Product(s): Laser collimator, LED flashlight, alt-azimuth mounts

IDAS

c/o Hutech Corporation; 23505 Crenshaw Boulevard #225, Torrance, CA 90505

Phone: (877) 289-2674, (310) 325-5511

Web site: www.sciencecenter.net/hutech

Product(s): Filters

iLanga, Inc.

12221 NE 82nd Lane, Kirkland, WA 98033

Phone: (425) 889-8273

Web site: www.ilangainc.com

Product(s): Astronomical software, including AstroPlanner

Imaginova

470 Park Avenue South, 9th Floor, New York, NY 10016

Phone: (212) 703-5800

Web site: www.imaginova.com

Product(s): *Starry Night* software

Insight Technology

445 Broadhead Street, Windsor, ON N9A-3W3 Canada

Phone: (519) 255-7644

Web site: www.thebeerchair.com

Product(s): Observing chair

Intes Micro Co. Ltd.

Shvernika str 4, 117036 Moscow, Russia

Phone: 95-126-9803

Web site: www.intes.su/defaulte.aspx

Product(s): Maksutov telescopes

Jim Fly

7806 Wildcreek Trail SE, Huntsville, AL 35802

Phone: (256) 882-2523

Web site: www.catseyecollimation.com

Product(s): Collimation tools, observing chair plans

Jim's Mobile, Inc. (JMI)

8550 West 14th Avenue, Lakewood, CO 80215

Phone: (303) 233-5353, (800) 247-0304

Web site: www.jmitelescopes.com

Product(s): Newtonian reflectors, accessories

JMB, Inc.

736 Oak Glen Circle, Fall Branch, TN 37656

Phone: (423) 348-8883

Web site: www.identi-view.com

Product(s): Solar filters

Johnsonian Designs

3466 E. County Road 20-C, Unit B-20, Loveland, CO 80537

Phone: (970) 219-6392

Web site: www.johnsonian.com

Product(s): Equatorial tracking tables

Just-Cheney Enterprises

3613 Alderwood Avenue, Bellingham, WA 98225

Phone: (360) 738-3766

Web site: www.dewshields.com

Product(s): Dew shields

Kendrick Astro Instruments

2920 Dundas Street West, Toronto, ON M6P 1Y8 Canada

Phone: (800) 393-5456, (416) 762-7946

Web site: www.kendrickastro.com

Product(s): Dew prevention system, portable observatories

Kokusai Kohki

Higashinokuchi Cho 1-3, Matsuo, Nishikyo Ku, Kyoto 615-8217 Japan

Phone: 81-75-394-2625

Web site: www.kkohki.com/English/Hello.html

Product(s): Eyepieces

Konus

Via Mirandola, 45-37026 Settimo di Pescantina (VR) Italy

Phone: 39-045-676-7670

Konus USA

7275 NW 87th Avenue, Miami, FL 33178

Phone: (305) 592-5500

Web site: www.konus.com

Product(s): Refractors, Newtonian reflectors

Kowa Optimed, Inc.

20001 South Vermont Avenue, Torrance, CA 90502

Phone: (310) 327-1913, (800) 966-5692

Web site: www.kowascope.com

Product(s): Binoculars

L. L. Bean

Freeport, ME 04033

Phone: (800) 441-5713 (USA and Canada), (800) 891-297 (UK), (207) 552-3027 (elsewhere)

Web site: www.llbean.com

Product(s): Cold-weather clothing and accessories

LaserMax, Inc.

3495 Winton Place, Building B, Rochester, NY 14623

Phone: (585) 272-5420, (800) 527-3703

Web site: www.oemlasers.com

Product(s): Laser collimators

Leupold and Stevens, Inc.

P.O. Box 688, Beaverton, OR 97075

Phone: (503) 526-1400

Web site: www.leupold.com

Product(s): Binoculars

LiteBox Telescopes

    1717 Kamamalu Avenue, Honolulu, HI 96813

    Phone: (808) 524-2450

    Web site: www.litebox-telescopes.com

    Product(s): Newtonian reflectors

Losmandy Astronomical Products (Hollywood General Machining)

    1033 North Sycamore Avenue, Los Angeles, CA 90038

    Phone: (323) 462-2855

    Web site: www.losmandy.com

    Product(s): Telescope mounts and accessories

Lumicon

    750 East Easy Street, Simi Valley, CA 93065

    Phone: (805) 520-0047, (800) 420-0255

    Web site: www.lumicon.com

    Product(s): Filters, astrophotography accessories, finderscopes

Lymax's Earth, Sky, and Astronomy

    P.O. Box 328, Grain Valley, MO 64029

    Phone: (888) 737-5050; (816) 737-5050

    Web site: www.lymax.com

    Product(s): Herald-Bobroff Astroatlas, observing accessories

Mag One Instruments

    16342 West Coachlight Drive, New Berlin, WI 53151

    Phone: (262) 785-0926

    Web site: www.mag1instruments.com

    Product(s): Newtonian reflectors

Manfrotto (Gruppo Manfrotto Srl)

    Via Sasso Rosso, 19; 36061 Bassano del Grappa (VI), Italy

    Phone: 39-042-455-5855

    Web site: www.manfrotto.com

    Product(s): Photographic tripods (see also Bogen)

MC Telescopes

    RR #3 Box 3745, Nicholson, PA 18446

    Phone: (570) 942-6838

    Web site: www.mctelescopes.com

    Product(s): Newtonian reflectors

Meade Instruments Corp.

    6001 Oak Canyon, Irvine, CA 92620

    Phone: (800) 626-3233; (949) 451-1450

    Web site: www.meade.com

    Product(s): Binoculars, telescopes, eyepieces, accessories, CCD imagers

Miyauchi

    177 Kanasaki, Minano-machi; Saitama 369-1621 Japan

    Phone: 81-49-462-4245

    Web site: www.miyauchi-opt.co.jp

    Product(s): Binoculars

Moonfish Group Optics S.L.

    C/ Vista Alegre 20, Matadepera 08290; Barcelona, Spain

    Web site: www.moonfishgroup.com

    Phone: 34-61-693-3291

    Product(s): Eyepieces, accessories

Mountain Instruments

    1213 South Auburn Street; Colfax, CA 95713

    Phone: (530) 346-9113

    Web site: www.mountaininstruments.com

    Product(s): Equatorial mounts

Murnaghan Instruments

    1781 Primrose Lane, West Palm Beach, FL 33414

    Phone: (561) 795-2201

    Web site: www.e-scopes.cc/ Murnaghan_Instruments_Corp56469.html

    Product(s): CCD flip mirrors

NatureWatch

    PO Box 22062, Charlottetown, Prince Edward Island C1A 9J2 Canada

    Phone: (902) 628-8095, (800) 693-8095

    Web site: www.naturewatchshop.com

    Product(s): Replacement wooden tripod legs

NightSky Scopes

    61432 Daspit Road, Lacombe, LA 70445

    Phone: (985) 882-9269

    Web site: www.nightskyscopes.com

    Product(s): Newtonian reflectors

Nikon

    Nikon Futaba Bldg., 3-25, Futaba 1-Chome, Shinagawa-Ku, Tokyo 142-0043 Japan

    Phone: 81-33-788-7697

    Nikon USA

    1300 Walt Whitman Road, Melville, NY 11747

    Phone: (800) 645-6689, (631) 547-4200

    Web site: www.ave.nikon.co.jp/index_e.htm; www.nikonusa.com

    Product(s): Binoculars, cameras

North Peace International
  Suite 810, 9707–110 Street, Capital Place, Edmonton, AB T5K 2L9 Canada
  Phone: (780) 944-0388
  Web site: www.telusplanet.net/public/northpc
  Product(s): Portable observatories

Nova Astronomics
  P.O. Box 31013, Halifax, NS B3K 5T9 Canada
  Phone: (902) 499-6196
  Web site: www.nova-astro.com
  Product(s): *Earth-Centered Universe* computer software

Nova Optical Systems
  14121 South Shaggy Mountain Road, Herriman, UT 84065
  Phone: (801) 446-1802
  Web site: www.nova-optical.com
  Product(s): Telescope mirrors

Novosibirsk Instrument-Making Plant (TAL Instruments)
  179/2 D. Kovalchuk Street, Novosibirsk, 630049 Russia
  Phone: 07-383-226-0765
  Web page: www.telescopes.ru
  Product(s): Reflectors, eyepieces, accessories

Oberwerk Corporation
  2440 Wildwood, Xenia, OH 45385
  Phone: (866) 244-2460, (937) 372-5409
  Web site: www.bigbinoculars.com
  Product(s): Binoculars, binocular accessories

Observa-Domes
  371 Commerce Park Drive, Jackson, MS 39213
  Phone: (601) 982-3333
  Web site: www.observa-dome.com
  Product(s): Observatory domes

Obsession Telescopes
  PO Box 804, Lake Mills, WI 53551
  Phone: (920) 648-2328
  Web site: www.obsessiontelescopes.com
  Product(s): Newtonian reflectors

Optec, Inc.
  199 Smith Street, Lowell, MI 49331
  Phone: (616) 897-9351, (888) 488-0381
  Web site: www.optecinc.com
  Product(s): Focal reducer/corrector

Optical Guidance Systems (OGS)
  2450 Huntingdon Pike, Huntingdon Valley, PA 19006
  Phone: (215) 947-5571
  Web site: www.opticalguidancesystems.com
  Product(s): Cassegrain reflectors

Optical Mechanics, Inc. (OMI)
  P.O. Box 2313, Iowa City, IA 52244
  Phone: (319) 351-3960
  Web site: www.opticalmechanics.com
  Product(s): Telescope mirrors, custom telescopes

Orion Optics (U.K.)
  Unit 21 Third Avenue, Crewe, Cheshire CW1 6XU UK
  Phone: 44(0)1270-500089
  Web site: www.orionoptics.co.uk
  Product(s): Newtonian reflectors, catadioptric telescopes

Orion Telescopes and Binoculars (USA)
  P.O. Box 1815, Santa Cruz, CA 95061
  Phone: (800) 447-1001, (831) 763-7000
  Web site: www.telescope.com
  Product(s): Binoculars, telescopes, eyepieces, accessories

Pacific Telescope Corporation (Sky-Watcher Telescopes)
  160-11880 Hammersmith Way, Richmond, B.C. V7A 5C8 Canada
  Phone: (604) 241-7027
  Web site: www.skywatchertelescope.com
  Product(s): Refractors, Newtonian reflectors, catadioptric telescopes, eyepieces, accessories

Parallax Instruments
  P.O. Box 327, Youngsville, NC 27596
  Phone: (919) 570-3140
  Web site: www.parallaxinstruments.com
  Product(s): Reflectors, mounts

Parks Optical
  750 East Easy Street, Simi Valley, CA 93065
  Phone: (805) 522-6722
  Web site: www.parksoptical.com
  Product(s): Binoculars, refractors, reflectors, accessories

Pegasus Optics
  121 N. CR 5601, Castroville, TX 78009
  Phone: (830) 538-9499
  Web site: www.pegasusoptics.com
  Product(s): Telescope mirrors

Pentax Corporation
2-36-9 Maeno-cho, Itabashi-ku, Tokyo 174-8639 Japan
Phone: 81-33-960-5151
Pentax Imaging Company
600 12th Street, Suite 300, Golden, CO 80401
Phone: (303) 728-0224, (800) 877-0155
Web sites:
www.pentax.co.jp/english/index.php;
www.pentaximaging.com
Product(s): Binoculars, telescopes, eyepieces

Performance Bike Shops
1 Performance Way, Chapel Hill, NC 27514
Phone: (800) 727-2453, (919) 933-9113
Web site: www.performancebike.com
Product(s): Cold-weather clothing and accessories (bike stuff, too!)

PixSoft, Inc.
208–3111 Portage Avenue, Winnipeg, MB R3K 0W4 Canada
Phone: (204) 885-4936
Web site: www.stargps.com
Product(s): GPS interface for Meade LX200-compatible non-GPS telescopes

Project Pluto
168 Ridge Road, Bowdoinham, ME 04008
Phone: (207) 666-5750, (800) 777-5886
Web site: www.projectpluto.com
Product(s): *Guide* computer software

Promaster
360 Tunxis Hill Road, Fairfield, CT 06825
Phone: (203) 336-0183
Web site: www.promaster.com
Product(s): Cameras

Questar Corporation
6204 Ingham Road, New Hope, PA 18938
Phone: (215) 862-5277
Web site: www.questar-corp.com
Product(s): Maksutov-Cassegrain catadioptric telescopes and accessories

RC Optical Systems
4025 E. Huntington Drive, Suite 105, Flagstaff, AZ 86004
Phone: (928) 526-5380
Web site: www.rcopticalsystems.com
Product(s): Cassegrain reflectors

R.E.I.
Sumner, WA 98352
Phone: (800) 426-4840, (253) 891-2500
Web site: www.rei.com
Product(s): Cold-weather clothing and accessories

R.F.S., Inc.
4935 Allison Street #7, Arvada, CO 80002
Phone: (303) 423-1264
Web site: www.buyastrostuff.com
Product(s): Observing chair

Rigel Systems
26850 Basswood, Rancho Palos Verdes, CA 90275
Phone: (310) 375-4149
Web site: www.rigelsys.com
Product(s): LED flashlights, collimation tools, QuikFinder, accessories

Round Table Platforms
1340 West Sturbridge, Hoffman Estates, IL 60195
Phone: (847) 202-1986
Web site: www.roundtableplatforms.com
Product(s): Equatorial tracking tables

SAC Imaging
P.O. Box 360982, Melbourne, FL 23936
Phone: (321) 259-6498
Web site: www.sac-imaging.com
Product(s): CCD imagers

Santa Barbara Instrument Group (SBIG)
147-A Castilian Drive, Santa Barbara, CA 93117
Phone: (805) 571-7244
Web site: www.sbig.com
Product(s): CCD cameras and accessories

Schmidling Productions, Inc.
18016 Church Road, Marengo, IL 60152
Phone: (815) 923-0031
Web site: http://schmidling.netfirms.com/ez-testr.htm
Product(s): Rónchi test eyepiece

Scientific American Book Club
P.O. Box 6375, Camp Hill, PA 17012
Phone: (717) 918-1065
Web site: www.sciambookclub.com
Product(s): Books

ScopeStuff
P.O. Box 3754, Cedar Park, TX 78630
Phone: (512) 259-9778
Web site: www.scopestuff.com
Product(s): Eyepieces, extension tubes, many
hard-to-find accessories

Sirius Observatories
33 Filmer Street, Clontarf QLD 4019, Australia
Phone: 61-07-3284-2111
Web site: www.siriusobservatories.com
Product(s): Observatory domes

Sirius Optics
17632 SE 102 Street, Renton, WA 98059
Phone: (425) 430-0555
Web site: www.siriusoptics.com
Product(s): Filters

Sky Domes Ltd. UK
Cockport Farm, Otterham, Camelford, Corn-
wall PL32 9SS UK
Phone: 44(0)845-644-8068
Web site: www.skydomes.co.uk
Product(s): Observatory domes

Sky Engineering
4630 N. University Drive, #329, Coral Springs,
FL 33067
Phone: (954) 345-8726
Web site: www.skyeng.com
Product(s): Sky Commander digital setting cir-
cles

Sky Instruments
M.P.O. 2037, Vancouver, BC V6B 3R6
Canada
Phone: (800) 648-4188, (604) 270-2330
Web site: www.antaresoptical.com
Product(s): Telescopes, eyepieces, finder-
scopes, filters

SkyMap Software
The Thompson Partnership, Brixham Business
Centre, 6a Fore Street; Brixham, Devon TQ5
8DS UK
Phone: 44(0)845-456-8412
Web site: www.skymap.com
Product(s): SkyMap Pro software

SkyShed
3404 Perth Road 180, RR #2; Staffa, ON N0K
1Y0 Canada
Phone (519) 345-0036

Web site: www.skyshed.com
Product(s): Roll-off roof observatories

Sky-Watcher Telescopes
see Pacific Telescope Corporation

Small Parts, Inc.
13980 N.W. 58th Court, P.O. Box 4650, Miami
Lakes, FL 33014
Phone: (305) 557-7955, (800) 220-4242
Web site: www.smallparts.com
Product(s): Nuts, bolts, fasteners, knobs, and
other assorted hardware

Software Bisque
912 12th Street, Golden, CO 80401
Phone: (800) 843-7599, (303) 278-4478
Web site: www.bisque.com
Product(s): *TheSky* computer software, Para-
mount equatorial mount

Star Instruments
555 Blackbird Roost #5, Flagstaff, AZ
86001
Phone: (928) 774-9177
Web site: www.star-instruments.com
Product(s): Cassegrain mirrors and other telescope optics

Starbound
68 Klaum Avenue, North Tonawanda, NY
14120
Phone: (716) 692-3671
Web site: none
Product(s): Observing chair

Starchair Engineering Pty Ltd
1 First Street, Gawler, S.A. 5118, Australia
Telephone: 61-08-8522-7272
Web site: www.starchair.com
Product(s): Starchair 3000 binocular chair

Stargazer Steve
1752 Rutherglen Crescent, Sudbury, ON P3A
2K3 Canada
Phone: (705) 566-1314
Web site: stargazer.isys.ca
Product(s): Newtonian reflectors

Starlight Xpress Ltd.
Foxley Green Farm; Ascot Road, Holyport, Berk-
shire, SL6 3LA UK
Phone: 44(0)162-877-7126
Web site: www.starlight-xpress.co.uk
Product(s): Starlight Xpress CCD cameras

Starmaster Telescopes
2160 Birch Road, Arcadia, KS 66711
Phone: (620) 638-4743
Web site: www.starmastertelescopes.com
Product(s): Newtonian reflectors, observing chair

Starsplitter Telescopes
3228 Rikkard Drive, Thousand Oaks, CA 91362
Phone: (805) 492-0489
Web site: www.starsplitter.com
Product(s): Newtonian reflectors

Steiner-Optik GmbH
P.O. Box 160241, 95448 Bayreuth, Germany
Phone: 49-09-2178-7916
Pioneer Marketing
97 Foster Road, Suite 5, Moorestown, NJ 08057
Phone: (609) 854-2424, (800) 257-7742
Web site: www.steiner.de, www.pioneer-research.com
Product(s): Binoculars

StellarCAT
1460 N Clanton Avenue, Sierra Vista, AZ 85635
Phone: (520) 432-4433
Web site: www.stellarcat.biz
Product(s): ServoCAT GoTo control

Stellarvue
11820 Kemper Road, Auburn, CA 95603
Phone: (530) 823-7796
Web site: www.stellarvue.com
Product(s): Refractors, binoculars

Stevens Optics
6516 SW Barnes Road, Portland, OR 97225
Phone: (503) 203-8603
Web site: www.stevensoptical.com
Products: Telescope mirrors

Swarovski Optik KG
Swarovskistr. 70; 6067 Absam Austria
Phone: 43-05-223-5110
Swarovski Optik North America
2 Slater Drive, Cranston, RI 02920
Phone: (401) 734-1800
Web site: www.swarovskioptik.at, www.swarovskioptik.com
Product(s): Binoculars

Swift Instruments, Inc.
2055 Gateway Place Suite 500, San Jose, CA 95110
Phone: (800) 523-4544, (408) 200-7500
Web site: www.swift-optics.com
Product(s): Binoculars

Synta Optical Technology Corporation
No. 89 Lane 4, Chia An W. Road, Long Tan Tao Yuan, Taiwan
No. 28 Huqiu Xi Road, Suzhou, China
Web site: none
Product(s): Refractors, Newtonian reflectors, eyepieces, accessories

T&T Binoculars Mounts
18 Strong Street Extension, East Haven, CT 06512
Phone: (203) 469-2845, (203) 272-1915
Web site: www.benjaminsweb.com/TandT
Product(s): Binocular mounts

Takahashi Seisakusho Ltd.
41-7 Oharacho, Itabashiku, Tokyo 174-0061 Japan
Phone: 81-03-3966-9491
Product(s): Refractors, reflectors, binoculars, mounts, eyepieces
U.S. Importer:
Land, Sea and Sky
3110 South Shepherd, Houston, TX 77098
Phone: (713) 529-3551
Web site: www.takahashiamerica.com
U.K. Importer:
True Technology Ltd
Woodpecker Cottage, Red Lane, Aldermaston, Berks, RG7 4PA UK
Phone: (44)-(0)118-970-0777
Web site: www.trutek-uk.com

Taurus Technologies
P.O. Box 14, Woodstown, NJ 08098
Phone: (856) 769-4509
Web site: www.taurus-tech.com
Product(s): Astrocameras, flip mirrors

Tech 2000
69 South Ridge Street, Monroeville, OH 44847
Phone: (419) 465-2997
Web site: homepages.accnorwalk.com/tddi/tech2000
Product(s): Dual-axis Dobsonian tracking system, laser collimator

Technical Innovations, Inc.
7851 Cessna Avenue, Gaithersburg, MD 20879

Phone: (301) 977-9000
Web site: www.homedome.com
Product(s): Observatory domes

Tectron Telescopes
5450 NW 52 Court, Chiefland, FL 32626
Phone: (352) 490-9101
Web site: www.amateurastronomy.com
Product(s): Collimation tools

Telescope Engineering Company
15730 West 6th Avenue, Golden, CO
80401
Phone: (303) 273-9322
Web site: www.telescopengineering.com
Product(s): Apochromatic refractors, Maksutov
telescopes

Tele Vue Optics
32 Elkay Drive, Chester, NY 10918
Phone: (854) 469-4551
Web site: www.televue.com
Product(s): Refractors, eyepieces

Telrad, Inc.
1701 Sheckler Cut-Off; Fallon, NV 89406
Phone: (775) 423-0747
Web site: none
Product(s): Telrad aiming device

Thousand Oaks Optical
P.O. Box 4813, Thousand Oaks, CA 91359
Phone: (805) 491-3642
Web site: www.thousandoaksoptical.com
Product(s): Solar filters, nebula filters

TL Systems
437 Drury Lane, Banning, CA 92220
Phone: (951) 922-8502
Web site:http://pw1.netcom.com/~tlsystem
Product(s): Equatorial tracking table kit

TMB Optical
P.O. Box 44331, Cleveland, OH 44144
Phone: (216) 524-1107
Web site: www.tmboptical.com
Product(s): Apochromatic refractors, eye-
pieces

True Technology Ltd
Woodpecker Cottage, Red Lane, Aldermas-
ton, Berks, RG7 4PA UK
Phone: 44(0)118-970-0777
Web site: www.trutek-uk.com
Product(s): CCD flip mirrors

TScopes
2081 Lakeville Road; Avon, NY 14414
Phone: (585) 226-2232
Web site: www.tscopes.com
Product(s): Newtonian reflectors

Tuma, Steven
1425 Greenwich Lane; Janesville, WI 53545
Phone: (608) 752-8366
Web site: www.deepsky2000.net
Product(s): *DeepSky* computer software

Universal Astronomics
6 River Court; Webster, MA 01570
Phone: (508) 943-5105
Web site: www.universalastronomics.com
Product(s): Binocular and telescope
mounts

Van Slyke Engineering
12815 Porcupine Lane, Colorado Springs, CO
80908
Phone: (719) 495-3828
Web site: www.observatory.org/vsengr.htm
Product(s): CCD flip mirrors, accessories

VERNONscope & Co.
5 Ithaca Road, Candor, NY 13743
Phone: (888) 303-3526, (607) 659-7000
Web site: www.vernonscope.com
Product(s): Brandon eyepieces, color filters,
Barlow lenses, accessories

Vixen Company, Ltd.
5-17 Higashitokorozawa, Tokorozawa, Saitama
359-0021 Japan
Phone: 81-42-944-4141
Vixen Optics
1010 Calle Cordillera, Suite 106, San
Clemente, CA 92673
Phone: (949) 429-6363
Web site: www.vixenoptics.com,
www.vixen.co.jp
Product(s): Refractors, Newtonian reflectors,
binoculars, eyepieces, telescope
mounts

Wildcard Innovations Pty. Ltd.
20 Kilmory Place Mount Kuring-Gai NSW 2080
Australia
Phone: 61-29-457-9049
Web site: www.wildcard-innovations.com.au
Product(s): Argo Navis digital setting circles

William Optics
  28 Fl., No. 29-5; Sec. 2, Jungjeng E. Road;
    Danshuei Jen, Taipei, 251 Taiwan
  Phone: 886-2-2809-3188
  William Optics USA
  4200 Avenida Sevilla, Cypress, CA 90630
  Phone: (714) 209-0388
  Web site: www.william-optics.com
  Product(s): Refractors, eyepieces, and accessories

Willmann-Bell, Inc.
  P.O. Box 35025, Richmond, VA 23235
  Phone: (804) 320-7016
  Web site: www.willbell.com
  Product(s): Books, software

Woden Optics
  1115 58th Street, Sacramento, CA 95819
  Phone: (916) 628-8673
  Web site: www.wodenoptics.com
  Product(s): Telescope mirrors

Zeiss AG (Carl Zeiss AG)
  Carl Zeiss Str. 22, 73447 Oberkochen Germany
  Phone: 49-736-4200
  Carl Zeiss, Inc.
  1 Zeiss Drive, Thornwood, NY 10594
  Phone: (914) 747-1800
  Web site: www.zeiss.de, www.zeiss.com
  Product(s): Binoculars

Dealers and Distributors

This is by no means an exhaustive list of dealers, but rather only a short inventory of retailers that sell astronomical equipment. Be sure to visit the *Star Ware* section of www.philharrington.net for a listing of astronomical retailers in all fifty states, as well as all Canadian provinces, and a wide variety of countries around the world. Many manufacturers also have complete lists of their dealers on their Web sites.

An important warning: if you are shopping by mail, *always* ask about shipping charges *before* ordering. Some offer exceptionally low prices only to charge the consumer exorbitant (and unpublished) shipping and handling costs. Not only will these hidden costs offset any savings, these dealers may end up being more expensive!

United States

Arizona
  Starizona
    5201 N. Oracle Road, Tucson, AZ 85704
    Phone: (520) 292-5010
    Web site: www.starizona.com
    Product line(s): Celestron, Tele Vue, Vixen
  Stellar Vision and Astronomy Shop
    1835 South Alvernon, #206; Tucson, AZ 85711
    Phone: (520) 571-0877
    Web site: www.azstellarvision.com
    Product line(s): Celestron, JMI, Meade, Orion Telescopes, Tele Vue, more

California
  Oceanside Photo and Telescope
    918 Mission Avenue, Oceanside, CA 92054
    Phone: (800) 483-6287, (760) 722-3343
    Web site: www.optcorp.com
    Product line(s): Borg, Celestron, Coronado, Discovery, Meade, Orion Telescopes, Stellarvue, Tele Vue, Vixen, William Optics, more
  Scope City
    730 East Easy Street, Simi Valley, CA 93065
    Phone: (800) 235-3344, (805) 522-6646
    Web site: www.scopecity.com

Product line(s): Carton, Celestron, Coronado, Fujinon, JMI, Konus, Kowa, Lumicon, Meade, Nikon, Parks, Pentax, Questar, Tele Vue, Telrad, Thousand Oaks, more

Woodland Hills Camera & Telescopes
5348 Topanga Canyon Boulevard, Woodland Hills, CA 91364
Phone: (818) 347-2270, (888) 427-8766
Web site: www.telescopes.net
Product line(s): Celestron, Coronado, Fujinon, Losmandy, Meade, SBIG, Tele Vue, Vixen, more

Colorado
S & S Optika
5174 South Broadway; Englewood, CO 80110
Phone: (303) 789-1089
Web site: www.sandsoptika.com
Product line(s): Celestron, DayStar, Pentax, Takahashi, Vixen, Zeiss, more

The Science Company
95 Lincoln Street, Denver, CO 80203-3922
Phone: (800) 372-6726
Web site: www.sciencecompany.com
Product line(s): Celestron, Konus, Meade

Florida
Internet Telescope Exchange
16222 133rd Drive N, Jupiter FL 33478
Phone: (800) 699-0906, (561) 282-3222
Web site: www.itetelescopes.com
Product line(s): Intes Micro, JMI, Losmandy, more

Georgia
Camera Bug, Ltd.
1799 Briarcliff Road, Atlanta, GA 30306
Phone: (877) 422-6284, (404) 873-4513
Web site: www.camerabug.com
Product line(s): Canon, Celestron, Meade, Minolta, Nikon, Pentax, Tele Vue, more

Illinois
20/20 Telescopes and Binoculars
3546 Ridge Road, Lansing, IL 60438
Phone: (887) 883-8883, (708) 474-2020
Product line(s): Celestron, Kendrick, Lumicon, Meade, Orion Telescopes, Tele Vue, more

OpticsPlanet.com, Inc.
15 E. Palatine Road, Unit 105, Prospect Heights, IL 60070
Phone: (888) 263-0356, (847) 574-6800
Web site: www.opticsplanet.net
Product line(s): Bushnell, Celestron, Coronado, Konus, Kowa, Lasermax, Meade, Nikon, Oberwerk, Pentax, Tele Vue, Vixen, more

Shutan Camera and Video
100 Fairway Drive, Vernon Hills, IL 60061
Phone: (800) 621-2248, (847) 367-4600
Web site: www.shutan.com
Product line(s): Celestron, Meade, Tele Vue, Vixen, more

Kansas
Science Education Center
2730 Boulevard Plaza, Wichita, KS 67211
Phone: (888) 597-2433
Web site: www.sciedcenter.com
Product line(s): Celestron, Meade, Tele Vue

Maryland
Company Seven
Box 2587, Montpelier, MD 20708
Phone: (301) 604-2500
Web site: www.company7.com
Product line(s): AstroSystems, Celestron, Fujinon, Kowa, Lumicon, Meade, Questar, Tele Vue, Universal Astronomics, more

Hands-On Optics
26437 Ridge Road, Damascus MD 20872
Phone: (866) 726-7371, (301) 482-0000
Web site: www.handsonoptics.com
Product line(s): Celestron, Coronado, Denkmeier, Kendrick, Lumicon, Meade, Stellarvue, Takahashi, Thousand Oaks, more

Michigan
Enerdyne, Inc.
223 St. Joseph Street, P.O. Box 366, Suttons Bay, MI 49682
Phone: (231) 271-6033
Web site: www.enerdynet.com
Product line(s): Celestron, Meade, Nikon, Orion Telescopes, Stellarvue, Vixen more

Rider's Hobby Shop
4035 Carpenter Road, Ypsilanti, MI
48197
Phone: (734) 477-7000
Web site: www.riders.com
Product line(s): Celestron, Meade, Orion
Telescopes, Tele Vue

Minnesota
Telescopes.com
200 6th Street, Proctor, MN 55810
Phone: (800) 303-5873, (218) 728-7953
Web site: www.telescopes.com
Product line(s): Barska, Baader, Celestron,
Coronado, Discovery, JMI, Kendrick,
Meade, Orion, Pentax, Tele Vue, Vixen,
more

Missouri
Lymax's Earth, Sky, and Astronomy
P.O. Box 328, Grain Valley, MO 64029
Phone: (888) 737-5050, (816) 737-5050
Web site: www.lymax.com
Product line(s): Celestron, Meade, Tele Vue,
more

Montana
Astro Stuff
1550 Amsterdam Road, Belgrade, MT
59714
Phone: (406) 388-1205
Web site: www.astrostuff.com
Product line(s): Celestron, JMB, Meade,
Orion Telescopes, Thousand Oaks Opti-
cal, more

New Hampshire
Astronomy Shoppe
3 Elm Street, Plaistow, NH 03865
Phone: (603) 382-0836
Web site: www.astronomy-shoppe.com
Product line(s): Celestron, Orion Telescopes,
Universal Astronomics, more

Rivers Camera Shop
454 Central Avenue, Dover, NH 03820
Phone: (800) 245-7963, (603) 742-4888
Web site: www.riverscamera.com
Product line(s): Celestron, Coronado,
Kendrick, Meade, Questar, Rigel, Star-
bound, Tele Vue, more

New Jersey
High Point Scientific, Inc.
442 Route 206, Montague, NJ 07827
Phone: (800) 266-9590, (973) 293-7200
Web site: www.highpointscientific.com
Product line(s): Celestron, Kendrick, Meade,
SBIG, Takahashi, Tele Vue, Vixen, more

Pisano Opticians
166 Wootton Street, Boonton, NJ 07005
Phone: (973) 402-2672
Web site: www.pisanoopticians.com
Product line(s): Celestron, Meade, JMI,
Orion Telescopes, more

New Mexico
New Mexico Astronomical
834 Gabaldon Road, Belen, NM 87002
Phone: (505) 864-2953
Web site: www.nmastronomical.com
Product line(s): Celestron, Discovery,
Meade, Orion Telescopes, Parks, Tele
Vue, Vixen more

New York
Adorama
42 West 18th Street, New York, NY 10011
Phone: (800) 223-2500, (212) 741-0466
Web site: www.adorama.com
Product line(s): Celestron, Coronado, Fuji-
non, Konus, Meade, Pentax, Takahashi,
Tele Vue, Thousand Oaks, Vixen, William,
more

Astrotec
1523 Lakeland Avenue, Bohemia, NY 11716
Phone: (631) 563-9009
Web site: www.astrotecs.com
Product line(s): Celestron, JMI, Kendrick,
Meade, Orion Telescopes, Tele Vue,
more

B&H Photo
420 Ninth Avenue, New York, NY 10001
Phone: (800) 221-5743, (212) 239-7765
Web site: www.bhphotovideo.com
Product line(s): Celestron, Fujinon, Meade,
Tele Vue, Vixen

Berger Brothers Camera Exchange
209 Broadway, Amityville, NY 11701
Phone: (800) 262-4160, (631) 264-4160

226 W. Jericho Turnpike (Route 25)
Syosset, NY 11791
Phone: (516) 496-1000
Web site: www.berger-bros.com
Product line(s): Canon, Celestron, Meade,
Pentax, more
Camera Concepts
33 East Main Street, Patchogue, NY 11772
Phone: (631) 475–1118
21 East Main Street, Riverhead, NY 11901
Phone: (631) 727-3283
Web site: www.cameraconcepts.com
Product line(s): AstroSystems, Apogee,
Celestron, Glatter, Kendrick, Meade,
Rigel, Tele Vue, Vixen, more
Continental Camera
5795 Transit Road, Depew, NY 14043
Phone: (716) 681-8038
Web site: www.continentalcamera.com
Product line(s): Celestron, Meade
Edmund Scientific Co.
60 Pearce Avenue, Tonawanda, NY 14150-
6711
Phone: (800) 728-6999
Web site: www.scientificsonline.com
Product line(s): Celestron, Edmund,
Meade, Orion Telescopes, Rigel, Tele
Vue
Focus Camera, Inc.
905 McDonald Avenue, Brooklyn, NY 11218
Phone: (888) 221-0828, (718) 437-8810
Web site: www.focuscamera.com
Product line(s): Canon, Celestron, Fujinon,
Meade, Nikon, Pentax, Tele Vue
Science & Hobby
Route 9 Columbia Plaza, Rensselaer, NY
12144
Phone: (800) 763-3212, (518) 477-7066
Web site: www.scienceandhobby.com
Product line(s): Celestron, Meade, Tele Vue
North Carolina
Camera Corner
2273 South Church Street, Burlington, NC
27215
Phone: (800) 868-2462, (336) 228-0251
Web site: www.cameracorner.com

Product line(s): Celestron, Discovery,
Meade, Tele Vue, more
Oklahoma
Astronomics
680 S.W. 24th Avenue, Norman, OK 73069
Phone: (800) 422-7876, (405) 364-0858
Web site: www.astronomics.com
Product line(s): Bushnell, Celestron, Coron-
ado, JMI, Konus, Meade, Questar, Stel-
larvue, Takahashi, Tele Vue, TMB, Vixen,
William, more
Pennsylvania
Helix Observing Accessories
P.O. Box 490, Gibsonia, PA 15044
Phone: (724) 316-0306
Web site: helix-mfg.com
Product line(s): Oberwerk, Stellarvue, more
Skies Unlimited
2922 Conestoga Road, Glenmoore, PA
19343
Phone: (888) 947-2673, (610) 321-9881
Web site: www.skiesunlimited.net
Product line(s): Celestron, Coronado,
Meade, Nikon, Tele Vue, Vixen, Zeiss,
more
Texas
Analytical Scientific
11049 Bandera Road, San Antonio, TX
78250
Phone: (800) 364-4848, (210) 684-7373
Web site: www.analyticalsci.com
Product line(s): Celestron, Meade, Tele Vue,
Telrad
Land, Sea, and Sky
3110 South Shepherd, Houston, TX 77098
Phone: (713) 529-3551
Web site: www.takahashiamerica.com
Product line(s): Celestron, Meade,
Miyauchi, Takahashi, Tele Vue
The Observatory
19009 Preston Road, Suite 114, Dallas TX
75252
Phone: (972) 248-1450
Web site: www.theobservatoryinc.com
Product line(s): Celestron, Fujinon, Meade,
Tele Vue, more

Washington

Anacortes Telescope & Wild Bird

9973 Padilla Heights Road, Anacortes, WA 98221

Phone: (800) 850-2001, (360) 588-9000

Web page: www.buytelescopes.com

Product line(s): Celestron, Denkmeier, Discovery, Fujinon, JMI, Kowa, Meade, Miyauchi, Oberwerk, Orion Telescopes, Questar, SBIG, Tele Vue, William Optics, Vixen, Zeiss, more

Wisconsin

Eagle Optics

2120 West Greenview Drive Suite #4, Middleton, WI 53562

Phone: (800) 289-1132, (608) 836-7172

Web site: www.eagleoptics.com

Product line(s): Bushnell, Celestron, Fujinon, Nikon, Pentax, more

Canada

Alberta

The Science Shop

316 Southgate Centre, Edmonton, AB T6H 4M6

Phone: (888) 434-0519, (780) 435-0519

Web site: www.thescienceshop.com

Product line(s): Celestron, JMI, Meade, Sky-Watcher, Sky Instruments, Tele Vue, Vixen, more

British Columbia

Island Eyepiece and Telescope Ltd.

P.O. Box 133, Mill Bay, BC V0R 2P0

Phone: (250) 743-6633

Web site: www.islandeyepiece.com

Product line(s): Celestron, Meade, Orion Telescopes, Sky Instruments, Sky-Watcher, Tele Vue, William, more

Vancouver Telescope Centre

1859 W. 4th Avenue, Vancouver, BC V6J 1M4

Phone: (877) 737-4303, (604) 737-4303

Web site: www.vancouvertelescope.com

Product line(s): Celestron, Intes Micro, Meade, Sky-Watcher, Tele Vue, William, more

Ontario

EfstonScience

3350 Dufferin Street, Toronto, ON M6A 3A4

Phone: (416) 787-4581, (888) 777-5255

Web site: www.telescopes.ca

Product line(s): Celestron, Coronado, Meade, Orion Telescopes, Pentax, Sky-Watcher, Tele Vue, more

Focus Scientific Ltd.

1489 Merivale Road North, Ottawa, ON K2E 5P3

Phone: (877) 815-1350, (613) 723-1350

Web page: www.focusscientific.com

Product line(s): Carton, Celestron, Meade, Nikon, Orion Telescopes, Sky Instruments, Sky-Watcher, more

Kendrick Astro Instruments

2920 Dundas Street West, Toronto, ON M6P 1Y8

Phone: (800) 393-5456, (416) 762-7946

Web site: www.kendrick-ai.com

Product(s): Baader, Celestron, JMI, Kendrick, Losmandy, Mag 1, Orion Telescopes, Pentax, Tele Vue, Vixen, more

Khan Scope Centre

3243 Dufferin Street, Toronto, ON M6A 2T2

Phone: (800) 580-7160, (416) 783-4140

Web site: www.khanscope.com

Product line(s): Antares, Celestron, JMI, Meade, SBIG, Sky-Watcher, Tele Vue, Thousand Oaks, William, more

O'Neil Photo & Optical, Inc.

356 William Street, London, ON N6B 3C7

Phone: (519) 679-8840

Web site: www.oneilphoto.on.ca

Product line(s): Celestron, Fujinon, Orion Telescopes, Sky-Watcher, Sky Instruments, Stellarvue, more

Perceptor

Brownsville Plaza, Suite 201, 17250 County Road 27, Schomberg, ON L0G 1T0

Phone: (877) 452-1610 (905) 939-2313

Product line(s): Carton, Celestron, Coronado, Meade, Orion Telescopes, Questar, SBIG, Tele Vue, Vixen, Zeiss, more

Sky Optics
> 4031 Fairview Street, Unit 216B, Burlington, ON L7L 2A4
> Phone: (877) 631-9944, (905) 631-9944
> Web site: www.skyoptics.net
> Product line(s): Celestron, Lumicon, Meade, Orion Telescopes, SBIG, Sky-Watcher, Sky Instruments, Tele Vue, more

Prince Edward Island

NatureWatch
> PO Box 22062, Charlottetown, PE C1A 9J2
> Phone: (800) 693-8095, (902) 628-8095
> Web site: www.naturewatchshop.com
> Product line(s): Celestron, Sky-Watcher

Quebec

Astronomie Plus (Lire La Nature Inc.)
> 1198 Chambly Road, Longueuil, QC J4J 3W6
> Phone: (450) 463-5072
> Web site: www.AstronomiePlus.com
> Product line(s): Apogee Instruments, Baader, Celestron, Meade, Nikon, Sky-Watcher, Pentax, SBIG, Tele Vue, Thousand Oaks, Zeiss, more

La Maison de l'Astronomie P.L. Inc.
> 8074 Saint Hubert, Montreal, QC H2R 2P3
> Phone: (514) 279-0063
> Web site: www.maisonastronomie.ca
> Product line(s): Celestron, Meade, Orion Telescopes, Sky-Watcher, Takahashi, Vixen, more

United Kingdom

Green Witch
> Unit 6 Dry Drayton Industries; Dry Drayton, Cambridge CB3 8AT
> Phone: 44(0)195-421-1288
> Web site: www.greenwich-observatory.co.uk
> Product line(s): Celestron, Meade, Orion Optics, Sky-Watcher, Tele Vue, more

Optical Vision Limited
> Unit 2b, Woolpit Business Park, Woolpit, Bury Street, Edmunds, Suffolk IP30 9UP
> Phone: 44(0)135-924-4200
> Web site: www.opticalvision.co.uk
> Product line(s): Intes Micro, Sky-Watcher, more

Pulsar Optical
> 1 Bridge Lane, Wimblington March, Cambridgeshire PE15 0RR
> Phone: 44(0)135-474-1443
> Web site: www.pulsar-optical.co.uk
> Product line(s): Celestron, Konus, Meade, Sky-Watcher, more

SCS Astro
> 1 Tone Hill, Wellington, Somerset TA21 0AU
> Phone: 44(0)182-366-5510
> Web site: www.scsastro.co.uk
> Product line(s): Apogee, Borg, Celestron, Kendrick, Lumicon, Meade, Orion Optics, Tele Vue, Vixen, more

Sherwoods Photo, Ltd.
> North Court House, Greenhill Farm, Morton Bagot, Studley, Warwickshire B80 7EL
> Phone: 44(0)152-785-7500
> Web site: www.sherwoods-photo.com
> Product line(s): Celestron, Fujinon, Meade, Orion Optics, Sky-Watcher, more

Telescope House (Broadhurst, Clarkson & Fuller Ltd.)
> Unit 6, Tunbridge Wells Trade Park, Longfield Road, Tunbridge Wells, Kent TN2 3QF
> Phone: 44(0)189-255-0100
> Web site: www.telescopehouse.co.uk
> Product line(s): Meade, Takahashi, Tele Vue, TMB, Vixen, William, more

True Technology Ltd
> Woodpecker Cottage, Red Lane, Aldermaston, Berks RG7 4PA
> Phone: 44(0)118-970-0777
> Web page: www.trutek-uk.com
> Product line(s): Astronomik, JMI, Starlight Xpress, Takahashi

Venturescope
> 11C The Wren Centre, Westbourne Road; Emsworth, Hampshire PO10 7RL
> Phone: 44(0)124-337-9322
> Web site: www.telescopesales.co.uk
> Product line(s): Celestron, Coronado, Fujinon, JMI, Kendrick, Miyauchi, Meade, Parks Optical, Tele Vue, Vixen, more

Other Countries

Australia

  Adelaide Optical Centre

    120 Grenfell Street, Adelaide, SA 5000

    Phone: 61(08) 8232-1050

    Web site: www.adelaideoptical.com.au

    Product line(s): Kendrick, Kowa, Meade, Vixen, William, more

  Astro Optical Supplies

    39 Hume Street, Crows Nest, NSW 2065

    Phone: 61(02) 9436-4360

    320 St. Kilda Road, St. Kilda, VIC 3182

    Phone: 61(03) 9593-9512

    Web site: www.astro-optical.com.au

    Product line(s): Celestron, Meade, Orion Telescopes, Vixen, more

  Astronomy & Electronics Centre

    P.O. Box 45, Cleve, SA 5640

    Phone: 61 (08) 8628-2435

    Web site: www.astronomy-electronics-centre.com.au

    Product line(s): Astro-Physics, Baader, Celestron, Miyauchi, Parks, Takahashi, Vixen, more

  Binocular and Telescope Shop

    55 York Street, Sydney, NSW 2000

    Phone: 61 (02) 9262-1344

    Web site: www.bintel.com.au

    Product line(s): Meade, Orion Telescopes, Sky-Watcher, more

Ireland (Republic of)

  Astronomy Ireland

    P.O. Box 2888, Dublin 1, Ireland

    Phone: (01) 847-0777

    Web site: www.astronomy.ie

    Product line(s): *Astronomy & Space* magazine, Canon, Celestron, Starlight XPress, more

New Zealand

  Skylab

    P.O. Box 12060, Christchurch, New Zealand

    Phone: 64-3-365-7700

    Web site: www.skylab.net.nz

    Product line(s): Celestron, Sky Instruments, more

South Africa

  Lynx Optics (Pty) Ltd

    P.O. Box 98078, Sloane Park, South Africa, 2152

    Phone: 27(0)11-792-6644

    Web site: www.lynx.co.za

    Product line(s): Kowa, Meade

Appendix D
An Astronomer's
Survival Guide

It's not always easy being an amateur astronomer. We are nocturnal creatures by nature, going outdoors when the rest of the world sleeps; braving cold, heat, bugs, and things that go bump in the night; always looking up when most everyone else is looking down (astronomers are the eternal optimists).

On page 400 is a checklist of things that I like to bring along for a night under the stars. They make the experience much more pleasurable, and the cold night a little warmer.

Astronomical

Telescope ____   Binoculars ____   Eyepieces ____   Filter(s) ____
Star atlas ____   Object list ____   Clipboard ____   Pen/pencil ____
Flashlight (red) ____   Flashlight (white) ____   Observing chair ____
Etc. _____

_____

Photographic

Camera(s) ____   Auxiliary lenses ____   Film ____   Memory card ____
Remote release ____   Tripod ____   Camera-to-telescope adapters ____
Etc. _____

_____

Miscellaneous

Sweatshirt ____   Long underwear ____   Heavy socks ____   Jacket ____
Winter coat ____   Gloves/mittens ____   Boots ____   Foot/hand warmers ____
Hat ____   Folding table ____   Chair ____   Insect repellent ____
Something warm to drink (nonalcoholic) ____   Food/snacks ____   Radio ____
Etc. _____

_____

Appendix E
Astronomical Resources

For those looking to expand their involvement with the hobby and science of astronomy, the following organizations are excellent ways to delve deeper into the universe.

Special Interest Societies

American Association of Variable Star Observers
   25 Birch Street, Cambridge, MA 02138
   Web site: www.aavso.org

American Meteor Society
   161 Vance Street, Chula Vista, CA 91910
   Web site: www.amsmeteors.org

Antique Telescope Society
   1878 Robinson Road, Dahlonega, GA 30533
   Web site: www.webari.com/oldscope

Association of Lunar and Planetary Observers
   305 Surrey Road, Savannah, GA 31410
   Web site: www.lpl.arizona.edu/alpo

Astronomical League
   9201 Ward Parkway Suite #100,
   Kansas City, MO 64114
   Web site: www.astroleague.org

Astronomical Society of the Pacific
   390 Ashton Avenue, San Francisco, CA 94112
   Web site: www.astrosociety.org

International Dark-Sky Association
   3225 N. First Avenue, Tucson, Arizona 85719
   Web site: www.darksky.org

International Meteor Organization
   161 Vance Street, Chula Vista, CA 91910
   Web site: www.imo.net

International Occultation Timing Association
   5403 Bluebird Trail, Stillwater, OK 74074-7600
   Web site: www.occultations.org

National Deep Sky Observer's Society
   1607 Washington Boulevard, Louisville, KY 40242
   Web site: http://users.erols.com/njastro/orgs/
   ndsos.htm

World Wide Web Resources

The Internet holds a whole universe of resources for the amateur astronomer. Trying to come up with a brief list is impossible, but here are a few Web sites that might be of interest. I'll lead off the list with my own Web site, www .philharrington.net, and its companion discussion group called, "Talking Telescopes," at http://groups.yahoo.com/group/telescopes. There, you will find updates to this book as they become available, including new product reviews, as well as online discussions of telescope pros and cons. Why not join us?

Telescopes and Equipment

Astro-Mart used equipment
  www.astromart.com
Astronomy Software Review Page
  www.seds.org/billa/astrosoftware.html
Cloudy Nights
  www.cloudynights.com
Ed Ting's Scope Reviews Home Page
  www.scopereviews.com
Excelsis Telescope Ratings
  www.excelsis.com/1.0/home.php

Maksutov Telescope Reviews
  www.lafterhall.com/makscopes.html
Northern Virginia Astronomy Club
  www.novac.com/resources/reviews
Cookbook CCD Camera Home Page
  http://wvi.com/~rberry
SkyNews Magazine
  http://skynewsmagazine.com

Sky Information

Abrams Planetarium
  www.pa.msu.edu/abrams
Astronomy magazine
  www.astronomy.com
Comet Observation Home Page
  www.cometobservation.com
Sky & Telescope magazine
  www.skytonight.com

Skymaps
  www.skymaps.com
Solar Eclipse Home Page
  http://sunearth.gsfc.nasa.gov/eclipse/eclipse
  .html
Spaceweather
  www.spaceweather.com

Astronomy-Related Newgroups and E-Mail Lists

A number of free newsgroups and electronic-mail lists exist on several topics of interest to the amateur astronomer. Newgroups are typically unmoderated and as such, can often lead to some truly spirited debates. No need to go through a formal joining process. Instead, jump right in!

Newsgroup	Topic
alt.sci.planetary	Discussion of planetary research and space missions
sci.astro	General discussion group on anything astronomical
sci.astro.amateur	General discussion group with a slant toward amateur observations and equipment
sci.astro.hubble	Discussions of the latest findings from the Hubble Space Telescope
sci.astro.planetarium	Planetarium professionals compare notes on shows and equipment
sci.astro.research	Researchers post questions and answers about various research-related topics

E-mail lists and discussion groups, on the other hand, are more specialized and are usually moderated by the list owner, who may or may not rule with a light touch. List owners often require you to register with the group, although the step is often more a formality than anything. Digest versions are often available, as well; consult the instructions that you receive after you subscribe initially.

By far the most popular forums are currently available through Yahoo! Groups. By going online to http://groups.yahoo.com and typing in "telescopes" in Yahoo's "Find a Group" search box will generate a list of more than 200 groups that are related to telescopes in some manner. Looking down that directory, it becomes pretty clear that many are devoted to one particular brand, or even one particular model, while others are more general in their coverage. "Talking Telescopes" falls under the latter. Other forums to visit when online include those managed by *Astronomy* magazine at www.astronomy.com. Other forum Web sites of interest include the discussion groups at www.cloudynights.com and www.astromart.com. Note that only those registered can post to most forums, although registration is free.

Appendix F
English/Metric Conversion

Most amateur astronomers in the United States speak of a telescope's aperture in terms of inches, while the rest of world uses millimeters or centimeters. The following table acts as a translation table to help convert telescope apertures from one system to another. Recall that there are 25.4 millimeters and 2.54 centimeters per inch.

Inches	Millimeters	Centimeters
2.8	70	7.0
3.1	80	8.0
3.9	100	10.0
4	102	10.2
4.5	114	11.4
5.1	130	13.0
6	152	15.2
8	203	20.3
10	254	25.4
12	305	30.5
12.5	318	31.8
14	356	35.6
15	381	38.1
16	406	40.6
18	457	45.7
20	508	50.8
24	610	61.0

Star Ware Reader Survey

To help prepare for the next edition of *Star Ware*, I'd like you to take a moment and complete the following survey. Photocopy it, fill it out, and send it to the address shown at the bottom. An online version is also available in the *Star Ware* section of my Web site, www.philharrington.net or via e-mail by writing to starware@philharrington.net. Feel free to reproduce it in your club newsletter as well.

Please answer all (or as many) of the questions below as possible. If you own more than one telescope, I'd like to hear about each. *Use as much room as you want!* I'll save all responses and reference them when it comes time to write the reviews. Please include your address should I need to get a hold of you for questions or clarifications. Rest assured that your address will not be given to anyone nor will any company ever see your survey.

Thanks in advance!

Your name: _____

Address: _____

City: _____ State/Country: _____ Zip: _____

E-mail: _____

How long have you been involved in astronomy?_____

Do you consider yourself a: Beginner   Intermediate   Advanced

TELESCOPE

How many telescopes do you own? _____

Telescope model: _____

How old is it? _____ Are you the original owner? _____

What do you like about it?

_____

_____

What don't you like about it?

_____

_____

Lived up to your expectations? _____ Buy it again? _____

Have you ever had to contact the company about a problem?

_____

Was it resolved? Explain.

_____

_____

EYEPIECES

What eyepieces do you own?

_____

How do they work? Any particular likes or dislikes? (Please list your impressions separately for each.)

_____

_____

ACCESSORIES

What accessories do you own? (Anything . . . binoculars, books, software, filters, finderscope, etc.)

_____

_____

_____

Any particular likes or dislikes? (Please list your impressions separately for each.)

_____

_____

_____

_____

Your vote for the best telescope of yesteryear (only models that are no longer made). _____

Anything you think should be included in the next edition?

_____

_____

Please mail this survey to Phil Harrington, c/o John Wiley & Sons, 111 River Street, Hoboken, NJ 07030, or respond by e-mail to starware@philharrington .net.

Index